Switchgear and Power System

CU01513121

Dedicated to the teachers who fuel their students'
imagination, enabling them to develop a new
outlook on life and a new kind of being

Contents

Part B: Power System Protection

Preface

Electrical equipment and appliances may get damaged by overloading, short circuiting or due to some other faults that may occur on the system. Incipient fault may become serious if the faulty equipment is not isolated immediately after the fault is detected. Due to economical reasons, the transmission level of voltage is kept as high as technically feasible. Switchgear and protection systems perform the functions of detection of the fault and isolation of the affected equipment or circuit while the operating voltage may be several thousands of volts. With the enhancement of voltage levels, power system engineers are challenged by several problems. This results in rise of faults. To take care of electrical faults, several switchgear instruments and apparatus like HRC fuses, isolators, earthing switches, contactors, reactors and circuit breakers are required to be installed.

In the present scenario, the deregulated energy market is under competitive environment. To compete the rivals, market driven power utilities are being forced to improve their productivity and efficiency. Knowing the fact that the existing power network and generating plants are operating closer to their limits, the system availability needs to be kept very high to ensure maximum continuity of electrical supply with minimum damage to life, equipment, and property. This is only possible, if a dedicated and smart control and protection system is installed with the power system.

Protection systems can get benefit from the enormous potential of existing communication networks. The new emerging technology "system protection schemes" (SPS) demands integrated communication networking between many terminals rather than point-to-point links.

There are very rare books available having the most modern topics like optical fibre-based relaying, microwave relaying, network protection, mains failure protection, discussion on the probable solutions of catastrophic failures, Power System Management and many more. These topics are made available in this book and presented in an illustrative manner, which makes this book entirely different from the other books available in the market.

Numerical problems solved with the help of MATLAB are given in a large number. This book is appropriate for the undergraduate students of electrical engineering as well as for postgraduate students specializing in Power Systems Engineering. It will also be useful to the research scholars and practising engineers.

The author sincerely hopes that the book shall earn wide appreciation from the faculty members and the students. The author shall welcome and gratefully acknowledge the suggestions and constructive criticism for improvement to the book.

Ravindra P. Singh

Part **A**

SWITCHGEAR

- ◆ **Definitions and Terminology**
- ◆ **Basics of Switchgear**
- ◆ **High Voltage Circuit-breakers**
- ◆ **Current-limiting Reactors**

Chapter 1

Definitions and Terminology

1.1
INTRODUCTION

Switchgear consists of circuit-breakers, isolators, switches, contactors, fuses, actuators, starters, potential transformers and current transformers. Following are the functions of the switchgear:

   (i) To monitor and measure the various power system quantities and parameters.
   (ii) To switch the healthy sections of the power system 'ON'.
   (iii) To switch the circuit 'OFF', whenever required.
   (iv) To isolate the faulty section of the power system network from the rest of the network.

Following are the essential requirements of the switchgear:

   (i) To fulfill maximum load during severe conditions, switchgear must be suitably rated.
   (ii) Fault detection without time delay is must so the monitoring equipment must have high sensitivity.
   (iii) To achieve desired reliability all the components of the switchgear must be ready to operate round the clock accurately.
   (iv) System must be capable to have correct discrimination.

Main circuit

Main circuit includes all the conducting components intended to be protected and controlled by the switchgear installed.

Auxiliary circuit

Signalling, indication and alarm circuits come under auxiliary circuits. These are the electrical circuits other than the main circuits and the control circuits.

Control circuit

Control circuit includes all the conducting parts outside the main circuit of the switchgear for opening and closing operations.

Ambient temperature

Ambient temperature is the temperature of atmosphere surrounding switchgear installations.

Ambient condition

Ambient condition includes standard temperature, pressure, pollution, humidity, rain, etc. at location of the switchgear installations.

Indoor switchgear

Switchgear installations residing inside the shed or building to protect the machineries and measuring instruments from rain, snow, pollution and sunrays are known as indoor switchgears. These are up to 11 kV ratings.

Outdoor switchgear

Switchgear installations residing outside under the open sky are known as outdoor switchgears. These are above 66 kV ratings.

Prospective current

Prospective current is the probable maximum current flow at the location of short circuit assuming negligibly small impedance of the circuit.

Rated breaking current

Under specific fault conditions, the maximum prospective current, which the switchgear carries is capable of interrupting at specified voltage.

Rated making current

At specified voltage and prescribed fault conditions rated making current is the peak value of the maximum current that switchgear is capable of making the circuit.

Short time rating

It is also known as short time withstand current. It is the current, which is carried by the switchgear in 'normally closed' position, under pre-specified service conditions.

Peak withstand current

Peak withstand current is the measure of mechanical strength of the equipment. Mechanical strength of the equipment is required to withstand the electrodynamic forces generated by the fault current. It is the value of peak current, which the switchgear can withstand momentarily in the closed position under specified conditions.

Power frequency withstand voltage

Power frequency withstand voltage is the rms value of the sinusoidal alternating voltage wave at power frequency, which the switchgear insulation can withstand under pre-specified testing conditions.

Opening time

Opening time is the time taken by circuit-breaker between the instant of opening command given and the instant when arcing contacts separated in all the three poles.

Arcing time

Arcing time is the duration between the instant of first arc initiation and the instant of arc extinction in all the poles.

Break time

Break time is the time interval from the beginning of the opening time to the end of the arcing time.

Closing time

Closing time is the time interval between the instant of initiation of the closing operation and the instant when the contacts touch all the poles of the switchgear.

Make time

Make time is the time interval between the instant of initiation of the closing operation and the instant when current starts to flow in the main circuit of the switchgear.

Dependent manual operation

If the speed of operation of the switchgear depends on the action of operator, the operation is known as dependent manual operation. It is recommended only for off-load switches and isolators.

Independent manual operation

If the speed of operation of the switchgear is independent of the action of operator, the operation is known as independent manual operation.

Dependent power operation

If the speed of operation of the switchgear depends on the electrical energy, i.e. electrical supply, the operation is known as dependent power operation.

Stored energy operation

Stored energy operation is an operation by means of which energy is stored in the mechanism itself prior to the initiation. However, storing process of the energy may also take place during the operation.

Operating mechanism

Operating mechanism of the switchgear includes the portion outside the main circuit that facilitates the motion to the 'contacts'.

Fixed trip mechanism

Fixed trip mechanism is the mechanism that cannot be released to open the switchgear except when it is in the closed position.

Trip-free mechanism

Trip-free mechanism is made available in the modern circuit-breakers and switches. It is the mechanism, which remains in open position, when the opening operation gets initiated just after the closing operation is initiated.

Chapter 2

Basics of Switchgear

2.1
INTRODUCTION

Switchgears are the equipment or systems used to regulate the electrical power network safely and efficiently, whether the network is under normal and healthy conditions or suffering from abnormalities. Fuses, isolators, contactors and circuit-breakers are the main components of the switchgears.

2.2
FUSES

A fuse is a device, which protects the electrical wires, cables and other electrical equipment against overloading and short circuit. It contains a fuse element, generally thin wire or strip mounted on insulated base. It is heated and destroyed while passing excessive current through it. It breaks the circuit by melting the fuse element, when the current flowing in the circuit exceeds a certain predetermined value. The fuse is placed in series with the equipment or its parts which are to be protected.

Fuse elements are made up of generally tin, lead, silver, zinc, aluminium, copper, etc. For the small current applications, an alloy of lead and tin in the ratio of 37 per cent and 63 per cent respectively is used. For the currents above 15 A, this alloy is not used because its size increases and after melting metal releases excessive volume of it. For the currents above 15 A silver is used as fuse material, but it is more costly. The advantage of using silver as fuse material is that it is not subjected to oxidation. Its oxide is unstable. The melting point of silver is 960°C and tin is 230°C.

2.2.1   Types of Fuses

They are of following two types:

 1. Rewirable fuse.
 2. Totally enclosed or cartridge fuse.

Rewirable fuse

This type of fuse is rewirable, i.e. the blown out fuse element can be replaced by a new element. The fuse element can be either open or semi-enclosed. Open type is rarely used. Semienclosed type is generally used for low-voltage applications in houses and installations for up to 440 V. This is the cheapest and simplest form of the protection.

Rewirable fuse is a set of two units. One is known as *base-unit* and the other is known as *fuse-carrier*. Fuse element can be fixed, removed and refixed in the fuse-carrier. Fuse-carrier is also known as *removable unit*. The base-unit is permanently fixed unit with the switchboard.

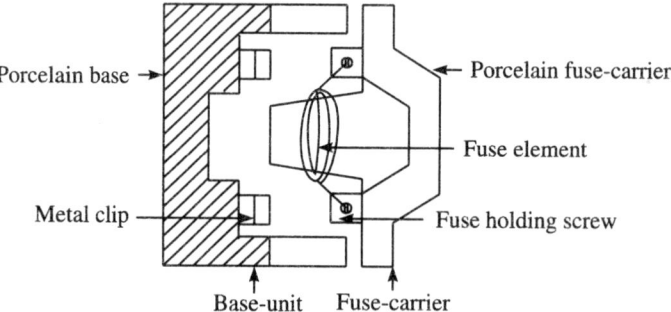

FIGURE 2.1   Constructional diagram of rewirable fuse.

Both the units are fixed on the porcelain chassis. The incoming wire is connected with the metal clip on one side of the porcelain base. The outgoing wire is connected with the other terminal of the porcelain base. These two terminals situated inside the base of the fuse-carrier are interconnected by fuse wire. This fuse wire is the weakest part of the circuit which melts during overloading or short circuit. After that a new fuse wire is connected between the terminals. A constructional diagram of rewirable fuse is shown in Figure 2.1. Table 2.1 below gives the values of rated current and the approximate fusing current for tin—copper fuse wires of different SWG and diameter.

TABLE 2.1

SWG	*Diameter* (cm)	*Rated current* (A)	*Approximate fusing current* (A)
20	0.09	34	70
25	0.05	15	30
30	0.031	8.5	13
35	0.021	5.0	8
40	0.012	1.5	3

Fusing factor = minimum fusing current/fuse rating (current)

Totally enclosed or cartridge fuse

The fuse element of this type is enclosed in a totally enclosed insulated container. It is provided with metal contacts on both the sides. Totally enclosed or cartridge fuses are of following two types:

(i) D-type cartridge fuse.

(ii) Link-type cartridge fuse or high-rupturing capacity (HRC) cartridge fuse.

D-type cartridge fuse unit is totally enclosed in an insulating container of tube shape. The container is filled up with power of arc quenching property, which is sealed at its ends with metallic cap. Fuse wire is fixed inside the container and cannot be removed or re-fixed. Whole unit is required to be replaced when fuse blows. These are used for about rating up to 800 A and voltage up to 660 V. A constructional diagram of D-type cartridge fuse is shown in Figure 2.2.

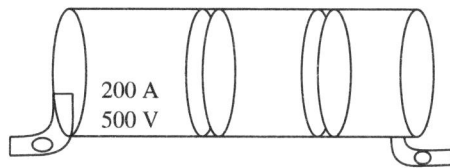

FIGURE 2.2   Constructional diagram of D-type cartridge fuse.

HRC fuse unit consists of a ceramic body usually of steatite, fuse element of silver, clean silica quartz, asbestos washers, porcelain plugs, brass end caps and copper tags. The space around the fuse element within the main body is filled with power, which has to promptly extinguish the arc produced during melting of the fuse element. The brass end caps and copper tags are generally electro-tinned. Special forged screw tightens the metal end caps with the ceramic body. End caps are welded with the contacts.

HRC fuses are uneconomical because whole unit is required to be replaced after the fuse blows. Another problem with HRC fuse is to identify the blown fuse among the number of similar fuses applied in the same board or feeder. However, by checking one by one the blown one can be identified. But HRC fuses are most reliable to withstand heavy stresses in clearing the fault. There is less temperature rise at even full load. Since HRC fuses are enclosed-type, therefore, there is no chance to damage nearby area. A constructional diagram of HRC cartridge fuse is shown in Figure 2.3.

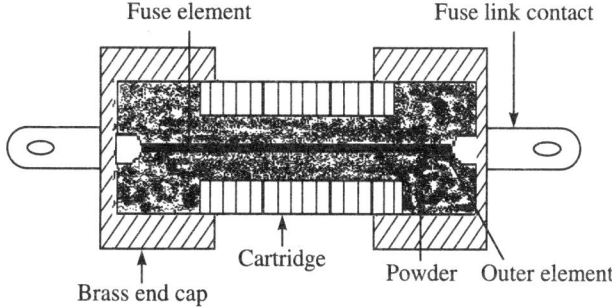

FIGURE 2.3   Constructional diagram of HRC cartridge fuse.

2.3

ISOLATORS

An isolator is a simple switchgear device with limited applications and basically used to visibly isolate a section of electric circuit or installation from the rest. It is also known as 'disconnector' or 'isolating switch'. Isolation is required during power source identification, changeover, busbars selection or maintenance. Isolators must have capabilities to carry continuous rated current during normal conditions. They are not automatic devices and do not have current-breaking capacity. Before operating the isolator, load is to be switched off; this is the reason why isolators do not have current-breaking capacity. However, considering the technical necessities isolators need to have current-breaking capabilities of the following:

(i) Breaking of cable charging currents.

(ii) Breaking of transformer magnetization currents, i.e. unloaded but charged transformer.

(iii) Breaking of line charging currents, i.e. disconnecting lines open at other end.

Isolators are available for each level of voltages up to 1000 kV and for continuous current carrying rating up to 4 kA. Isolators have also been developed for making current capacity up to 80 kA and for thermal capacity of 50 kA for one second.

2.3.1   Types of Isolators

Isolators are of following types:

1. Knife blade indoor isolator
2. Sliding contact indoor isolator
3. Centre post-rotating outdoor isolator
4. Double pole-rotating outdoor isolator
5. Pantograph outdoors isolator
6. Rocking or tilting isolator
7. Single break isolator

Knife blade indoor isolator

Knife blade indoor isolators are operated manually. They are suitable up to 12 kV applications. In these types of isolators, two types of contacts, viz. fixed contacts and moving contacts are fitted. Knife-type contact arms are used to serve the purpose of moving contact. A shaft carrying the moving contacts is rotated by a lever. As operating lever moves upwards it drives shaft, which rotates the knife contact arms to engage them with U-shaped fixed contacts. Springs are used to maintain the contact pressure.

They are also used as loads break switches, which are operated during small loading conditions like energized transformer carrying magnetizing current. While using load break switches, they are provided with highly abrasion-resistant arcing contacts in addition to the main contacts on both moving and fixed sides of each phase. Figure 2.4 shows the knife blade indoor isolator.

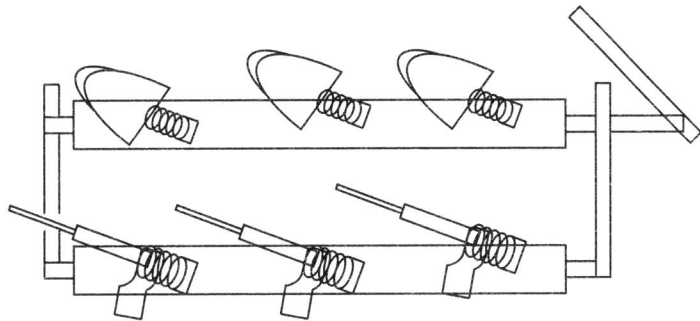

FIGURE 2.4   Knife blade indoor isolator.

Sliding contact indoor isolator

This is a compact size isolator having following qualities:

(i) Long mechanical life

(ii) High fault current capacity

(iii) Less maintenance

Sliding contact indoor isolators are capable to be suitably applied up to 24 kV systems. The main components of these isolators such as fixed contacts, moving arms set in moving assembly, connection pads, insulating jacket, transfer contact, arc extinction chamber, arcing ring, tubular contact, insulated arm, secondary contact and its rod, metal cap, base frame, cast iron insulator and discharge opening are shown in Figure 2.5.

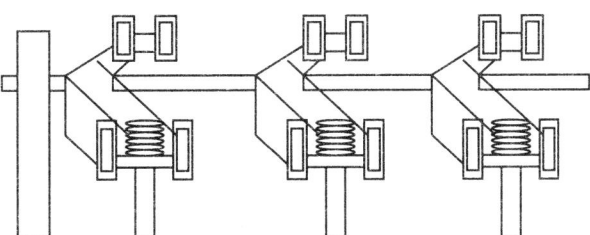

FIGURE 2.5   Sliding contact indoor isolator.

In this type six ribbed cast resin insulators are fixed on the base frame. Three-pole assembly is fixed with these ribbed insulators. Tubular contact enclosed inside the insulating jacket and internal secondary contact rod with its sliding contact and spring set make the moving assembly for the isolator. During the normal condition such as in the closed position of the switch current continues to flow through the connecting pad, fixed contact, tubular contact and transfer contact.

During the opening process, the tubular contact withdraws from the fixed contact but the secondary contact rod remains in the grip of the secondary contact. An arc is created between the transfer contact and the arcing ring. While moving, contact assembly travels down the arc chamber and the lower tip of the tubular contact leaves the transfer contact. Arc gets lengthened with further travel of the moving contact downwards. Surrounding materials around the arcing ring extinguish the arc before moving contact completes the opening.

Centre post-rotating outdoors isolator

This is a compact size isolator which requires less space. This is also known as 'double break-type isolator'. Its operating mechanism is simple because only one part for each pole needs to rotate in this type of isolator.

In this type of isolator, three insulator stacks are fixed on each pole. Insulator stacks are supported by solid steel frame. Moving contact assembly is fixed on the central insulator. The fixed contact assemblies and terminal pads are fixed with two extreme insulators.

The merits of centre post-rotating-type outdoor isolators are:

(i) During normally closed position, a high-pressure contact ensures elimination of malfunctioning of the isolators.

(ii) Contact pressure in this isolator gets released before taking place the blades movement in the opening direction.

(iii) In order to achieve gang operation, all the three poles are interconnected. A down rod arranges simultaneous operation of all the three poles.

Figure 2.6 depicts this type of isolator.

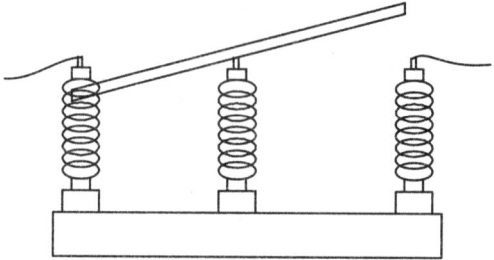

FIGURE 2.6   Centre post-rotating outdoors isolator (open circuit).

Double pole-rotating outdoor isolator

In this type of insulator, each insulator post carries a contact assembly consisting of a terminal pad connected to high voltage conductors. Contact plate and a switchblade are also fixed with contact assembly. Rotating bearing assemblies (two in number) are attached with base frame. Base frame interconnects L-beam. Another switchblade and contact sphere are also fixed with the contact assembly. The two bearings move in opposite direction by motor driving system. Either contact roller or flexible connector is connected with the terminal pads for keeping the HV conductors stationary. The operating mechanism is mounted at the bottom of the frame. The design is such that sufficient contact pressure is maintained and contacts are wiped as well as lubricated during each closing operation. Double pole-rotating outdoor isolators are available in modular pattern and could be used in single, double or triple pole versions. Three poles are generally mounted side by side, however, these can also be mounted in linear formation. Figure 2.7 depicts this type of isolator.

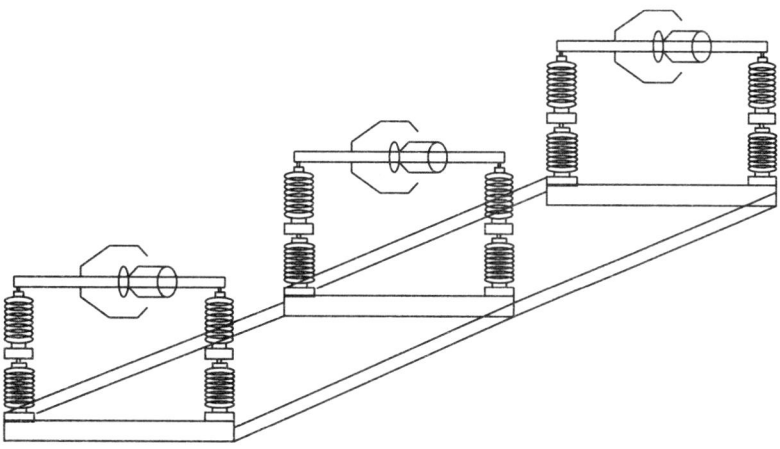

FIGURE 2.7   Double pole-rotating outdoor isolator.

Pantograph outdoors isolator

The constructional profile of this type of isolator is similar to that of the pantograph of an electric locomotive. A three-phase isolator consists of three identical poles. An insulator post supports the pantograph assembly and drives housing of each pole. A pantograph assembly consists of:

(i) Scissors-shaped copper tube linkage

(ii) Conducting joints

(iii) Pantograph contact

When an isolator moves into 'close' position, the pantograph contact move vertically towards the overhead conductor. Adjustment of the position of the overhead conductor should be done in such a manner that in fully closed condition, the tip of the pantograph contacts rise above the tip of the stir-up contacts. Figure 2.8 depicts this type of isolator.

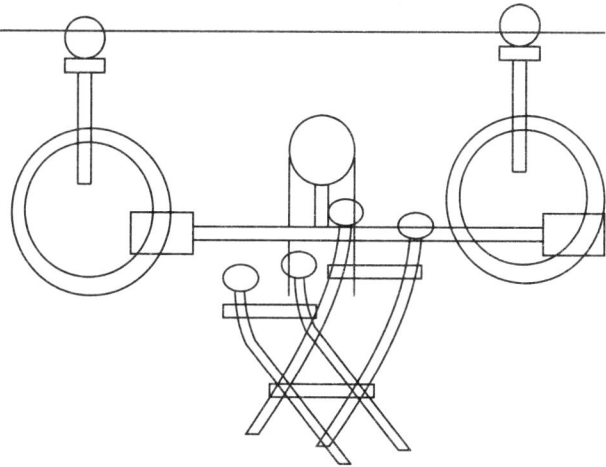

FIGURE 2.8   Pantograph outdoors isolator.

In the fully closed condition, the pantograph contacts firmly grip the stir-up contacts suspended from the overhead conductors. A rotating insulator with some insulating strength of main support insulator transmits the actuation motion to the pantograph drive housing where angular movement of the insulator housing takes place. Since this isolator occupies less floor area hence is widely used in the switchyards.

Rocking or tilting isolator

Rocking or tilting isolators are simplest outdoor isolators and are available for up to 33 kV, 1000 A rating. They are made up by integrating the combination of isolators and dropout fuses. These isolators are known as either rocking-type isolators or tilting isolators. Three isolators are fixed with each pole on the base frame. The fixed contact is attached with top insulator and the other insulator is 'rocking insulator', carries moving contact. Moving contact is connected by flexible braids to the fuse holder, which is supported by a linkage to avoid being hung while isolator is open. Two socket contacts fitted on the bottom insulators hold the dropout fuse. With the help of a hook-stick from the ground, electrician can remove or place the dropout fuse. They are economical, simple, and easy for maintenance and operation.

Single-break isolator

It is suitable for lower voltage applications, i.e. 11 kV and below. Two insulator posts support each pole. One pole is fixed while other is kept free to rotate. Fixed insulator carries fixed contact and rotating insulator carries switchblade and contact plate. Contact plate closes and opens the isolator. Single-break isolators are applied to:

(i) Neutral circuit isolator of the transformer.

(ii) Bus switching of the station earthing.

(iii) Switching of the earthing resistor, etc.

Figure 2.9 shows a single-break isolator.

FIGURE 2.9 Single-break isolator.

2.4
EARTHING SWITCHES

Earthing switches are required in the switchyards and substations. These are also called earthing devices. The main functions of the earthing switches are:

(i) To earth the feeder cables at both the ends of the transmission line while maintenance of any equipment of the substation is undergoing.

(ii) To discharge the static charges due to earth capacitance of the energized long transmission overhead lines or cables.

Figure 2.10 shows the schematic diagram of an earthing switch. During switching 'ON' the far end of the contact arm (earthing blade) engages firmly with the fixed contact. Fixed contact is connected to the earthing circuitry. The earthing system of the installation is connected with the other end of the contact arm. Three earthing blades joined together to make the earthing switch. Earthing switches associated with the isolators are mounted on the isolator near the foot of the pole insulator that carries the earthed terminal. Earthing blade is engaged with the contact assembly. Operating mechanisms of the main isolators and the earthing switches are same.

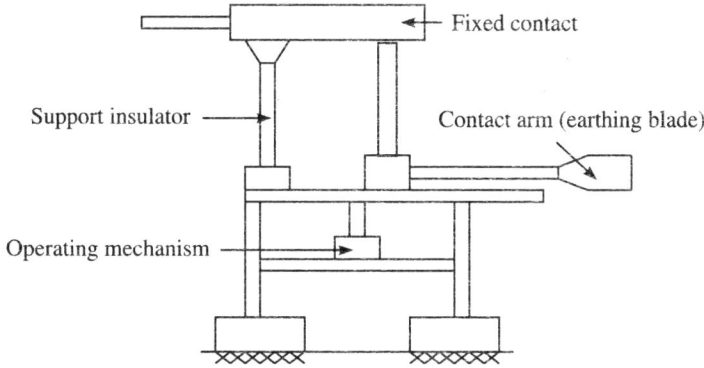

FIGURE 2.10   An earthing switch.

2.5
CONTACTORS

A contactor is a mechanical switching device which is capable of making, carrying, and breaking electric current under normal circuit conditions including operating overload conditions.

2.5.1   Types of Contactors

They are of following two types:

1. Push-button-type contactor
2. Electromagnetic-type contactor

Push-button-type contactor

This provides control of an equipment make or break of an electric circuit by pressing a button.

As long as the push-button is held pressed, it remains pressed. Spring inside the push-button brings the contact back in its initial position as and when the pressure on the push-button is released.

Two sets of contacts, one 'NO' (normally open) and the second 'NC' (normally closed) are generally used. Both the sets are used in the circuit, where one is kept in 'NO' state and the other in 'NC' state.

For alternating current, push-buttons are rated from 2 A to 3 A, 500 V. The contact tips are made up of silver that can withstand several switching operations. Push-button-type contactors are used for starting, stopping, reversing, getting multi-speed of motors and other appliances. Figure 2.11(a) and (b) shows the combination of two sets of push-buttons.

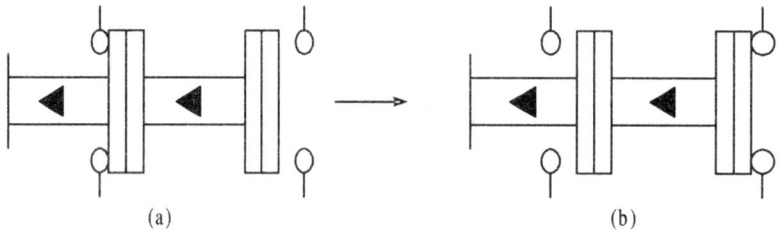

(a)   (b)

FIGURE 2.11   (a) Original position of one 'NO' and the other 'NC' push-button. (b) Push-button in pressed condition.

Electromagnetic-type contactor

This is made to operate by electromagnetic force rather than by hand pressure for making and breaking an electric circuit and its operating speed does not depend on the operator. In this contactor, a coil is connected across a voltage source as per rating of the contactor. For making and breaking of the three-phase supply, three main contacts, one for each phase, are used. The supply source energizes the coil. The main working component, i.e. the contactor coil is available for different voltage ratings of ac, dc or combination of both. Ratings are selected as per requirement. Figure 2.12 shows an electromagnetic-type contactor with a core.

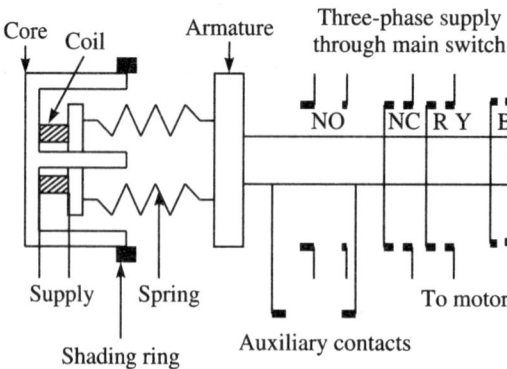

FIGURE 2.12   Electromagnetic-type contactor.

Ratings:

  (i) Rated voltage

  (ii) Rated operational voltage

  (iii) Rated insulation voltage

  (iv) Rated current

  (v) Rated thermal current

  (vi) Rated operational current

  (vii) Rated duty and service conditions

2.6
CIRCUIT-BREAKERS

A circuit-breaker is an automatic device capable of making and breaking an electric circuit under normal and abnormal conditions in the event of overloading or short circuits. A circuit-breaker makes and breaks the electrical circuits as and when required. During opening the circuit, separation of the circuit-breaker contacts is required. Separation of the contact must happen in an insulating fluid, i.e. air, oil or gas. Insulating fluid is used to:

  (i) Extinguish the arc produced due to the fault current interruption.

  (ii) Provide insulation between the contacts and from each contact to the earth.

2.6.1 Insulating Fluids

Fluids suitable for an arc extinction of a circuit-breaker depend on the size and rating of the circuit-breaker. Commonly used insulating fluids in the circuit-breakers are:

  (i) Normal air at atmospheric pressure

  (ii) Compressed air

  (iii) Hydrogen producing oil

  (iv) Ultra-high vacuum

  (v) Sulphur hexafluoride gas (SF_6)

  Gases are most preferred insulating fluids for fast-operating circuit-breakers.

2.6.2 Properties of Insulating and Arc Quenching Materials

  (i) Should have high dielectric strength.

  (ii) Should be thermally and chemically stable.

  (iii) Should be non-inflammable.

  (iv) Should have high thermal conductivity to keep properly the contacts, conductors cool and to help quick arc quenching.

  (v) Should have low dissociation temperature.

(vi) Should not produce carbon during arcing or any conducting products.

(vii) Should be commercially available.

Air is preferred for most of the applications because it can be compressed to extremely high pressure at room temperature. Under compression, its dielectric strength becomes better than the electronegative gases.

Sulphur hexafluoride gas (SF_6) has better insulating and arc quenching properties over air at equal pressure. But the circuit-breakers using SF_6 at 14 kg/cm^2 or more pressure are required to have heaters installed in high-pressure reservoir. Heaters are required to heat up the SF_6 above the pressure 14 kg/cm^2 in order to avoid to be liquefied. The dielectric strength of SF_6 at normal pressure and temperature is about three times that of air. SF_6 also has dielectric strength comparable to the transformer oil at twice of atmospheric pressure.

2.6.3   Initiation of Arc in Circuit-breakers

During separation of the circuit-breaker contacts in the presence of load current or fault current, current carrying contacts generate arc. Generation of arc can be explained as follows.

A voltage gradient of 10^3 kV/cm is enough to initiate the field emission. Availability of this very high voltage gradient for a fraction of microsecond is required to produce an arc through the process of field emission. When separation of the circuit-breaker contacts takes place, at beginning very small gap appears which causes to set up very high voltage gradient. This voltage gradient causes ionization of the medium between the contacts. Initially, a large number of electrons are liberated from the cathode and travel towards anode. While reaching anode, these electrons on the way collide with the molecules and atoms present in the medium and create other electrons, and these in turn take energy from the field, and thus multiply. So, this way discharge occurs due to the emission of the electrons and in the presence of high current, discharge attains sufficiently high temperature to cause thermal ionization. So, it is fact that initially an arc is initiated due to field effect and then maintained due to thermal ionization.

In the presence of the current above 100 A, the constituent gases of air circuit-breakers such as nitrogen, oxygen and copper vapour get dissociated into atoms and the arc gets initiated. The oxygen gas remains dissociated even in the presence of the current of the order of 1 A.

The gases hydrogen, acetylene, methane and ethylene are liberated and the oil of circuit-breakers get heated and consequently decompose the oil.

2.6.4   Arc Interruption

The arc between the circuit-breaker contacts consists of ionized gas particles. An arc interruption is possible, if the contact gap is deionized. Deionization process of the contact gap can be achieved by the following ways:

(i) High pressure

(ii) Forced convention and turbulence

(iii) Arc splitting

Deionizing by applying high pressure depends on the principle of application of gas blast directing along the discharge. Quenching of the discharge and then effective cooling can be achieved by applying gases at very high pressure that is known as *gas blasting*.

Since, gases in the oil circuit-breaker are generated at high pressure, turbulence near the surface of the arc are used in the process of deionization for extinction of the arc.

The third method of deionization leading to extinction of the arc uses two methods, which are:

- Lengthening the arc by forcing into splitters.
- Splitting the arc into smaller arcs.

2.6.5   Current Interruption

The main function of the switchgear is to interrupt the current in either normal condition or in abnormal conditions. As and when the operating mechanism of the breaker gets tripping signal from the auxiliary circuit, the operating mechanism gets activated and the circuit-breaker contacts start separating. During contact separation an arc is produced. This arc essentially requires to be quenched.

Since ac current flows naturally, this provides current zeros at every half cycle. At current zero, its interruption is easier than dc current. So, designer exploits the zero current position. Circuit-breaker contacts open the circuit when current goes to zero value and does not rise again.

At the time of starting the contact separation, a substantial percentage of the system voltage appears between the contacts. Voltage stress generated across the initially made gap breaks down the insulating medium and the electrons are liberated. The liberated electrons move towards positive terminals resulting in an electric arc. The arc elongates as contact go farther. Now arc will get extinguished, if the voltage required across the arc to maintain it is more than the system voltage, otherwise the arc will persist. It is obvious that building-up of the dietetic strength across the contact gap will be highest at current zero. If the dielectric strength of the gap exceeds the voltage stress, the arc cannot restrike. So, by the time current zero occurs, the deionization in the gap completes and current interruption is recommended at the moment when current zero occurs. If this is not achieved at first half cycle of the current zero, the same forces again come in action so interrupt the current forever at one of the subsequent current zero.

2.6.6   Effect of Power Factor

As moving contact starts separating the fixed contact of the circuit-breaker, voltage starts building up between the contacts. The instantaneous value of this voltage depends upon the power factor of the faulted network. If the power factor is poor, at the current zero, voltage remains high, means voltage stress between the contacts is high and the chances of restrike the arc raise. And, if the power factor of the network is good means the current and voltage are in phase or closer at current zero. The voltage magnitude also remains low, which generates low voltage stress between the contacts and less chance to restrike the arc. This is the reason, while designing and testing the circuit-breaker for its interruption performance, the assumed worse case may be the current and voltage with 90°-phase difference at current zero.

An arc voltage does not follow the linear relationship with the circuit current. This is the reason why at the time of opening the circuit-breaker contacts, an arc is produced and the current does not remain sinusoidal.

2.6.7   Parallel Capacitances

Some capacitance in parallel with the circuit-breaker always exists due to:

(i) Capacitance to earth of the line and connection on the generator side of the breaker. This capacitance cannot be avoided when earth fault is on the load side.

(ii) Capacitance between each contact and metallic body of the breaker.

The line capacitance and self-capacitance of the circuit-breaker as shown in circuit diagram [Figure 2.13(a) and (b)] are effectively in parallel with the arcs. The contact gap voltage after current zero and the arc current just before the current zero are influenced by these capacitances.

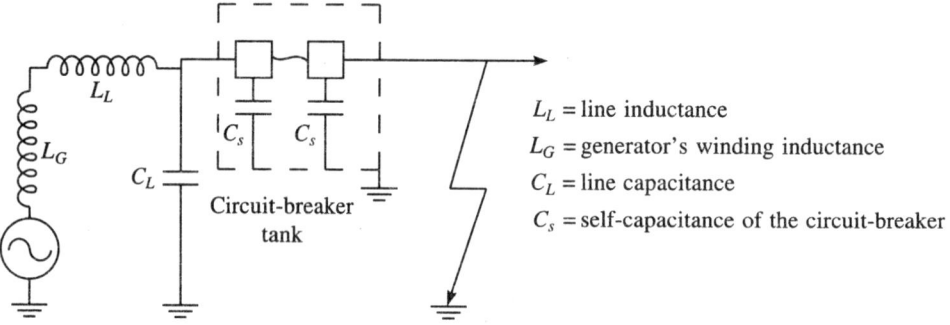

L_L = line inductance
L_G = generator's winding inductance
C_L = line capacitance
C_s = self-capacitance of the circuit-breaker

FIGURE 2.13(a)   Actual circuit configuration.

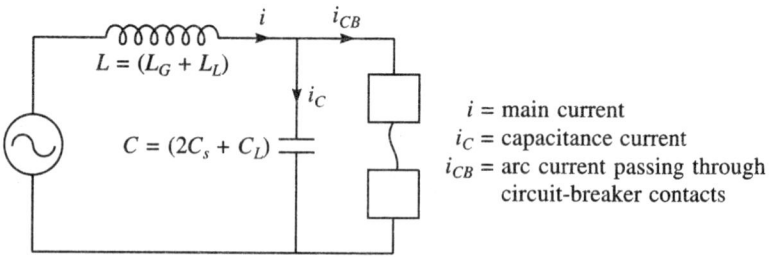

i = main current
i_C = capacitance current
i_{CB} = arc current passing through circuit-breaker contacts

FIGURE 2.13(b)   Simplified circuit configuration.

If the inductive reactance of L counterbalances the capacitive reactance of C, the power factor of the network is unity, which makes the duty of the circuit-breaker little easier.

2.6.8   Arc Interruption Theories

An arc interruption between the circuit-breaker contacts can be done by the following two methods:

1. Deionizing method also known as 'high resistance method'. This method includes cooling, lengthening and splitting of the arc.
2. Current zero method also known as 'low resistance method'.

The phenomenon of arc extinction is described by the following two theories:

1. Energy balance theory
2. Voltage race theory

Energy balance theory

According to this theory, if the rate of heat generated between the contacts is less than the rate of heat dissipation, the arc tends to be extinguished. In the reverse situation, when the rate of heat generated between the contacts is more than the rate of heat dissipation, the arc tends to restrike. Heat generated between the contacts is maximum between the two limits, first at the initial state, when the contacts are about to open (the restriking voltage is zero) and after fully opening the contacts when a very high resistance appears between the contacts, which, in turn, develops enough dielectric strength to check the restriking of the voltage. Heat generated between these two limits is suppressed and removed by cooling, lengthening and splitting the arc at a faster rate than the heat generated by the arc discharges. The arc thus gets extinguished.

Voltage race theory

An arc extinction depends on the complete deionization at current zero. Ionization of the medium between the contacts depends on the voltage appeared between the contacts, known as *restriking voltage*. For loss/less system, restriking voltage v at any instant t is expressed as:

$$v = V \left[1 - \cos \frac{t}{\sqrt{LC}} \right]$$

where

V = value of the voltage at the instant of interruption

L = series inductance of the line up to fault point

C = shunt capacitance of the line up to fault point

If the fault is near the source, means values of L and C are small, from the expression above; one can understand that for lower values of L and C, the value of restriking voltage v will be higher. The value of restriking voltage v is also function of the system voltage at the instant of arc interruption v. If the power factor of the system is poor say highly lagging, the voltage V will correspond to the peak system voltage. But it does not always happen. Fault can occur at any instant and the fault can have any degree of asymmetry. As circuit-breaker interrupts at current zero, the recovery voltage oscillates around the instantaneous value of the supply. At current zero, if the power factor is unity, the system voltage will also be at its zero and the chance of restriking voltage will become nil. Restriking voltage, arc voltage, fault current and recovery voltage along with the system voltage are shown in Figure 2.14.

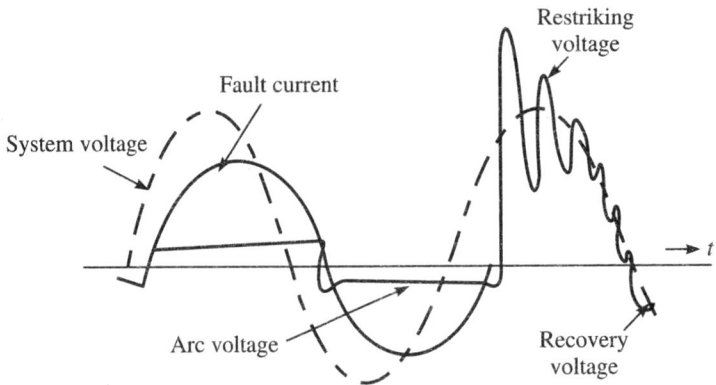

FIGURE 2.14   Arc voltage, restriking voltage and recovery voltage.

2.6.9   Current Chopping

Current chopping is the breaking of fault current before it passes through the neural zero. This happens when low-inductive current is interrupted. The example may be the current due to no load magnetizing current of a transformer or the current passing through a circuit-breaker producing varying degree of deionizing force.

Figure 2.15(a), (b) and (c) helps in describing the phenomenon of current chopping. If we look Figure 2.15(a) and (b), we see that the arc current decreases to zero in normal fashion. It is because as the temperature increases the arc resistance increases. In the beginning, arc voltage is small (G and H are the beginning points in the figure). Now, in the event of the appearance of disproportionate high deionizing force, at certain value of arc current, the arc destabilizes collapsing arc current to zero before natural zero [see the instant u in Figure 2.15(a)]. This is first chop. Referring the equivalent circuit [Figure 2.15(c)], inductance L opposes the rate of fall of current and hence the current through L cannot cease to flow instantly. This current gets directed towards capacitance C. Now, capacitance voltage, parallel to circuit-breaker contacts rises with a rate, such as $dv/dt = i_{arc}/C$. Since, this voltage rise is parallel to the contacts of the circuit-breaker, it means this rising voltage appears across the separating contacts and escalates the rate of fall of the current flowing through the L.

Since the rate of voltage rise is inversely proportional to the capacitance, the low value of the capacitance may cause excessive voltage rise, which may in turn makes the circuit-breaker to restrike at the low-system voltage. So, the involvement of low inductance and capacitance may cause frequent current chopping, finally lead to dangerous resonance conditions in the system. Dangerous resonance conditions are undesirable which cause harmful overvoltages. By decreasing the value of parallel capacitance, the maximum value of instantaneous current to be chopped can be decreased. For smaller magnetizing current of the transformer, the rate of rise of voltage on current collapse can be damped by the effects of eddy currents and hysteresis loss.

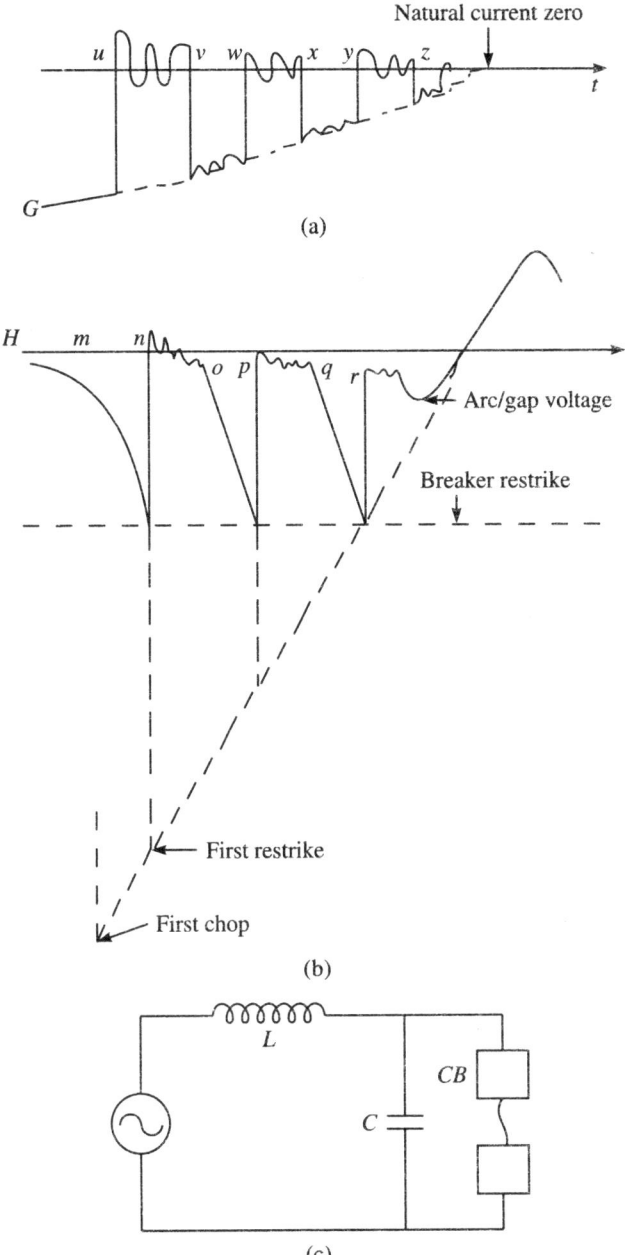

FIGURE 2.15(a), (b) and (c) Current chopping.

If i_a is the instantaneous value of the arc current, where chopping takes place, the prospective value of the voltage to which the capacitor will be charged, will be:

$$V = i_a \sqrt{\frac{L}{C}}$$

2.6.10   Resistance Switching

Very high overvoltage due to current chopping should be avoided, otherwise it may endanger the system. A resistance across the contacts of the breaker, as shown in Figure 2.16, can be connected to reduce the dangerously high voltages.

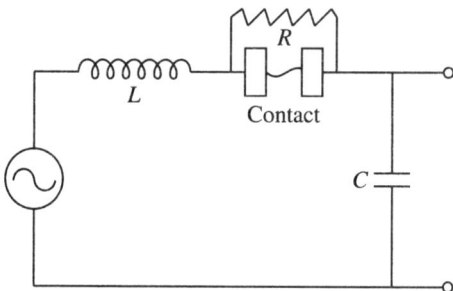

FIGURE 2.16   Resistance switching.

Introduction of the resistance R in the circuit helps the circuit-breaker in the following ways:

(i) It reduces duties of the breaker by reducing the rate of rise of the restriking voltage.

(ii) Resistor R helps in distributing the transient recovery voltage uniformly across the gaps of a multibreak circuit-breaker.

(iii) It reduces transient voltages during switching out inductive or capacitive loads.

The parallel resistance R with the circuit-breaker damps the restriking voltage and prevents it from its oscillations. It is known as *critical resistance*. The value of the critical resistance will be:

$$R = \left(\frac{1}{2}\right)\sqrt{\frac{1}{C}} \qquad \text{when } C \text{ is in parallel}$$

$$= 2\sqrt{\frac{1}{C}} \qquad \text{when } C \text{ is in series}$$

The damped frequency of the oscillation of the RLC circuit:

$$f_d = \frac{1}{2\pi}\sqrt{\frac{1}{LC} - \frac{1}{4R^2C^2}}$$

2.6.11   Capacitance Current Breaking

Capacitance between the line conductor and the earth wire of the long unloaded transmission line results in small capacitive current. This current encounters with voltage during breaking the circuit. Considering the line capacitance, the equivalent circuit is shown in Figure 2.17.

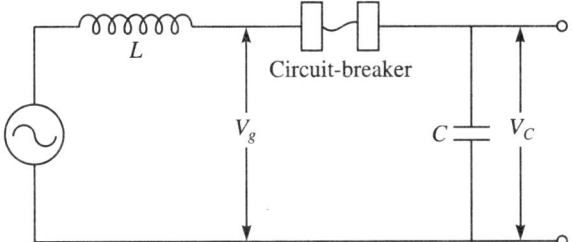

FIGURE 2.17 An equivalent circuit for capacitance current breaking.

Here, in both the contacts of the circuit-breaker, the generator side contact remains at generator voltage (V_g) and the other contact at line voltage say V_C.

The voltage difference between the two contacts of the circuit-breaker may cause considerable current flow between the contacts and obstruct the deionization of the medium. So, to avoid striking of the gap, the breaker resistance must be sufficiently increased.

2.6.12 Restriking Voltage

Restriking voltage is a phenomenon corresponding to the post-current zero situations. At current zero, as the arc gets quenched a voltage appears across the contact gap. The appearance of transient voltage across the breaker contacts tries to restrike the arc and the voltage is known as *Restriking Voltage*. An illustration of the waveforms of the different voltages involved during arc extinction in the circuit-breaker is shown in Figure 2.18. The trend of voltage rise from current zero to the definite value is dependent on the power factor of the circuit and presence of inductance and capacitance. A circuit in the absence of the capacitance may cause instantaneous voltage rise. There will be no time available for building-up of the dielectric strength. In this situation, restriking the arc cannot be avoided.

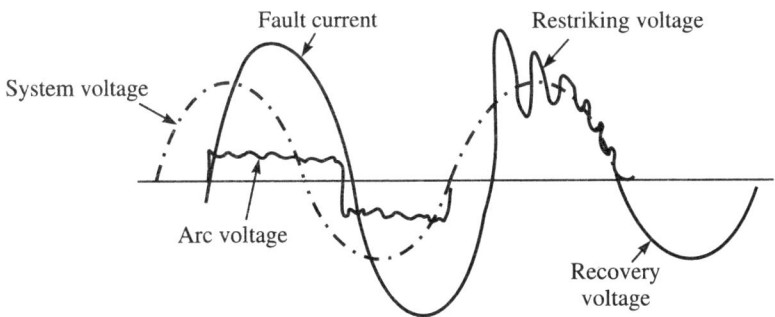

FIGURE 2.18 Restriking voltage waveforms.

Rate of rise of restriking voltage (RRRV)

Figure 2.19(a) and (b) shows an equivalent circuit of a circuit-breaker connected in the system under fault.

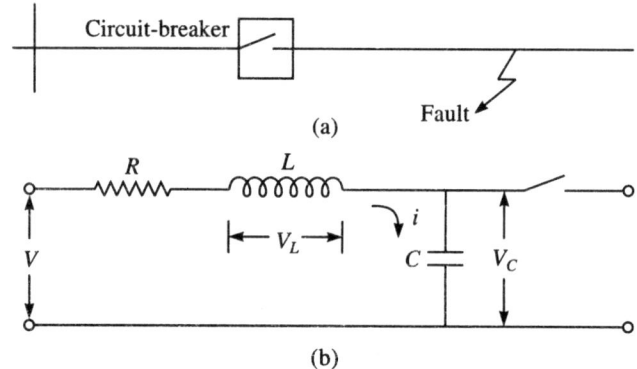

FIGURE 2.19(a) and (b)   An equivalent circuit of a circuit-breaker, connected in the system.

As the circuit-breaker contacts close, the current flows through the resistance and inductance (L) of the line. If the circuit-breaker contacts get fully open, the current gets diverted through the capacitance (C) of the line, as shown in Figure 2.19(b). Oscillations in the current and voltage waves take place due to the presence of inductance (L) and capacitance (C). Voltage across capacitance (C) is restriking voltage. For the transmission lines $R << L$; resistance of the line is neglected during analysis. So, the systems steady state voltage:

$$v = V_m \sin \omega t = V_L + V_C = L\left(\frac{di}{dt}\right) + V_C$$

and

$$i = \frac{dq}{dt} = \frac{d(CV_C)}{dt}$$

Again

$$\frac{di}{dt} = \frac{d^2(CV_C)}{dt^2} = C\left(\frac{d^2V_C}{dt^2}\right)$$

So,

$$v = V_m \sin \omega t = LC\left(\frac{d^2V_C}{dt^2}\right) + V_C$$

Solution of the above equation gives the restriking voltage V_C:

$$V_C = V_m (1 - \cos \omega t) \tag{2.1}$$

Now, the rate of rise of restriking voltage (RRRV):

$$\frac{dV_C}{dt} = \frac{d[V_m(1 - \cos \omega t)]}{dt} = \omega V_m \sin \omega t$$

or

$$\frac{dV_C}{dt} = \omega V_m \sin \omega t \tag{2.2}$$

From Eqs. (2.1) and (2.2), the maximum value of restriking voltage:

$$(V_C)_{max} = 2V_m$$

and the maximum value of the rate of rise of restriking voltage:

$$(RRRV)_{\text{max}} = \omega V_m = \frac{V_m}{\sqrt{LC}}$$

This is maximum when

$$\frac{t}{\sqrt{LC}} = \frac{\pi}{2}$$

or

$$t = \frac{\pi\sqrt{LC}}{2}$$

Here, natural frequency of oscillations:

$$f_n = \frac{1}{2\pi\sqrt{LC}}$$

Effect of resistance switching on RRRV: Parallel resistance across the circuit-breaker contacts, indirectly helps in increasing the arc resistance. It is because parallel resistance R decreases the current in the arc and hence ionization rate gets decreased increasing the arc resistance. Decreased ionization process helps in reducing the maximum restriking voltage and RRRV.

Recovery voltage

Power frequency rms voltage, which appears across the circuit-breaker contacts after the transient oscillations die out and final extinction of arc has resulted in all the poles, is known as *Recovery Voltage.*

Instantaneous recovery voltage at the instant of arc extinction is known as *Active Recovery Voltage.*

Example 1   An instantaneous value of magnetizing current of 5.5 A in a 132 kV transmission line having line to ground capacitance of 0.015 µF and inductance of 5 H is required to be interrupted. Determine:

(i) The voltage across the contacts of the circuit-breaker at the time of current interruption.

(ii) The resistance to be inserted across the contacts in order to avoid restriking voltage.

Solution

(i) Voltage appearing across the circuit-breaker contacts at the time of current interruption:

$$V = i_a \sqrt{\frac{L}{C}}$$

$$= \frac{5.5\sqrt{5}}{0.015 \times 10^{-6}}$$

$$= 5.5 \times 1.8 \times 10^4$$

$$= 99 \text{ kV}$$

(ii) Since capacitance is in parallel, so to avoid restriking voltage, the required resistance across the circuit-breaker contacts:

$$R = \frac{1}{2}\sqrt{\frac{L}{C}}$$

$$= \frac{0.5\sqrt{5}}{0.015 \times 10^{-6}}$$

$$= 0.5 \times 1.8 \times 10^{4}$$

$$= 9 \text{ k}\Omega$$

Answer: (i) 99 kV, (ii) 9 kΩ

Note: Using the following MATLAB programming, we can solve the above problem.

```
clear;
clc;
% -------data given----------
ia=5.5;
% ia=instantaneous value of magnetizing current
C=0.015*(10^-6);
% C= Line to ground capacitance
L=5;
% L= transmission line inductance
% -----------solution----------
V=ia*sqrt(L/C)
% V= Voltage appearing across the circuit-breaker contacts
% at the time of current interruption
R=(1/2)*sqrt(L/C)
% R= Resistance to be inserted across the contacts
% in order to avoid restriking voltage
```

Example 2   For a 132 kV system, calculate:
   (i) The frequency of oscillations.
   (ii) The maximum value of restriking voltage.
   (iii) The maximum value of RRRV.

The inductance and capacitance of the transmission line up to fault point are 0.02 H and 0.025 μF respectively.

Solution

(i) The frequency of oscillations:

$$f_n = \frac{1}{2\pi\sqrt{LC}}$$

$$= \frac{1}{2\pi\sqrt{0.02 \times 0.025 \times 10^{-6}}}$$

$$= \frac{1}{2\pi \times 22.3 \times 10^{-6}}$$

$$= 7.14 \text{ kHz}$$

(ii) The maximum value of restriking voltage:

$$2V_m = 2 \times \sqrt{2} \times \frac{132}{\sqrt{3}}$$

$$= 215.8 \text{ kV}$$

(iii) The maximum value of RRRV:

$$\omega V_m = 2\pi f_n V_m$$

$$= 2\pi \times 7140 \times \sqrt{2} \times \frac{132}{\sqrt{3}}$$

$$= 4.84 \text{ kV/}\mu\text{s}$$

Answer: (i) 7.14 kHz, (ii) 215.8 kV, (iii) 4.84 kV/μs

Note: Using the following MATLAB programming, we can solve the above problem.

```
clear
clc
% -------data given---------
L=0.02;
C=0.025*(10^-6);
% L and C are the inductance and capacitance of the
  transmission line up to
% fault point
V=132*(10^3);
% V=System voltage (line to line)
% ----------solution-----------
Vm=sqrt(2)*V/sqrt(3);
% Vm= maximum system voltage (phase to phase)
fn=1/(2*pi*(sqrt(L*C)))
% fn=frequency of oscillation
Vr=2*Vm
% Vr=maximum value of restriking voltage
W=2*pi*fn;
RRRVmax=W*Vm
% RRRVmax= Maximum value of RRRV
```

Example 3 For a 132 kV, 50 Hz transmission line, the reactance and capacitance up to the fault point from the location of the circuit-breaker are 10 Ω and 0.35 μF respectively and 500 Ω switching resistor is connected externally. Calculate:

(i) The damped frequency of oscillations.

(ii) The critical resistance.

(iii) The switching resistance, if the damped frequency of oscillation is one third of the natural frequency of oscillation.

Solution

(i) Damped frequency of oscillations:

$$f_d = \frac{1}{2\pi}\sqrt{\frac{1}{LC} - \frac{1}{4R^2C^2}}$$

and

$$L = \frac{X_L}{2\pi f} = \frac{10}{2\pi \times 50} = 0.032 \text{ H}$$

So,

$$f_d = \frac{1}{2\pi}\sqrt{\frac{1}{0.032 \times 0.035 \times 10^{-6}} - \frac{1}{4 \times 500^2 \times (0.035 \times 10^{-6})^2}}$$

$$= 1.393 \text{ kHz}$$

(ii) Critical resistance:

$$R = 0.5\sqrt{\frac{L}{C}} = 0.5\sqrt{\frac{0.032}{0.035 \times 10^{-6}}}$$

$$= 0.478 \text{ k}\Omega$$

(iii) Switching resistance:

Given:

$$\text{Damped frequency of oscillations} = \frac{\text{Natural frequency of oscillations}}{3}$$

$$= \frac{1}{3} \times \frac{1}{2\pi} \times \sqrt{\frac{1}{LC}}$$

$$= \frac{1}{6\pi}\sqrt{\frac{1}{0.032 \times 0.035 \times 10^{-6}}}$$

$$= 1.59 \text{ kHz}$$

So,

$$\frac{1}{2\pi}\sqrt{\frac{1}{LC} - \frac{1}{4R^2C^2}} = 1.59 \times 10^3$$

$$\frac{1}{2\pi}\sqrt{\frac{1}{0.032 \times 0.035 \times 10^{-6}} - \frac{1}{4 \times R^2 \times (0.035 \times 10^{-6})^2}}$$

$$= 1.59 \times 10^3$$

Hence,

$$R = 507.61 \ \Omega$$

Answer: (i) 1.393 kHz, (ii) 0.478 kΩ, (iii) 507.61 Ω

Note: Using the following MATLAB programming, we can solve the above problem.

```
clear;
clc;
% -------data given--------
Xl=10;
C=0.035*(10^-6);
% Xl & C are the reactance and capacitance up to the fault
   point
%from the location of the circuit-breaker respectively
V=132*(10^3);
Rs=500;
f=50;
% Rs is the switching resistor, f and V are the system
   frequency and voltage respectively
% -----SOLUTION-------
% (i)Damped frequency of oscillations
L=Xl/(2*pi*f);
fd=(1/(2*pi))*sqrt((1/(L*C))-(1/(4*Rs^2*C^2)))
% (ii)Critical resistance
Rc=0.5*sqrt(L/C)
% (iii)Switching resistance (Rs1), if damped frequency
% of oscillation is one third of the natural frequency of
   oscillation
fn=(1/(2*pi))*sqrt(1/(L*C));
% fn is the natural frequency of oscillation
fd1=fn/3;
Rs1=(1/(2*C))*sqrt(L*C/(1-(4*pi^2*fd1^2*L*C)))
```

Example 4   Determine the RRRV for the circuit-breaker installed on a 400 kV, 3ϕ, 50 Hz system. Following data were recorded, when a short circuit grounded fault occurred. Given:

Recovery voltage = 0.97 of full line value

Power factor of the fault = 0.45

Natural frequency for symmetrical breaking current = 16 kHz

Solution   The peak value of line to neutral voltage = $\sqrt{2} \times \dfrac{400}{\sqrt{3}}$ = 327 kV

And, recovery voltage = 0.97 × 327 = 317 kV

Active recovery voltage = $V_m \sin \phi$ = 317 × 0.893

$$= 283.1 \ kV$$

where $\phi = \cos^{-1} 0.45 = 63.26°$ and hence $\sin \phi = 0.893$

Now, the maximum restriking voltage $(v) = 2 \times$ active recovery voltage

$$= 2 \times 283.1$$

$$= 566.2 \text{ kV}$$

So, $\text{RRRV} = 2f_n V$

$$= 2 \times 16 \times 10^3 \times 566.2 \text{ kV/s}$$

$$= 18.118 \text{ kV/}\mu\text{s}$$

Answer: 18.118 kV/μs

Note: Using the following MATLAB programming, we can solve the above problem.

```
clear
clc
%---------data given-----
V=400;
f=50;
pf=0.45;
% V and f are the system voltage and frequency respectively
  and
% pf is the power factor of the fault
fn=16*10^3;
% fn=Natural frequency for symmetrical breaking current
% ---------SOLUTION----------------------
Vm=sqrt(2)*V/sqrt(3);
% Vm is the peak value of line to neutral voltage
Vr=0.97*Vm;
% Vr is the recovery voltage= 0.97*Vm, given
phi=acos(pf);
% phi is the phase angle
Vra=Vr*sin(phi);
% Vra is the active recovery voltage
Vrm=2*Vra;
%Vrm is the the maximum restriking voltage
RRRV=2*fn*Vrm
```

Example 5 A 3-pole, 132 kV circuit-breaker gave the following results, after conducting the short circuit test:

(i) The recovery voltage = 0.95 times full line value.

(ii) Power factor of the fault = 0.45.

(iii) Natural frequency of oscillations of restriking voltage = 16 kHz.

(iv) Multiplying factor = 1.5 for isolated 3-phase fault.

Assume breaking current is a symmetrical current and fault does not involve ground through the grounded neutral. Calculate the average rate of rise of restriking voltage.

Solution Peak voltage of the line to ground $(V_{L-G})_{\text{peak}} = \sqrt{2} \times \dfrac{132}{\sqrt{3}} = 107.77$ kV

Instantaneous value of recovery voltage $(V_r) = KV_m \sin \phi$

Here,

$$K = 1.5$$

$$V_m = 0.95 \times 107.77 = 102.39 \text{ kV}$$

and

$$\sin \phi = \sin (\cos^{-1} 0.45) = 0.893$$

So,

$$V_r = 1.5 \times 102.39 \times 0.893$$

$$= 137.15 \text{ kV}$$

And natural frequency of oscillations

$$f_n = \frac{1}{2\pi} \times \sqrt{\frac{1}{LC}} = 16 \times 10^3$$

So,

$$\frac{1}{\pi} \times \sqrt{\frac{1}{LC}} = 2 \times 16 \times 10^3 = 32 \times 10^3$$

Now,

$$(\text{RRRV})_{\text{average}} = \frac{2 \times V_r}{\pi \times \sqrt{(LC)}} = 2 \times 137.15 \times 32 \times 10^3 \text{ kV/s}$$

$$= 8.778 \text{ kV/}\mu\text{s}$$

Answer: 8.778 kV/μs

Note: Using the following MATLAB programming, we can solve the above problem.

```
clear
clc
%---------data given---------
V=132*10^3;
fn=16*10^3;
% V and fn are the system voltage and natural frequency of
   oscillation respectively
pf=0.45;
k=1.5;
% k and pf are the multiplying factor and power factor
   respectively
% -------------solution---------
phi=acos(pf);
% phi is the phase angle
Vmg=sqrt(2)*V/sqrt(3);
% Vmg is the Peak voltage of the line to ground
Vm=0.95*Vmg;
% Vm is the recovery voltage
Vr=k*Vm*sin(phi);
% Instantaneous value of recovery voltage
RRRV=2*Vr*2*fn
```

2.6.13 Current Zero Pause

Performing detailed analysis and thereafter having proper information about types of circuit-breakers, magnitude of the fault and circuit conditions facilitates to understand the little bit complicated phenomenon of around current zero (any instant near zero). Generally, at current zero say at m in Figure 2.20, breakdown of the medium takes place and current rises to full value. At the next current zero, i.e. after completion of the half-cycle say at instant n, the current is successfully supposed to be interrupted. Due to more active deionizing effect before current zero (the instant n), fast current decay takes place. Parallel capacitance plays role in lengthening and decaying current zero. The gap tries to build up dielectric strength after current zero, but does not get success in the presence of parallel capacitance. *So, the phenomenon of small current flow over a rather long duration is known as Current Zero Pause.*

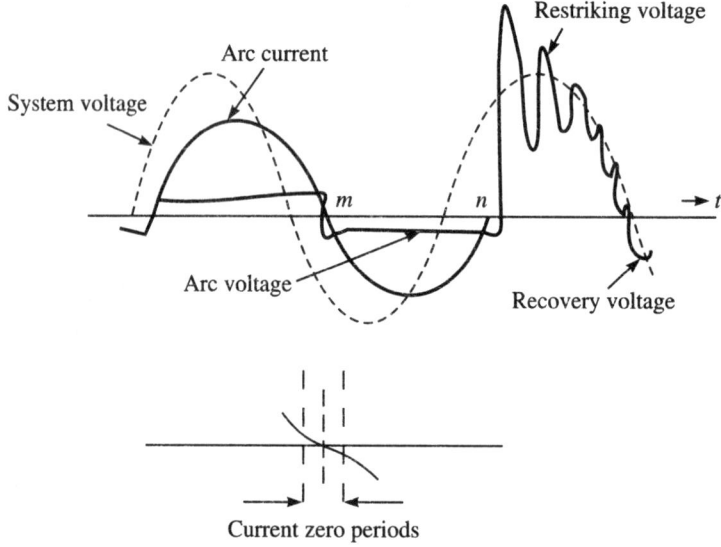

FIGURE 2.20   Waveforms of the current zero pause.

The rate of ionization and deionization processes depends on dielectric strength of the medium of contact gap which is known as *residual column* at current zero. The rising voltage after current zero tends to feed energy to the residual column. This energy can be lessening by cooling the residual column by rapidly lengthening the arc, by air blasting or by convention. So, in other words, it can be said that if the cooling of the column is dominantly effective, i.e. energy from the column is effectively removed, the fault current will be successfully interrupted.

Spark breakdown takes place during excessive ionization in the residual column after current zero, and if the rate of input energy is more than the rate at which it is removed, temperature of the residual column will continuously increase and thermal breakdown will take place.

2.6.14   Main Parts of Circuit-breakers

(i) Poles with interrupter, porcelain support, arc quenching medium, etc.

(ii) Operating mechanism

(iii) Support structure, and

(iv) Control cabinet

2.6.15   Main Types of Circuit-breakers

(i) Miniature circuit-breaker

(ii) Air circuit-breaker

(iii) Oil circuit-breaker

(iv) Minimum oil circuit-breaker

(v) Air blast circuit-breaker

(vi) Sulphur hexafluoride gas (SF_6) circuit-breaker

(vii) Vacuum circuit-breaker

2.6.16   Rating of Alternating Current Circuit-breaker

Each ac circuit-breaker should have the following specified values:

(i)	Rated voltage	240/415 V (ac), 220 V (dc)
(ii)	Rated normal current	0.5 A to 63 A
(iii)	Protection class	P 30 as per IS 2147
(iv)	Rated frequency	50 Hz to 60 Hz
(v)	No. of poles	three
(vi)	Rated breaking current	9 kA at 240/415 V
(vii)	Permissible ambient temperature	5 + 55°C
(viii)	Connecting conductor size	as per requirement, say up to 35 mm^2
(ix)	Electrical endurance	more than 50,000 operating cycles
(x)	Mechanical endurance	more than one lakh operating cycles

2.6.17   Miniature Circuit-breaker (MCB)

Miniature circuit-breaker provides overload and short circuit protection and works as a circuit-breaker. It also serves the purpose of fuse units during normal operating conditions. But it does not require servicing or rewiring as in the case of fuse units. MCB gives better service for protection because this operates just on 5 per cent to 15 per cent excessive overloads. Fuse wire of the fuse units melt on 50 per cent to 100 per cent load or on even more. It is much faster than fuses. So, MCB is more reliable and easy to handle.

Miniature circuit-breakers are basically electromechanical device. They operate (disconnect) the circuit, if the circuit current reaches the predetermined value. MCBs are the better choice to be fitted on the consumer's distribution board. MCBs operate accurately and efficiently during overloading and short circuit conditions both. Fuses operate in mostly short circuit conditions and in excessive overloading conditions. Looking on the operating knob that moves automatically to the 'OFF' position can identify the tripped MCB. A single pole MCB unit is shown in Figure 2.21 and cross-sectional view of a MCB is shown in Figure 2.22. An arc produced due to the current interruption is broken down into small pockets of arcs in the arc-chute stack in order to reduce its impact.

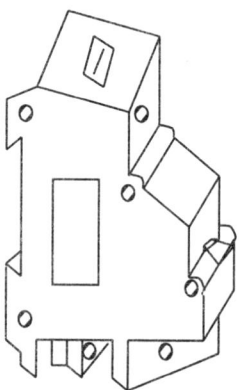

FIGURE 2.21 A single pole MCB unit.

The arc produced at the time of separation of the contacts actually moves very fast under the influence of magnetic field and is directed towards the arc-chute stack. Increase in the number of arc-chute plates reduces the plate voltage and quickly extinguishes the arc. A non-melting material (silver graphite) is welded with contact, which makes the contact tips. Properties of the housing material of the MCB are: it is flame-retardant and water non-absorbent, have high-melting temperature; high dielectric strength and low coefficient of linear thermal expansion. So, high strength plastic is used as housing material.

Angular vents are provided to prevent any ingress dust and to make the electrical contact dust-free.

The moving parts are encased in a sheet metal casing. Plastic casing cannot withstand several operations of the MCB.

Uses of miniature circuit-breakers

Miniature circuit-breakers are being used extensively, replacing the conventional rewirable fuses. They are required in homes, shops, distribution boards, offices, power loads, etc.

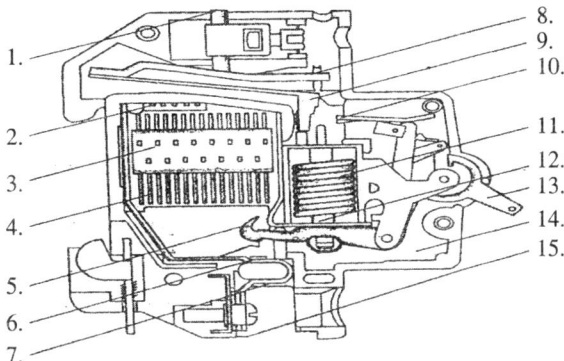

1. Outgoing terminal, 2. Angular vents, 3. Arc-chute holder, 4. Arc-chute, 5. Moving contact, 6. Silver graphite contact tip, 7. Fixed contact, 8. Bimetal carrier, 9. Bimetal strip, 10. Trip lever, 11. Solenoid, 12. Plunger, 13. Operating knob, 14. Toggle, 15. Incoming terminal

FIGURE 2.22   Cross-sectional view of MCB.

Types of miniature circuit-breakers

They are categorized into following two series:

(i) *L-series MCBs:*   They are suitable for low and steady currents like geysers, bulbs, electric iron, oven, heaters, meters, cables, wires, etc.

(ii) *G-series MCBs:*   They are suitable for high inrush current peaks. They are applied for protecting the air conditioners, motors, halogen lamps, machine tools, etc.

2.6.18   Residual Current Circuit-breaker (RCCB)

The other type of circuit-breaker is RCCB (Residual Current Circuit-breaker), which is also known as ELCB (Earth Leakage Circuit-breaker).

Two wires, one for phase and the other for neutral, are used to carry current in electrical appliances. Leakage current flows if the metallic part of the appliance comes in direct contact with the naked part of the phase wire that may be due to the insulation failure or accident. In the case, if the metallic part of the appliance comes in touch with an earthed object, the leakage current passes to the ground. Absence of proper earthing and shielding of this path, may cause fire hazards and undesirable overloading. The earth leakage is the main cause of electric shock. So, this type of leakage and its after-effects may be avoided by using the RCCB, popularly known as ELCB. Its construction is more or less same as described earlier and is useful in protecting the following:

(i) Protection against fire.

(ii) Protection against indirect contact.

(iii) Protection against electrocution to human and domestic cattle.

(iv) Also protects from energy loss otherwise undetected leakage current causes continuous energy loss that may be in small amount but longer duration energy loss becomes considerable enough.

EXERCISES

1. With the help of a neat sketch, describe the pantograph-type isolator.

2. What are the main functions of isolators? Explain the different types of isolators.

3. How are isolators different from load break switches?

4. How are isolators different from earthing switches?

5. Describe the tilting-type isolator with the help of a neat sketch.

6. Write in brief about current-breaking capacity of isolators.

7. Write short note on 'Single-break Isolator'.

8. Describe in brief the 'Current Zero Pause'.

9. Explain the phenomenon of 'Current Chopping'.

10. How power factor influences current interruption in circuit-breakers.

11. Describe the causes and after-effects of restriking voltage.

12. Describe the properties of insulating and arc quenching materials.

13. Explain push-button-type contactor and electromagnetic contactor.

14. Write short note on the following:
    (i)   Transient restriking voltage
    (ii)   HRC fuses
    (iii) Earthing switch
    (iv)   RCCB or ELCB
    (v)   MCB
    (vi)   RRRV

15. A three-phase alternator of rating 66 kV, 50 Hz has an earthed neutral. Per phase inductance and capacitance of the system are 7 mH and 0.01 μF respectively. Following observations are achieved by performing the short circuit test:
    Pf of the fault = 0.25
    Fault current symmetrical recovery voltage = 90% of full line voltage
    Fault is isolated from the ground.
    Calculate the RRRV.
    Answer: 5.34 kV/μs

16. A circuit-breaker is designed to interrupt a transformer magnetizing current (rms) of 11 A, chops the current at an instantaneous value of 7 A. The values of inductance and capacitance of the circuit are 35.2 H and 0.0023 μF respectively, and the entire inductive energy is transferred to the capacitance. Calculate the value of voltage that appears across the contacts of the circuit-breaker.
    Answer: 866 kV

17. Following observations were achieved on a single-frequency transient during short circuit test on a circuit-breaker:

Time to reach the peak restriking voltage = 40 μs

Peak restriking voltage = 100 kV

Calculate (i) the average RRRV, and (ii) the frequency of oscillation.

Answer: (i) 2.5 kV/μs, (ii) 12.5 kHz

Chapter 3

High Voltage Circuit-breakers

3.1

INTRODUCTION

Electrical equipment and appliances are damaged by overloading, short circuiting and by some other faults. Incipient fault may become serious if the faulty equipment does not get isolated immediately the fault is detected. Switchgear and protection systems perform the functions of detection of the fault and isolation of the affected equipment or circuit.

High voltage circuit-breakers in association with the proper protective schemes are performing very complex and tough operations of switching circuits. Without these breakers, power system network could not have reached at present prestigious level of supplying bulk power safely, with extreme precision, speed and reliability.

In the beginning, knife switches and mercury switches were performing the switching operations. Rewirable fuses were used for protection purpose. With the growth of power transmission systems, these ordinary switches and fuses could not handle the large power transmission. As an alternative, first *oil circuit-breaker* (OCB) was developed in Europe in the early 20th century. Oil circuit-breakers were enclosed in a tank filled with transformer oil. Transformer oil was required to provide insulation as well as arc extinction medium.

Oil circuit-breakers work satisfactorily but have demerits also like fire hazards, therefore, they are not recommended for EHV applications. *Air blast circuit-breakers* were developed in 1925 and are popular for EHV applications due to elimination of fire hazards. The most advanced circuit-breakers are SF_6 circuit-breakers. They are in operation since 1945. Vacuum circuit-breakers are the latest developed circuit-breakers which are in use for UHV.

Following are the types of circuit-breakers which are operational in high voltage substations:

    (i) Bulk oil circuit-breaker

    (ii) Minimum oil circuit-breaker

    (iii) Air circuit-breaker

(iv) Air blast circuit-breaker

(v) Vacuum circuit-breaker

(vi) SF$_6$ circuit-breaker

(vii) HVDC circuit-breaker

3.2
BULK OIL CIRCUIT-BREAKER (BOCB)

Bulk oil circuit-breakers are also known as *tank circuit-breakers*. For insulation and arc quenching purposes, mineral oil is used. They are available for 500 V to 220 kV applications.

3.2.1   Advantages of Bulk Oil Circuit-breaker

(i) Reliable and relatively free from current chopping phenomenon.

(ii) Simple and robust construction.

(iii) Relatively cheaper, maintenance easy.

(iv) Unaffected by air pollution.

(v) Available for a large range of current and voltage ratings.

(vi) Does not need support facilities like compressed air or high pressure insulating medium.

3.2.2   Limitations of Bulk Oil Circuit-breaker

(i) Not suitable above 220 kV.

(ii) Arcing problems as arcing by-products are explosive and not suitable for auto-reclosing duty.

(iii) Requires frequent maintenance.

(iv) For extra high voltages, volume of the breaker becomes economically unfeasible.

(v) Longer breaking time in comparison to other types of breakers.

(vi) Not suitable for frequent switching applications such as furnaces, railway drives, etc.

(vii) Requires huge amount of oil that is deteriorated due to arcing, atmospheric moisture, heating, etc. Top-up oil requirement is expensive.

The moving contact starts separating from the fixed contact and the appearance of the system voltage across the gap creates high voltage stress that decomposes the mineral oil and causes current flow between the contacts. This sudden creation of the current causes arc generation. The arc heats up the oil resulting vaporization. This action causes decomposition of the oil into the by-products such as ethane, methane, hydrogen in the form of gas bubbles. If there is continuous arcing, the size of the gas bubbles increases. These gas bubbles push the oil in all the directions and exert very heavy stresses on the tank wall.

This is required to cool the arc and deionize the gas very rapidly so that current could be interrupted at next current zero. Lengthening the arc also helps in reducing the voltage stress, which, in turn, reduces the voltage restrike. If the successful building-up of the dielectric strength exceeds the restriking voltage, the current gets interrupted.

Pressure chambers, explosion pot and de-ion grid are used to control the arc in the oil circuit-breaker. These are the methods for arc quenching already discussed in the earlier sections.

Oil circuit-breakers are classified into 'indoor circuit-breakers' and 'outdoor circuit-breakers'. Indoor circuit-breakers are employed in completely enclosed building. They are recommended up to 33 kV. Outdoor circuit-breakers fulfill any rated requirement and employed in weatherproof housings.

3.2.3 Construction of Bulk Oil Circuit-breaker

Oil tank, bushings, arc-control device, fixed-contact assembly, moving- contact assembly and operating mechanism are the main parts of an oil circuit-breaker.

All the internal parts of the oil circuit-breakers are housed in a tank. Mineral oil is filled in the tank. Tank absorbs and dissipates energy released during current interruption. Due to heat dissipation oil and gas movement cause stress. To withstand the stress, tank design must be robust and top-plate strong. Tank size depends on phase-to-phase and phase-to-body clearances. Modern arc-control devices facilitate the confinement of the gas pressure in pressure chamber. Gas pressure gets damped as the turbulence travels out from pressure chamber.

Figure 3.1 shows sketch of a bulk oil circuit-breaker. There are generally four types of bushings:

   (i) Plain single dielectric bushing

  (ii) Condenser bushing

 (iii) Oil-filled bushing

  (iv) Dielectric fluid-filled condenser bushing

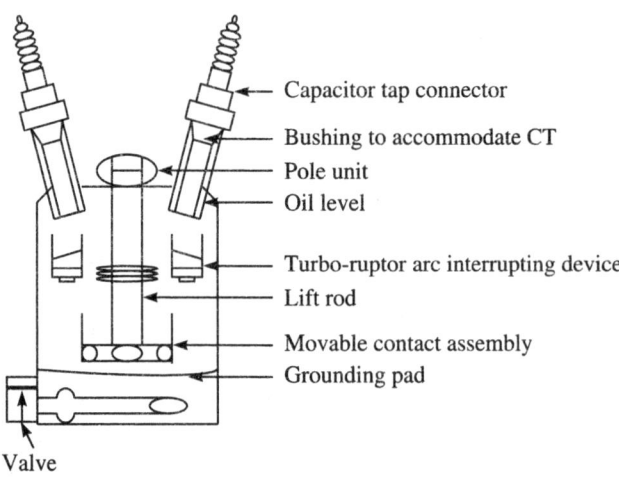

FIGURE 3.1 Bulk oil circuit-breaker.

Bushing is insulation across a high voltage conductor passing through a frame or metal sheet at earth potential. Overvoltages are always possible in the system. Lower sections of the bushings are kept in the oil to prevent internal explosion so that flashover near lower ends does not take place.

The insulation of the plain single dielectric bushing is made up of cast resin or bakelite paper. The concept of condenser bushing is to apply homogeneous and uniform field strength between the conductors and the grounded materials. Equal voltage sharing takes place, when a voltage is applied across a set of series, connected capacitors are graded properly. But the condenser bushings are expensive. Another type of bushing is liquid dielectric (mineral oil)-filled bushing. It is cheaper than the condenser-type bushing. Transformer oil (mineral oil) having sufficient dielectric strength or sulphur hexafluoride gas is filled in the space between conductor and outer porcelain. This prevents impurity formation. It is used for high voltage applications.

For ultra-high voltage applications, bushings consisting characteristics of both the condenser bushing and the oil-filled bushing are applied. These are known as *dielectric fluid-filled condenser bushings*. In this type of bushings unlike condenser bushings, several condenser-type coaxial cores are generally connected in series, which facilitate easy construction and better cooling of the dielectric fluid.

3.3
MINIMUM OIL CIRCUIT-BREAKER (MOCB)

Minimum oil circuit-breakers are also known as *small oil volume breakers*, which contain low quantity of oil in their tanks. Risk of fire is more in bulk oil circuit-breakers and to reduce the risk of fire, minimum oil circuit-breakers were developed. In MOCB mineral oil (only 10 per cent of the bulk oil circuit-breakers) is used for arc extinction and solid post insulators are used to provide insulation. This way use of oil gets minimized. MOCB is used widely up to 1000 kV, 4000 A with 50 kA short circuit breaking capacities. Figure 3.2 shows sketch of a minimum oil circuit-breaker.

MOCB for indoor applications is applicable up to 33 kV. Indoor-type MOCB contains entire interior in an either fibre-glass hollow insulator or porcelain insulator chamber. Minimum oil circuit-breakers have single compartment or chamber per pole. Each pole assembly is mounted on cast iron made breaker structure with two supporting insulators.

Multiple chambers are made in outdoor MOCB construction. The insulating post provides insulation comprising the hollow post insulators mounted one above another. Arc extinction chambers are designed according to the need of single break or double break provisions. So, minimum oil circuit-breakers are generally of three types namely indoor type, single break outdoor type and multibreak outdoor type.

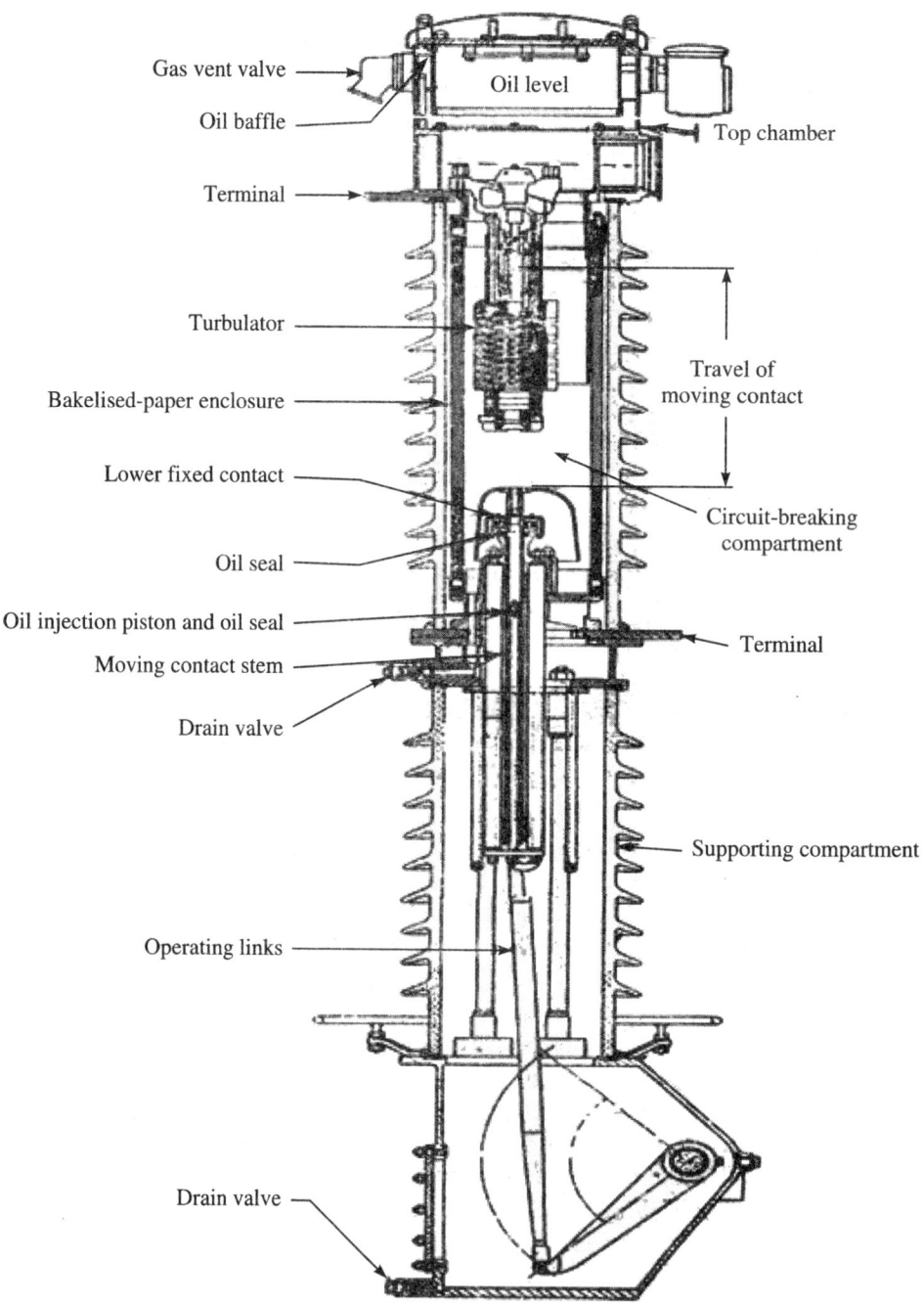

FIGURE 3.2 Minimum oil circuit-breaker.

3.3.1 Advantages of Minimum Oil Circuit-breaker

(i) Less fire risk as contains small amount of oil.

(ii) Low cost of oil replacement.

(iii) Applicable for high voltage and current applications.

(iv) Compact size and comparatively cheaper maintenance and inventories.

(v) Simpler and less expensive maintenance.

(vi) Less installation cost as compared to air blast circuit-breakers.

(vii) Performs well for all duties.

(viii) Useful in small installations as no additional facilities are required.

3.3.2 Limitations of Minimum Oil Circuit-breaker

(i) Longer operating time as compared to air blast circuit-breakers and SF_6 circuit-breakers.

(ii) Risk of fire hazard due to presence of oil.

(iii) Need of regular monitoring of the loss of oil volume and the deposition of impurities.

3.4
AIR CIRCUIT-BREAKER (ACB)

An arc interruption in oil takes place due to hydrogen gas generated by decomposition of oil. Arc interruption properties of hydrogen gas are much better than the clean air but clean air has the following distinct *merits* over oil while interrupting the arc:

(i) Elimination of fire hazards.

(ii) Elimination of maintenance due to presence of oil as in OCB.

(iii) Cheaper as compared to OCB because no top-up of oil is required.

(iv) Absence of heavy pressure due to movement of oil and gas.

Using the effective principles of arc control and proper use of air pressure overcome relatively inferior quality of arc extinction. The popular type of air circuit-breaker (ACB) is *Arc-chute air circuit-breaker*. In such type of ACB, arc is extinguished by applying the principle of lengthening and increasing the power loss of the arc. Figure 3.3(a) and (b) shows how the arc discharge moves upward with the help of the effects of electromagnetic and thermal both. These arc discharges are then directed towards the chute. The chute is consisted of splitters and baffles. Baffles improve cooling while splitters lengthen the arc.

Demerit of the ACB is that if the electromagnetic fields are weak, such as at low current, the breaker becomes less efficient. At low current, the movement of the arc towards chute becomes slower and high speed interruption is not achieved.

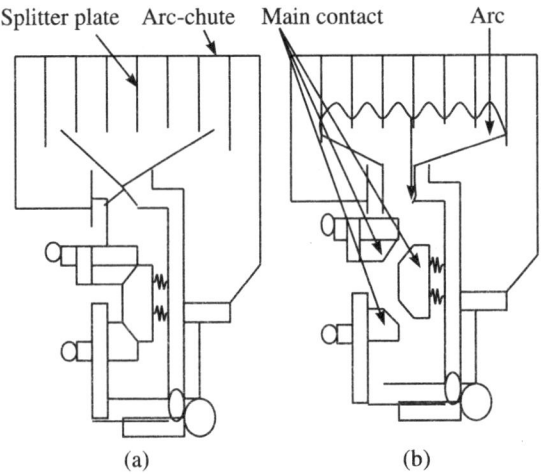

Splitter plate   Arc-chute   Main contact   Arc

(a)   (b)

FIGURE 3.3   (a) Air circuit-breaker in normal state, (b) Air circuit-breaker in operation.

3.5
AIR BLAST CIRCUIT-BREAKER (ABCB)

Air blast circuit-breaker is different from *air-break circuit-breaker*. Air-break circuit-breaker works on the principle of 'current limiting' or 'lengthening of arc' whereas ABCB works on the principle of current interruption by directing a blast of air at very high velocity and pressure towards arcing region.

A reservoir containing dry and clean air is required for ABCB circuit-breakers. To supply air, a compressed air plant with standby compressor is essential for ABCB. It is useful for voltages above 100 kV and for high fault current interruption duties.

In ABCB, contacts are surrounded by air at normal pressure during closed position. When tripping signal is received by the ABCB, high pressure air opens the blast valve and enters the 'interruption chamber' where current interruption takes place. A mechanism for contacts opening opens the moving contact with the pressure of air. Air pressure causes the opening of moving contact against spring pressure. Nozzles direct high pressure air towards arc which sweeps across the arc and quenches it at current zero as per prespecified conditions. Air pressure in the main receiver, which is maintained at 42 kgf/cm^2 for 132 kV and at 210 kgf/cm^2 for 400 kV, ABCB. Constructional view showing different parts of air blast circuit-breaker is shown in Figure 3.4.

3.5.1   Construction of Air Blast Circuit-breaker

A spring holds the fixed and moving contacts and arc chamber is fixed with the air reservoir. Air entering the arc chamber exerts pressure on the moving contacts that move when air pressure exceeds the spring force. Nozzle is made strong enough to carry the air with sonic velocity. Optimization of the contact gap compatible with the maximum breaking capacity is necessary. Arc is kept in high velocity blast of air which converges into nozzle throat.

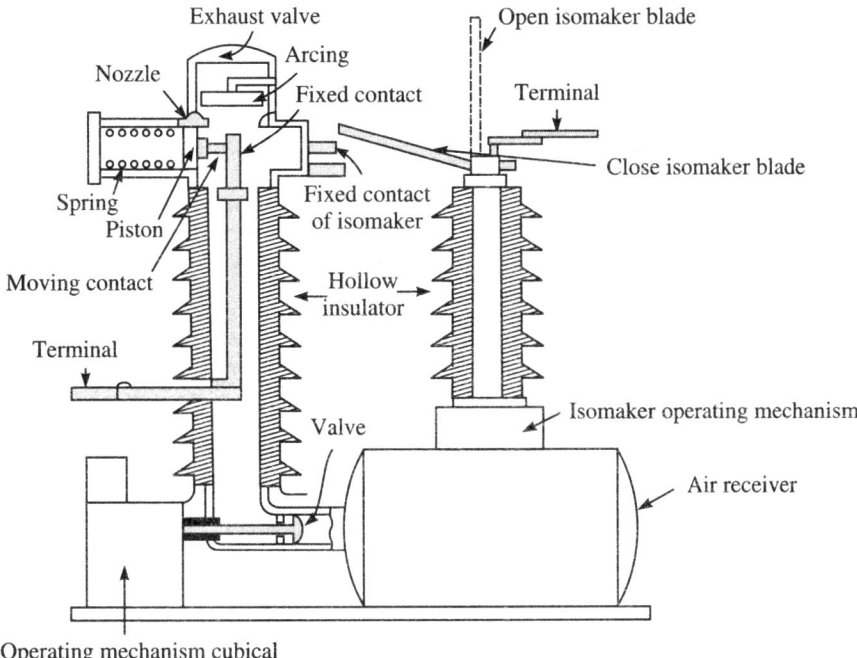

FIGURE 3.4 Air blast circuit-breaker.

An auxiliary switch for isolating the contact gap is also used. This isolating switch is necessary because small contact gap after interruption constitutes inadequate clearance for the normal system voltage. Isolating switch is arranged to open immediately after fault interruption which provides required insulation clearance. Air pressure on the moving contact should be continuously maintained till the isolator fully opens the circuit. As air pressure in the chamber falls below the spring pressure moving contact returns to engage the fixed contact.

Arcing time depends on the value of breaking current. For higher breaking current, arcing time is smaller.

3.5.2 Advantages of Air Blast Circuit-breaker

(i) Energy required for arc extinction is supplied from high air pressure and is independent of the current to be interrupted.

(ii) Arc energy is less so ABCB is suitable for frequent operations.

(iii) No risk of fire.

(iv) Blast removes the arcing products and no expenses required like oil replacement in the OCB.

(v) Operating time is very small so contacts deterioration is less.

(vi) Due to rapid growth of dielectric strength, small contact gap is required, which reduces the size of the device.

3.5.3   Limitations of Air Blast Circuit-breaker

   (i) Air has comparatively inferior cooling and arc extinguishing properties.

  (ii) ABCB is sensitive to variations in the RRRV.

 (iii) Compressor plant which supplies air needs maintenance.

3.6
VACUUM CIRCUIT-BREAKER (VCB)

A system measured in mm Hg and maintained below the atmospheric pressure is known as *vacuum system*. One torr of pressure is equal to 1 mm Hg pressure and 760 mm Hg pressure at 0°C is one standard atmospheric pressure. The pattern of current growth prior to breakdown in the vacuum is different from general class of low and high pressure arcs.

   Research and development in the field of plasma physics, metallurgy, high precision machines and computers helped the engineers and researchers to develop vacuum interrupters in the sixties. After that vacuum circuit-breakers became reality. At present, up to 33 kV rated vacuum circuit-breakers are being manufactured.

   The *advantages* of using vacuum as an arc interrupting medium are:

   (i) Dielectric strength of high vacuum is estimated eight times greater than air and four times that of SF_6 at standard atmospheric pressure for same contact gap.

  (ii) The contact gap in vacuum recovers very fast after arc interruption as compared to other insulating medium.

 (iii) In a vacuum, arc energy dissipation is extremely small as compared to oil or SF_6. Small amount of arc energy release causes less erosion to the contacts and helps in minimizing the maintenance.

 (iv) Vacuum circuit-breaker needs lighter contacts with shorter contact travel. There is no resistance to the moving parts because in vacuum, there is no material.

Arc interruption in vacuum

Vacuum of the order of 10^{-6} to 10^{-7} bar is applied in the vacuum circuit-breaker. Initiation of the metal vapour discharge takes place during opening of the contacts in such high vacuum. This is the fact that arc has insulating property, but in vacuum, arc has conductive property. This means arc in vacuum has very low resistance compared to those in other mediums. Arc in vacuum is supported by vaporization of metal because there is hardly any air particle present in the vacuum. At the current zero when arc extinguishes the conductive metal vapour recondenses on the contact surfaces in few microseconds.

3.6.1   Construction of Vacuum Circuit-breaker

It is well known that the arc in vacuum behaves differently from the other mediums, so the requirements of the contact materials for the vacuum circuit-breakers are:

(i) High cold and hot hardness to prevent wear and tear during normal opening and closing operations.

(ii) Good thermal conductivity to assist rapid cooling of arc-roots.

(iii) Good electrical conductivity.

(iv) High specific gravity.

(v) Sufficient mechanical strength.

(vi) Resistant to develop film on the contact surface.

(vii) Very low gaseous content to contact material.

(viii) Good antiwelding properties.

(ix) Uniform and low contact erosion.

(x) Generate low vapour pressure.

In the vacuum circuit-breakers, vacuum interrupter is housed inside a metallic tank. The metallic tank must be at earth potential. In order to insulate the live parts, solid insulators and SF_6 at pressure of 2–3 bars are employed.

The pole of vacuum circuit-breakers comprises earth metal tank. Tank is filled with SF_6 to provide insulation strength. Tank houses vacuum interrupter, incoming and outgoing bushings and interrupter drive. Requirement of clearances between live parts and tank is much smaller as compared to oil or air because of the presence of SF_6 gas. Interrupter drive converts the rotary movement of the coupling rod into linear travel. Two bushing conductors are connected with contact terminals. Current transformers are placed on both the bushings. Flexible-type weather seal is provided on each bushing. Current transformers are placed and removed by removing this flexible seal without disturbing the pole enclosure.

Kiosk for vacuum circuit-breaker are made and fabricated by using the sheet of the steel. Epoxy powder coating on the sheet makes it weatherproof. Operating kiosk houses operating mechanism, fuses, drive linkages, control switches, auxiliary switches and other ancillary equipments. Inspection windows are provided to enable to view the indicators. Access door is hinged at the top. Constructional view of a vacuum circuit-breaker is shown in Figure 3.5.

3.6.2   Advantages of Vacuum Circuit-breaker

(i) Absence of arc quenching medium and simple working mechanism makes vacuum circuit-breaker practically maintenance-free.

(ii) Low running cost, as no auxiliary set-up is required.

(iii) Light in weight, so less vibrations and pressure on foundation.

(iv) Does not create pollution of any kind.

(v) Preferred for autoreclosing duty because the maximum arc duration in the VCBs is less, such as 15–30 milliseconds.

(vi) Small in size, suitable for indoor substations.

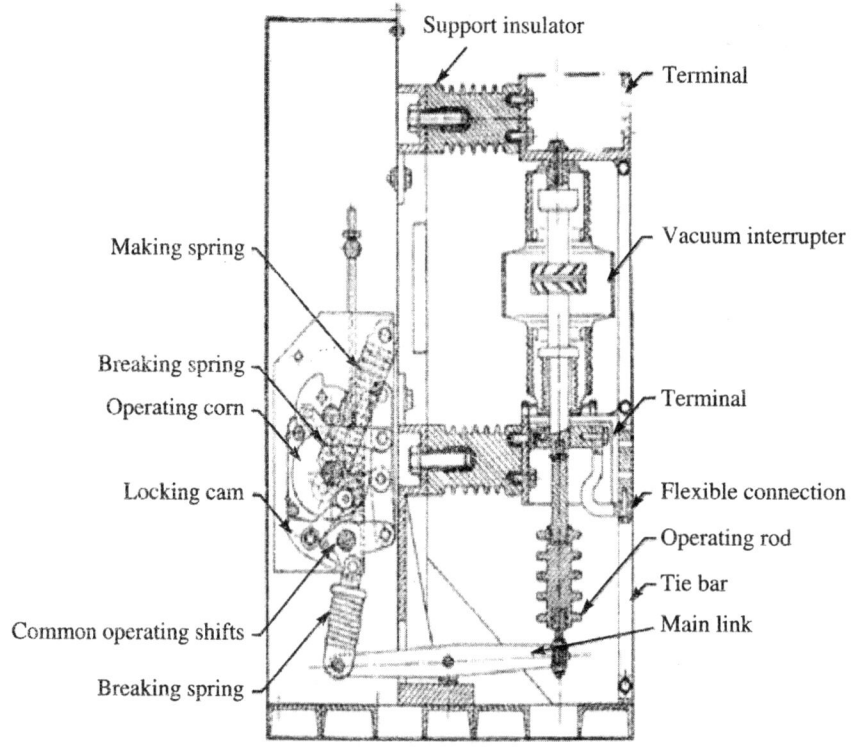

Support insulator
Terminal
Making spring
Vacuum interrupter
Breaking spring
Operating corn
Terminal
Locking cam
Flexible connection
Operating rod
Tie bar
Common operating shifts
Main link
Breaking spring

FIGURE 3.5   Vacuum circuit-breaker.

3.6.3   Limitations of Vacuum Circuit-breaker

(i) Installation cost of VCB is high.

(ii) The main working component, the vacuum interrupter is unrepairable and needs to be replaced with a new unit.

(iii) Generally, current chopping takes place at very low values of current and current chopping leads to overvoltage.

3.7
SULPHUR HEXAFLUORIDE GAS (SF₆) CIRCUIT-BREAKER

Because of the electronegative property SF_6 has an excellent insulating strength. Its molecules easily absorb free electrons in the arc path to form negatively charged ions. Negatively charged ions are heavier than free electrons and move slowly. This is the reason why the column of gas between the contacts becomes bad conductor of electricity in the presence of negatively charged ions. SF_6 is heavy, chemically inert, non-toxic, non-inflammable, odourless and colourless gas. It is chemically and physically stable up to 500°C at normal atmospheric pressure. Since, SF_6

has low thermal conductivity, the heat transfer to adjoining areas during arcing is slow and limited. It helps in keeping the surrounding temperature below the temperature required for ionization. This way arc gets constricted and arc interruption becomes easy. The dielectric strength of SF_6 at standard atmospheric pressure is two to three times greater than that of air and increases with pressure.

Arc interruption in SF$_6$

In the arc interruption chamber, filled with SF_6 at pressure of 3 kgf/cm^2, current interruption takes place. Arc strikes between the contacts, when they separate. Simultaneously, decomposition of SF_6 into sulphur and fluorine atoms takes place. Experimental evidences show that around 3000°K, ionization of the gas begins which are captured by the fluorine atoms to form negative ions. Fluorine ion is 185 times heavier than the electrons. Because of high weight, fluorine ion moves much slower so that each replacement of an electron by a negative ion reduces the current flow rate by a factor of 185.

With the application of pressure of SF_6 jets at 15 kgf/cm^2 at the arc, current interruption is achieved. Heat generated by the arc is taken away by the pressurized gas. Pressurized gas jet is generated by admitting gas under high pressure from the received or by specially designed pressurizing devices.

3.7.1   Construction of SF$_6$ Circuit-breaker

A vital part of the SF_6 circuit-breaker is internal gas reservoir, which is made available for each pole, and is connected to a common high pressure tank. Each pole comprises a low pressure tank which contains two series connected interrupters and two current bushings for external connections. The bushings accommodate secondary of the current transformers at their bases. Interrupter units are supported by the lower portion of the bushings. SF_6 at 3 kgf/cm^2 pressure is filled in the space around interrupter units. High pressure tank is connected with the compressed gas system of the substation through a set of pipes, valves, filters, etc. Pneumatic operating mechanism is fitted inside a cubicle mounted at one end of the breaker. The gas equipment meant for creating double pressure SF_6 circuit are placed in other cubicle located between the two poles.

Sketch of the complete assembly of a SF_6 circuit-breaker is shown in Figure 3.6. In these breakers, gas pressure is maintained at high-level so careful scaling is required to prevent gas leakage at joints. Alarms and a set of lockout switches are installed properly to give warning. At the moment gas pressure drops below certain values.

Safety devices take care of the fault in the control circuit, which does not allow the compressor to build up excessive pressure in the high pressure reservoir. Temperature is required to be maintained around 20°C. A heater is fitted in the high pressure chamber to maintain the temperature. A thermostat switches 'ON', when the ambient temperature falls below 16°C.

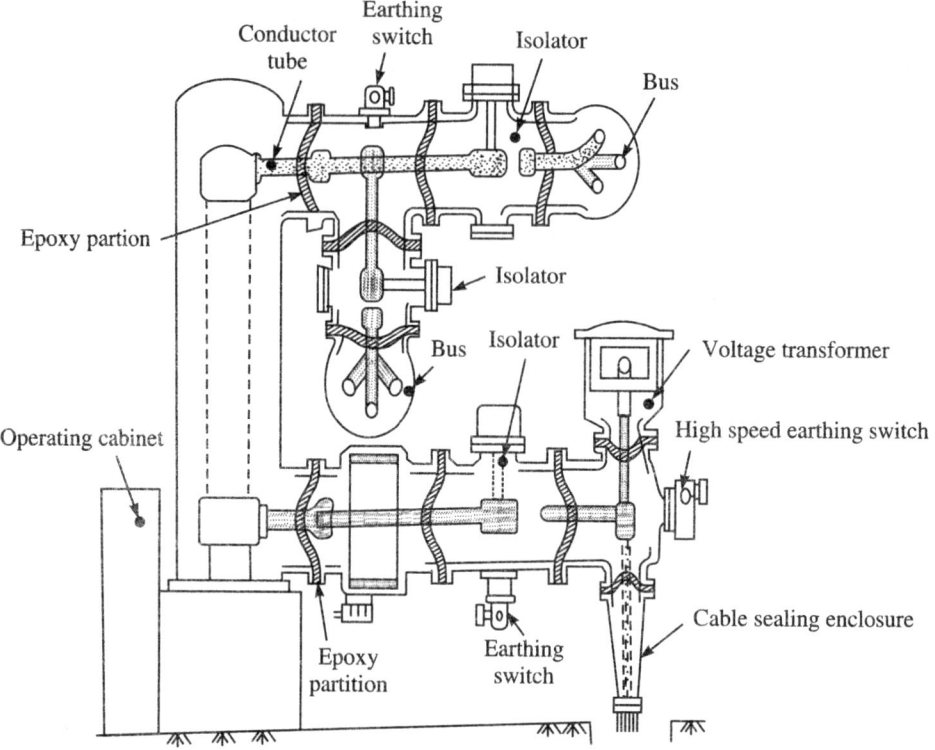

FIGURE 3.6 SF$_6$ circuit-breaker.

3.7.2 Operation of SF$_6$ Circuit-breaker

Accelerating spring pulls away the moving contact just after receiving the trip signal. Simultaneously, blast valve gets opened and gas with high pressure enters the arc region from the internal gas reservoir. High pressure gas passes through nozzles and intersects the arc and then escapes into the low pressure tank. The moment current gets interrupted, the blast valve closes immediately. Pumping back of the excess gas to the high pressure tank also takes place. Both the interrupter units operate in unison and moving contacts are mounted on the ladder assembly made up of insulating rod, this moving contacts ladder assembly moves by an operating mechanism.

3.7.3 Advantages of SF$_6$ Circuit-breaker

(i) Absolutely free from hazards because SF$_6$ has properties of extreme inertness at standard temperature and pressure.

(ii) Provides minimum current chopping tendency.

(iii) Offers very short break-time and arcing period.

(iv) Due to high dielectric strength of SF_6, the clearance requirement is drastically reduced. Fast building-up of the dielectric strength makes SF_6 circuit-breakers suitable for rapid and multi-shots reclose.

(v) Performance is independent of the atmospheric conditions.

(vi) Less contact erosion because of low-arcing time. Breaker gives noiseless operation because of low-velocity operation.

(vii) The closed circuit gas cycle eliminates the moisture problem.

(viii) SF_6 is non-toxic and has no danger to the environment even during its leakage.

3.7.4 Limitations of SF_6 Circuit-breaker

(i) Replacement of SF_6 is expensive and time consuming.

(ii) Economical only above 66 kV.

(iii) Very expensive for smaller substations.

3.8
HIGH VOLTAGE DIRECT CURRENT (HVDC) CIRCUIT-BREAKER

HVDC transmission system has proved to be most cost-effective technology for transmitting bulk power from generating end to the users end.

Direct current is unidirectional current and flows without natural zeros. High direct current is difficult to be interrupted at high voltages of present transmission lines of the complex networks. Direct current circuit-breaking is not achieved actually by breaking the short circuit currents but achieved by interrupting the load current in circuits at high potential with respect to ground. Direct current switches have been developed which interrupt the fault current by creating artificial zero of the fault current. Artificial zero may be created through the contacts of the switch by the oscillatory discharge of a capacitor. Circuit of the switch needs to be designed in such a way that the crest value of the oscillatory current is greater than the direct current to be interrupted.

Circuit of the switch is shown in Figure 3.7. For the normal operation, contacts M and B are normally closed contacts and contact A is normally open contact. Since A is normally open contact, capacitor C gets charged through the resistance R. Resistance R is a high-valued resistor to limit the current. As signal reaches the switch to interrupt the fault current (I_d), operating mechanism closes contact A and opens contact B simultaneously. Oscillations get initiated immediately after the contact M opens and current gets interrupted at its zero value. After interrupting the fault current, contact A gets opened and B closed.

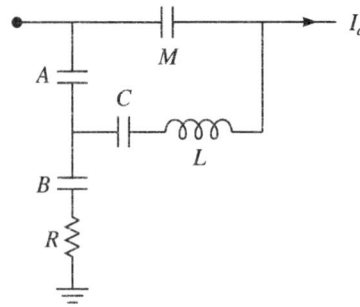

FIGURE 3.7   Circuit diagram of a HVDC switch.

3.9

AUTORECLOSING

Depending on the application, autoreclosing is described either by single-phase or by three-phase. Reclosing after predetermined time, when line to ground fault takes place, is known as *single-phase autoreclosing*. During *three-phase autoreclosing* all the three phases, which are independent of the types of fault, are opened after predetermined time.

3.10

CIRCUIT-BREAKER TESTING

To confirm quality and qualifying the eligibility for application with certain specifications, following performance testings of the circuit-breakers are required.

Short circuit tests are conducted to certify the ratings of the circuit-breakers. These tests are conducted either in laboratories or in the field. Laboratory testings are very common. Specially designed high voltage generators are used to feed power during laboratory testing. The advantages of the laboratory testing are:

 (i) Test voltage and currents may be reproduced.

 (ii) Indirect methods can also be used.

 (iii) Test parameters like current, voltage, power factor, frequency, restriking voltage, etc. can be accurately controlled.

 (iv) Laboratory testings are suitable in the research and development.

On the other side, field testing represents actual conditions but cannot be reproduced repeatedly, hence is not suitable for research and development. Also desired test conditions such as RRRV, recovery voltage, etc. cannot be achieved conveniently.

3.10.1   Test Equipment

Operation of the testing equipment is intermittent, so generator is of special design to deliver power up to 2000 MVA. To withstand and tackle the effects of high electromagnetic forces, rigid bracings of the conductors and coil-ends are used. During design of the generator reactance is chosen at its minimum value.

Resistors and reactors are suitably used to control the power factor. Capacitors are used to control the rate of re-striking voltage and the line-charging currents.

For high voltage testing applications, generally banks of single-phase transformers are employed. A transformer of normal rating of 30 MVA may require delivering 475 MVA under short circuit conditions.

Two circuit-breakers, first master circuit-breakers and the second test circuit-breakers are used. Master circuit-breakers always have higher rating than the test circuit-breakers.

High air pressure is required to ensure that make-switch operates properly. Before the short circuit current reaches its peak, the circuit-breaker contacts must be fully closed.

3.10.2 Test Method

A short-timing sequence of several operations is the part of testing procedure. Measuring instruments of the testing plant such as voltmeter, ammeter, power factor metre, frequency metre, impulse voltage measuring set, etc. are first required to be adjusted properly.

In the next step, measuring circuit is required to be made correctly checked and verified. Oscillograph loops need to be suitably calibrated.

Sequence operations of 0.2 second time duration are managed by the drum switch. Following sequence operations are required to be performed for testing the breaking capacity of the circuit-breaker.

Step 1: Switch 'OFF' the motor driving, the generator after achieving pre-specified speed.

Step 2: Switch 'ON' the impulse excitation.

Step 3: Close master circuit-breaker.

Step 4: Switch 'ON' the powerscope to get oscillograph.

Step 5: Close the make switch.

Step 6: Open the circuit-breaker under test.

Step 7: Open the master circuit-breaker.

Step 8: Switch 'OFF' the exciter circuit.

Schematic diagram of the test equipment is shown in Figure 3.8(a) and the equivalent circuit is shown in Figure 3.8(b).

Short-time current test is performed by passing the current through the circuit-breaker under the test for short duration of one second. Oscillogram on the powerscope shows the magnitude and nature of the current.

Making capacity test is performed first by closing the master to make switches. After that, circuit-breaker under the test is closed to apply the short circuit current. The making current is observed and determined by the powerscope.

For breaking capacity testing, the master circuit-breaker and the testing circuit-breaker are first closed, and by closing the making switch, short circuit current is applied. The testing circuit-breaker is opened at pre-specified moment and this breaking current is observed and measured by powerscope.

Example 1   Determine the rated normal current, breaking current, making current, and the short-time rating current for a circuit-breaker, rated as 1000 A, 2000 MVA, 66 kV, for 3 seconds of a three-phase oil circuit-breaker.

Solution

Rated normal current = 1000 A (given)

$$\text{Breaking current} = \frac{2000}{\sqrt{3} \times 66} = 17.5 \text{ kA}$$

Making current = 2.5 times of breaking current

$$= 2.5 \times 17.5 = 43.75 \text{ kA}$$

Short-time rating = 17.5 kA for 3.0 s

                                     Answer: 1000 A, 17.5 kA, 43.75 kA, 17.5 kA for 3.0 s

Note: Using the following MATLAB programming, we can solve the above problem.

```
clear
clc
Ir=1000;
%Ir=rated current
P=2000*(10^6);
% P= Power rating
V=66*(10^3);
% V= voltage
t=3;
%t=time
%---------solution-----------
Ib=P/sqrt(3)*V)
% Ib=Breaking current
Im=2.5*Ib
% Im=Making current =2.5 times of breaking current
% Answer: Rated normal current=1000A, Breaking current=1.
  7495e + 004A, Making current = 4.3739e+004 A
% Short time rating= 17.5 kA for 3.0 s
```

Example 2   One transformer is connected with an alternator rated as 10 MVA, 13.8 kV, $X''_d = 10\%$, $X'_d = 15\%$ and $X_d = 100\%$. Three-phase short circuit between the breaker and the transformer occurred, when alternator was operating at no load and rated voltage. Calculate the following for circuit-breaker:

   (i) Steady state short circuit current.
  (ii) Initial symmetric rms current.
 (iii) Maximum dc component of the short circuit current.
 (iv) Momentary current rating.
  (v) Current to be interrupted.
 (vi) Interrupting kVA.

Solution   Given:

Steady state reactance $X_d = 100\%$

Steady state short circuit MVA = 10 MVA

    (i) Steady state short circuit current:

$$= \frac{10 \times 10^6}{\sqrt{3} \times 13.8 \times 10^3}$$

$$= 418.37 \text{ A}$$

   (ii) Initial symmetric rms current:

Current corresponding to the sub-transient state at 10% reactance = Initial symmetric rms current

So, short circuit MVA $= \dfrac{10}{10/100} = 100$ MVA

And short circuit rms current = Initial symmetric rms current

$$= \frac{100}{\sqrt{3} \times 13.8} = 4184 \text{ A}$$

  (iii) Maximum dc component of the short circuit current:

$$= \text{Peak value of the sub-transient current}$$

$$= \sqrt{2} \times 4184 = 5917 \text{ A}$$

  (iv) Momentary current rating:

$$= 1.6 \text{ times of the short circuit rms current}$$

$$= 1.6 \times 4184$$

$$= 6694 \text{ A}$$

   (v) Current to be interrupted:

$$= 1.2 \text{ times of the symmetrical breaking current of a three-cycled breaker}$$

$$= 1.2 \times 4184$$

$$= 5021 \text{ A}$$

  (vi) Interrupting kVA:

$$= \sqrt{3} \times 13.8 \times 5021$$

$$= 120 \text{ MVA}$$

        Answer: 418.37 A, 4184 A, 5917 A, 6694 A, 5021 A, 120 MVA

Note: Using the following MATLAB programming, we can solve the above problem.

```
clear
clc
P=10*10^6;
V=13.8*10^3;
Xdd=10;
%Xdd= sub transient reactance
```

```
Xd=15;
% Xd=Transient reactance
X=100;
% X= Steady state reactance
%------------------SOLUTION----------------
Is=P/sqrt(3)*V)
% Is= steady state current
Psc=P/Xdd/100)
%Psc= short circuit MVA
Ist=Psc/sqrt(3)*V)
%Ist=Current corresponding to the sub-transient state
Iirms=Ist
%Iirms= Initial symmetric rms current
%Current corresponding to the sub-transient state at 10%
    reactance
% = Initial symmetric rms current
Imdc=sqrt(2)*Iirms
%Imdc= Maximum dc component of the short circuit current
Imr=1.6*Iirms
%Imr=Momentary current rating
Iir=1.2*Iirms
%Iir= Current to be interrupted
KVAir=sqrt(3)*(V/(10^3))*Iir
%KVAir=Interrupting kVA
%ANSWER: Is = 418.3698, Ist=4.1837e+003, Iirms=4.1837e+003,
    Imdc =5.9166e+003
% Imr =6.6939e+003, Iir =5.0204e+003, KVAir =1.2000e+005
```

EXERCISES

1. With the help of a neat sketch, describe the various parts of a bulk oil circuit-breaker (BOCB).

2. Explain the current interruption and arc control in a BOCB.

3. Compare MOCB with BOCB.

4. List the advantages and limitations of the minimum oil circuit-breaker (MOCB).

5. Write the merits and demerits of BOCB.

6. Describe MOCB giving a neat sketch.

7. What are the important advantages and disadvantages of air blast circuit-breakers (ABCB)?

8. Describe in detail the process of current interruption and arc control in a typical ABCB.

9. Explain the properties of SF_6 which make it suitable to use as insulating and interrupting medium.

10. With the help of a simple sectional view, describe the different components of a SF_6 circuit-breaker.

11. Compare ABCB with SF_6 circuit-breaker.

12. Discuss in detail about the current interruption and arc control in vacuum circuit-breaker (VCB).

13. Describe the various components of a VCB.

14. Discuss the merits and limitations of a VCB.

15. Compare SF_6 circuit-breaker with VCB.

16. Write short note on HVDC circuit-breaking.

17. Why is current interruption easier in an ac circuit than in a dc circuit?

18. Write in brief about air circuit-breaker.

19. What do you understand by autoreclosing? Describe in brief.

20. Why are circuit-breakers designed to have a short-time rating?

21. Why circuit-breaker testing is required? Explain in detail.

22. Describe important common components of circuit-breakers.

23. A generating station 'A' has short circuit capacity of 1000 MVA. Another station 'B' has short circuit capacity of 650 MVA. They are operating at 11 kV. Find the short MVA, if they are interconnected by a cable of 0.5 Ohm reactance per phase.

Answer: 1176.5 MVA

Chapter 4

Current-limiting Reactors

4.1
INTRODUCTION

Current-limiting reactors comprise high inductances, which are brought in circuit to compensate the reactive power. Compensating the reactive power means improving the power factor of the network and bringing the system voltage within the specified limit.

It can be seen below, how inductive reactance helps reducing the short circuit current. As seen in Figure 4.1, X_i is the internal reactance of the generator and X is the reactance of the current-limiting reactor. Let, short circuit fault occurs on the line at point F, as shown in Figure 4.1.

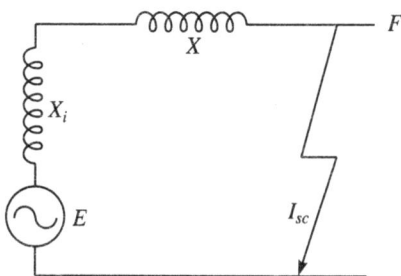

FIGURE 4.1 Short circuit fault.

Short circuit current, if the reactor is in the circuit, $I_{sc_1} = \dfrac{E}{X_i + X}$

Short circuit current without reactor, $I_{sc_2} = \dfrac{E}{X_i}$

Further, assuming, $X_i = X$

$$I_{sc_1} = 2I_{sc_2}$$

So, it is clear that injecting reactance of the reactor equal to the internal reactance of the generator, fault current could be reduced to half of its value.

4.2
TYPES OF REACTORS

Commercially, following four types of reactors are being developed:

   (i) Dry air-cored reactor
  (ii) Oil-immersed, non-magnetically-shielded reactor
(iii) Oil-immersed, magnetically-shielded reactor
(iv) Saturated reactor

While designing a reactor, the basic requirement must be kept in mind that the reactor core does not get saturated even if the fault current reaches the highest value. In the reactors, fluxes flow all around the coil with high flux density near the coil reducing as distance increases from the coil. Shield made up of ferromagnetic material is generally provided so that fluxes confine to it. Shielding reduces the heating of external metals created by induced current because it forces the fluxes to be confined in the shield.

4.2.1 Dry Air-cored Reactor

Dry air-cored reactor is a simplest type of reactor. It is unshielded and executed in the open. Cooling of this reactor is done either by natural air circulation or by forced air circulation. A wooden frame which holds the cylindrical coil and the reactor is supported on a frame of synthetic resin glass reinforced.

Permeability of the air is very small and hence inductance created by the reactor is also small. Special care is required while installing the reactors because they are not shielded. These reactors are required to be installed in specially designed rooms free from the metallic loops and reinforcement. These reactors are applied up to 33 kV for supplying small reactive powers. An air-cored reactor is shown in Figure 4.2.

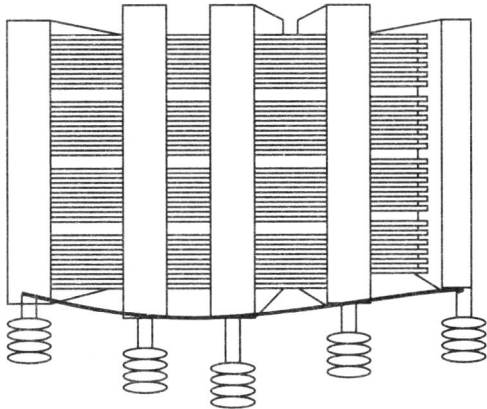

FIGURE 4.2 Air-cored reactor.

4.2.2   Oil-immersed, Non-magnetically-shielded Reactor

In this reactor, a tank filled with transformer oil carries the reactor's coil, which is without iron core. A cylindrical-shaped aluminium enclosure separates coil and the tank. Aluminium enclosure works as non-magnetic shield. It can be applied for both indoor and outdoor applications. It is manufactured for larger voltage and current rating to supply relatively higher reactive power.

The aluminium cylinder used to shield the fluxes is designed in such a way that the induced flux in it opposes the main flux generated by the current coil. These two fluxes should be equal outside the enclosure so that they oppose and cancel each other.

4.2.3   Oil-immersed, Magnetically-shielded Reactor

In this type of reactor, coils with core are put inside the transformer oil, filled in a tank. Laminated silicon steel made core is fixed between the coil and the steel tank. Oil-immersed, magnetically-shielded reactors may be air-cored or metal-cored having one or more air gaps. Fixing the laminated silicon steel made core between the coil and the steel tank provides path to the magnetic fluxes and stray fluxes get minimized. Stray fluxes are possible in small amount, if the core is made up of iron with gaps. No separate shield is provided in these type of reactors. Similar cooling schemes of transformers are applied in these reactors. These types of reactors can be designed and applied up to any kVAR rating. Oil-immersed, magnetically-shielded reactor is shown in Figure 4.3.

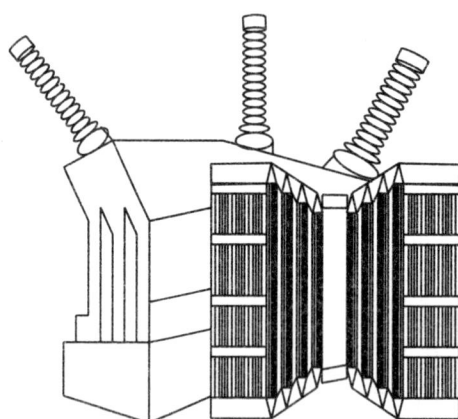

FIGURE 4.3   Oil-immersed, magnetically-shielded reactor.

4.2.4   Saturated Reactor

Saturated reactor is being used in short circuit limiting coupling circuit. This is made by having closed magnetic core like transformer. Core is made up with smaller cross-sections, for example core of a saturated reactor may be designed to have only 20% to 25% cross-section of series reactor. So, the saturated reactors become fully saturated at the rated voltage. In some applications, they also get saturated below the rated voltage.

4.3
REACTANCE CALCULATION

Series reactance in the form of high voltage reactors is applied in the circuit to limit the fault current to the desired value. Equivalent circuit diagram is shown in Figure 4.4.

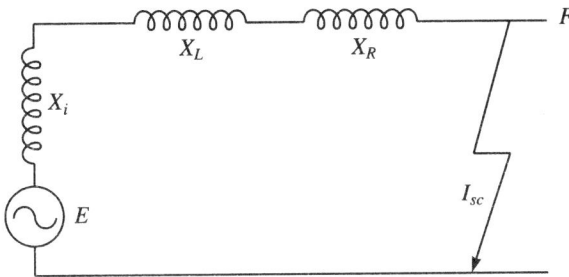

FIGURE 4.4   Equivalent circuit diagram.

Let,

I = full load current (ampere)

E = rated voltage (volt)

X_R = reactance of the reactor (ohm)

X_L = line reactance (ohm)

X_i = generator reactance (ohm)

I_{sc} = fault current (ampere)

The percentage reactances:

$$\% \ X_R = \frac{IX_R}{E} \times 100$$

$$\% \ X_i = \frac{IX_i}{E} \times 100$$

$$\% \ X_L = \frac{IX_L}{E} \times 100$$

and short circuit current, i.e. fault current:

$$I_{sc} = \frac{E}{X_i + X_L + X_R}$$

Putting the values of X_i, X_L and X_R in the terms of % reactances:

$$I_{sc} = \frac{I}{\%X_i + \%X_L + \%X_R} \times 100$$

Now, the short circuit power:

$$(\text{kVA})_{sc} = \frac{\text{Base kVA} \times 100}{\%X_i + \%X_L + \%X_R}$$

$$= \frac{\text{Base kVA} \times 100}{\%X_t}$$

$\% X_t$ = resultant reactance up to the fault point.

4.4
APPLICATIONS OF REACTORS

   (i) They are basically used to limit the short circuit currents. Since, old installations suffer from frequent faults, current-limiting reactors are essentially required to be installed there.
  (ii) Capacitance of the long transmission lines becomes considerably high, if the line is unloaded or lightly loaded due to the line charging. The receiving end voltage becomes higher than the sending end voltage. Reactor is switched 'ON' to inject inductive reactance which compensates capacitive reactance.
 (iii) During expansion of the industrial plants, additional transformers are installed to enhance the capacity of the electrical power supply. Enhancement of the capacity results rise in fault levels, so current-limiting reactors are required.
  (iv) Generators are required to be added up to the common busbars, if the generating capacity is to be enhanced. Connecting additional generators increase the possibility of fault level. Current-limiting reactors are needed to be installed.
   (v) In the same way as described above, if the switching or receiving stations are extended by adding up new incoming lines, the overall fault level increases and the current-limiting reactor is required.
  (vi) Current-limiting reactors are also used in short circuit test plants and arc furnace plants.

4.5
SHORT CIRCUIT, CURRENT-LIMITING COUPLING CIRCUIT (SCCLCC)

There is undesirable voltage drop across the series reactance of the reactor while reactor limits the short circuit current. Voltage drop requires to be reduced. Introducing SCCLCC in series with the series reactor as shown in Figure 4.5 reduces the voltage drop up to zero depending on the design requirements.

   In this circuit, a capacitor (C) offering reactance equal to that of the series reactor (R) is installed. SR_1 and SR_2 are the saturated reactors which are so designed that at normal operating voltage across capacitor (C), they do not get saturated and offer enough impedance to check the current through these reactors. Load current (I_1) flows through capacitance (C). Resistor (R) and capacitor (*C*) are turned in order to keep voltages at *A* and *B* at zero level. In the presence of fault in either sides of the SCCLCC, the voltage across *R* and *C* increases with the rise in voltage across *C*, simultaneously reactors SR_1 and SR_2 get saturated. Saturated reactors offer sharp decrease in reactances and capacitor (*C*) gets shunted. Now reactor (*R*) withstands the network voltage limiting fault current.

FC is known as filter circuit which is used to eliminate harmonics generated by the saturated reactors. SCCLCC returns to its normal state after fault interruption by action of the switchgear. During this reversion process auxiliary saturated reactor (SR_1) gets desaturated first. With desaturation of SR_1, damping resistor (R_D) comes in series with, main saturated reactor (SR_2). Now, parallel circuit gets detuned rapidly and SR_2 desaturates.

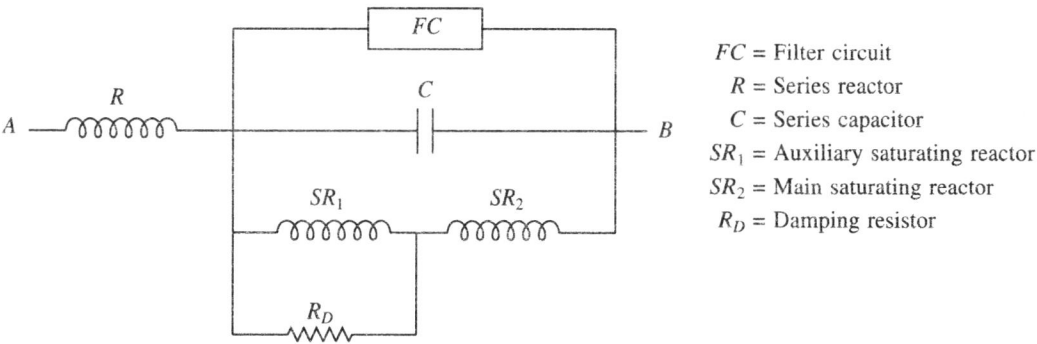

FC = Filter circuit
R = Series reactor
C = Series capacitor
SR_1 = Auxiliary saturating reactor
SR_2 = Main saturating reactor
R_D = Damping resistor

FIGURE 4.5   Short circuit, current-limiting coupling circuit.

Example 1   A single phase power system network is shown in Figure 4.6. A short circuit occurs at point *F*. The voltage at point *F* is 66 kV. Calculate:

(i) The fault current, without series reactor.

(ii) The fault current, with series reactor.

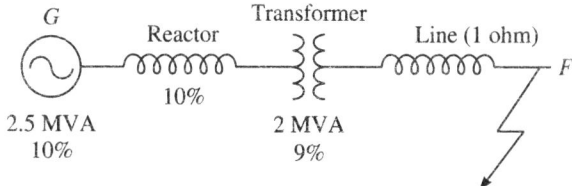

FIGURE 4.6   Power system network.

Solution

Impedance diagram of the given network:

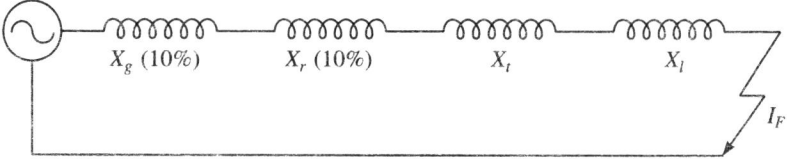

Here,

X_g = generator reactance
X_r = reactor reactance
X_t = transformer reactance
X_l = line reactance
I_F = short circuit fault current

Assume, the base kVA be 2.5 MVA and kVA 66 kV.
Now,

$$\text{Base current} = \frac{\text{Base kVA}}{\text{Base kV}} = \frac{2500}{66} = 37.88 \text{ A}$$

$$\text{Base impedance} = \frac{\text{Base voltage}}{\text{Base current}} = \frac{66000}{37.88}$$

$$= 1742 \text{ ohm}$$

Transformer reactance at new base:

$$\%X_t = 9\% \times \frac{2500}{2000} = 11.25\%$$

Reactance of the transmission line up to fault point F:

$$\%X_l = \frac{\text{Ohmic reactance}}{\text{Base impedance}} \times 100 = \frac{1 \times 100}{1742}$$

$$= 0.057\%$$

(i) Total reactance up to fault point without series reactor:

$$\%X_T = 10\% + 11.25\% + 0.057\%$$

$$= 21.307\%$$

Now, the fault current without reactor:

$$I_F = \frac{\text{Full load current, i.e. base current}}{\%\text{ reactance up to fault point}} \times 100$$

$$= \frac{37.88 \times 100}{21.307} = 177.78 \text{ A}$$

(ii) Total reactance up to fault point with series reactor:

$$\%X_{TR} = 10\% + 10\% + 11.25\% + 0.057\%$$

$$= 31.307\%$$

Now, the fault current with reactor:

$$I_{FR} = \frac{37.88 \times 100}{31.307} = 120.99 \text{ A}$$

Note: There is about 32% reduction in fault current after using the reactor.

Answer: (i) 177.78 A, (ii) 120.99 A

Note: Using the following MATLAB programming, we can solve the above problem.

```
clear
clc
KVAb=2.5*10^3;
%KVAb= Base KVA
KVb=66;
% Vb= Base voltage
Xt=9;
% Xt=Transformer reactance
KVAt=2*10^3;
% KVAt=Transformer KVA
Xl=1;
%Xl= Line reactance
X=10;
%X=%Reactor reactance
Xc=10;
% Xc=%Reactor reactance
%----------------SOLUTION-------------------
Ib=(KVAb/KVb);
% Ib= Base current
Zb=(KVb*10^3)/Ib;
% Zb= base impedance
Xt1=Xt*(KVAb/KVAt);
%Xt1= Transformer %reactance at new base
Xl1=(Xl/Zb)*100;
%Xl1= %Reactance of the transmission line up to fault point
XT=Xt1+Xl1+X;
% XT=Total %reactance up to fault point without series
   reactor
IF=(Ib/XT)*100
% IF=fault current without reactor
XTR=Xt1+Xl1+X+Xc;
% XTR=Total %reactance up to fault point with series reactor
IFR=(Ib/XTR)*100
% IFR=fault current with reactor
IFr = ((IF-IFR)/IF)*100;
% IFR=%reduction in fault current after using reactor
%ANSWER: IF = 177.7730, IFR =120.9899,
% %reduction in fault current after using reactor=32%
```

Example 2   Three generators each rated at 75 MVA with 20% reactance are installed in a generating station. While designing the busbars to limit the busbar fault kVA up to 750 MVA, estimate the value of bus reactance. Reactance diagram is shown in Figure 4.7.

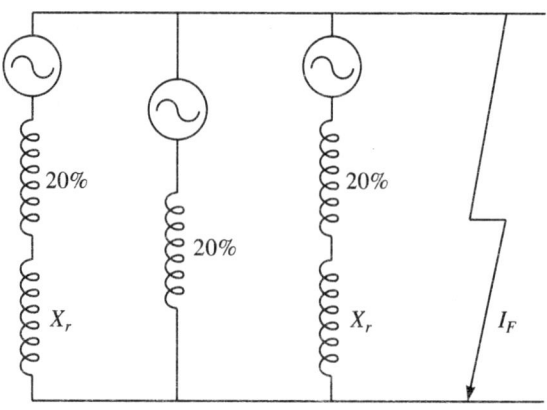

FIGURE 4.7   Reactance diagram.

Solution   Thevenin's equivalent of the reactance diagram:

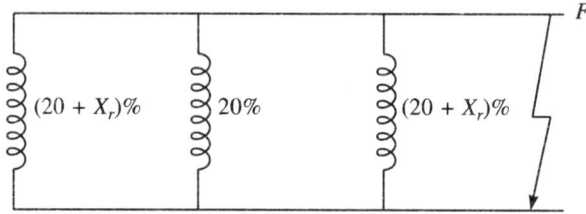

Let the base kVA is 75 MVA and reactance at F, X_e:

$$\frac{1}{X_e} = \frac{1}{20} + \frac{1}{20 + X_r} + \frac{1}{20 + X_r}$$

$$= \frac{60 + X_r}{400 + 20X_r} \tag{1}$$

Now,   Short circuit MVA $= \dfrac{\text{Base MVA} \times 100}{\%X_e}$

$$750 = \frac{75 \times 100}{\%X_e}$$

or $\%X_e = 10$ \hfill (2)

Putting the value of X_e in Eq. (1):

$$\frac{1}{10} = \frac{60 + X_r}{400 + 20X_r}$$

$$400 + 20X_r = 600 + 10X_r$$

$$10X_r = 200$$

$$X_r = 20\% \text{ on the kVA base of 75 MVA}$$

Answer: 20% on the kVA base of 75 MVA

Note: Using the following MATLAB programming, we can solve the above problem.

```
clear
clc
%----------data given----------
MVAb=75;
% MVAb= Base MVA
MVAsc=750;
% MVAsc= short circuit MVA
X=20;
% X= %Reactor reactance
%----------------SOLUTION-------------------
Xe=(MVAb*100)/MVAsc;
% Xe=total %reactance at the fault point
% (1/Xe) = (1/X) + (1/(X+Xr)) + (1/(X+Xr))
% (1/Xe) = (3*X+Xr)/ (X^2+X*Xr)
Xr = (3*X*Xe-X^2)/(X-Xe)
% Xr = % bus reactance added with X to limit busbar fault MVA
  up to 750 MVA
% ANSWER: Xr=20, So 20% bus reactance is required to limit
  busbar fault kVA up to 750 MVA on the kVA base of 75 MVA
```

EXERCISES

1. Describe in detail the need of current-limiting reactors.

2. Discuss the requirements of reactors for generators, feeders and busbars.

3. Explain dry air-cored reactors, mentioning their applications.

4. Discuss oil-immersed, non-magnetically-shielded reactor.

5. Explain oil-immersed, magnetically-shielded reactor.

6. Write notes on:
   (a) Saturated reactor
   (b) SCCLCC (short circuit, current-limiting coupling circuit)

7. A circuit-breaker of short circuit rating 150 MVA is installed in a feeder circuit. Calculate the value of series reactor *R* for the system as shown in the following figure. Assume grid as infinite bus.

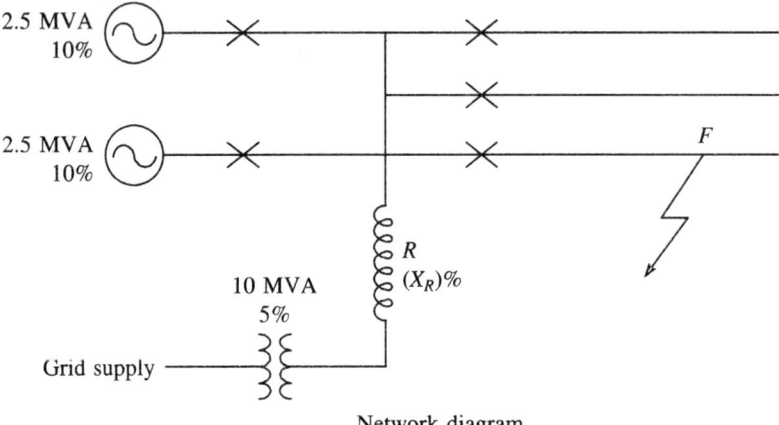

Network diagram

Answer: 5% based on 500 kVAR

8. A station operating at 33 kV is divided into two sections A and B interconnected through a reactor X. Section A has three generators each 15 MVA having 15% reactance. Section B is fed from a grid through a 75 MVA transformer with 8% reactance. The circuit-breakers have rupturing capacity of 750 MVA. Find the reactance of reactor X to prevent the breakers being overloaded if a three phase fault occurs on outgoing feeder connected to A.

Answer: 1.26 ohm

Part B

POWER SYSTEM PROTECTION

Chapter 5

Introduction

Today's highly competitive environment necessitates greater efficiency in electric power industry because of growth of electric energy consumption and the consequent power sector reforms and the entry of private entrepreneurs. Thus, the main problem that the power industry faces today concerns reliability. Ensuring reliable power supply means providing steady and uninterrupted power supply to all electricity consumers. In order to achieve this objective, all the five major zones of a modern power system (described in Section 5.1) need to be properly and effectively protected.

5.1
STRUCTURE OF A POWER SYSTEM

The modern power system is a complex interconnected network divided into the following five major zones:

1. **Generation**, one of the essential components of the power system, uses alternators of 11 kV and 33 kV rating. Modern generating units use ac generators with rotating rectifiers known as *brushless excitation systems*. Since these modern alternators do not require a commutator, this feature enables ac generators to generate a typically high voltage of 33 kV unlike the older generators, which could generate a voltage of only 11 kV. The power rating of the plant generators varies from 50 MW to 1500 MW. Steam turbines operate at high speeds of 3600 or 1800 rpm for 60 Hz supply, and 3000 or 1500 rpm for 50 Hz supply, respectively. Because of this high operating speed of the steam turbines, the generators used have a cylindrical rotor. Hydraulic turbines operate at a lower speed because of less pressure/thrust generated by water and hence the generators used in this case have a salient pole type rotor.

2. **Transformers** are used to step up voltages at the generating end and to step them down at the receiving end. Generally, because of the insulation requirements and the need to

maintain practicality in design, the maximum generated voltage that can be achieved ranges from 11 kV to 33 kV. In a modern utility system, the power generated may undergo four or five transformations as it travels from the generator to the ultimate user.

3. ***Transmission and sub-transmission*** pertain to that section of the power system wherein the voltage level remains at its maximum level. For economic reasons, the transmission level of the voltage is stepped up from the generated voltage and has presently reached more than 1000 kV. Typically, the sub-transmission voltage level ranges from 33 kV/66 kV to 132 kV. In order to maintain the transmission line voltage level within permissible limits, capacitor banks and reactor banks are installed at the sub-stations. Power is also supplied to large industries from the sub-transmission system in most cases.

4. The ***distribution system*** is that part of power system, which connects distribution sub-stations to the consumer's service mains. The primary distribution lines are of 11 kV and 33 kV. The secondary distribution network reduces the voltage up to 400/230 V to facilitate utilisation by residential and commercial consumers. Small industrial customers may also be served directly by primary feeders.

5. ***Loads*** on power systems are mainly of residential, commercial and industrial types. Residential and commercial loads mainly consist of lighting, heating and cooling. These loads are independent of the frequency and consume negligibly small reactive power. Industrial loads are of both types, namely active and reactive.

An example of an interconnected power system network is shown in Figure 5.1.

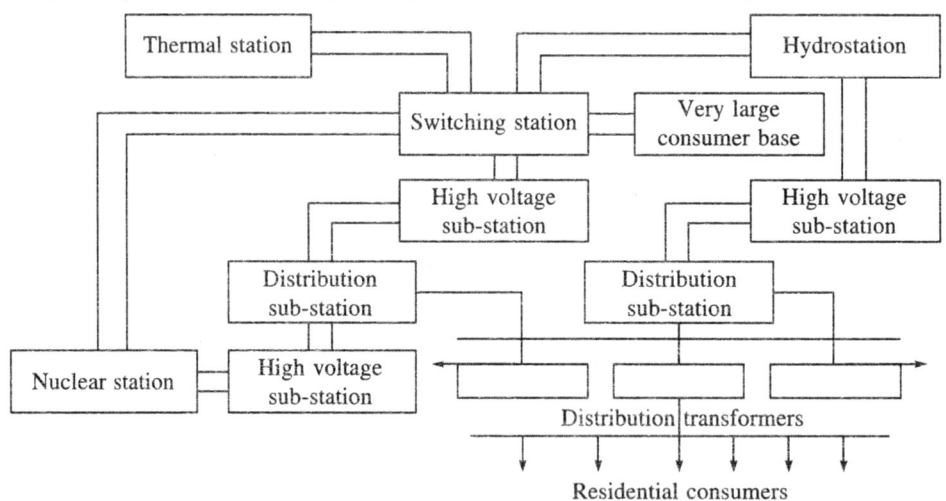

FIGURE 5.1   Basic components of an interconnected power system network.

5.2
WHAT IS THE NEED FOR POWER SYSTEM PROTECTION?

The power system network in any country or region constitutes the backbone of its economic development and is necessary for improving the quality of human life. This necessitates the effective, reliable and efficient operation of the power system. For economic reasons, the components of a power system, i.e. alternators, transformers, reactors, circuit-breakers, busbars, transmission lines, distribution lines, etc. are all generally designed for normal operation of the network. In the event of an abnormality in any section of the power system network, it is essential to isolate the faulty section from the healthy sections otherwise the spread of the abnormal current and overvoltage will adversely affect the healthy sections and may result in a major blackout.

So, if any abnormality takes place in any section of the power network, the affected component should immediately be isolated from the rest of the network in order to restore power in the remaining sections. The isolation of the faulty section/component from the rest of the sections is undertaken by the protection scheme, which includes protective relays and switchgears, i.e. circuit-breakers. It is thus imperative to design an appropriate protection scheme for the proposed power system at the planning stage itself. The primary function of the relay is to operate only if a fault occurs and not under normal operating conditions. A protective relay senses the fault, identifies the location of the fault and sends a command signal to the appropriate circuit-breaker (a circuit interrupting device which operates by closing its trip coil) which disconnects the faulted section.

Under abnormal conditions, changes occur in the basic electrical quantities like current, voltage, frequency and phase angle. In order to detect abnormal conditions in any section of the power system, a relay utilises one or more of these electrical quantities. Abnormal conditions in the power systems include:

- Short circuits in the transmission or distribution lines
- Overvoltages due to switching or lightning
- Overspeeding of generators or motors
- Loss of excitation of the machines
- Overheating of the stator and rotor of the machines
- Insulation breakdown between the intercoils of the windings
- Low oil levels in the transformers and circuit-breakers.

It is thus evident that even if all possible precautions are taken in the power system design, there is a need to design a proper power protection system as well, in order to detect intolerable or undesirable conditions such as above within an assigned area. The protective relay system so designed must therefore be reliable and fast in operation, besides being maintainable.

5.3
DEVELOPMENT OF PROTECTIVE RELAYS

In the beginning, small generators with smaller electrical networks were used. For automatic protection, fuses were used to isolate the faulted sections. But the fuses had to be replaced before the restoration of the supply could be effected. This demerit of the fuses was overcome by the induction of the attracted armature type electromagnetic relays in 1902. These relays were based on the overcurrent principle. With development and progress, however, a need for *selectivity* (a relay system that provides maximum possible service continuity with minimum system disconnection) was felt. In order to fulfil this requirement for selectivity, induction disc type inverse time current relays came into practice in late 1909. The concept of directional discrimination of faults was incorporated in these relays. The pilot wires were used for conveying information from one end to the other end of the line. The development of distance relays in the form of impedance started in 1923. Later on, the induction type mho relays with very high precision came into practice. After that, polarised dc relays with better accuracy and sensitivity were developed in 1939.

The first static relays using thermionic valves were designed in the early 1940s. Later, the advent of transistor circuits opened the door for the development of many new protection concepts like the phase comparator, block spike comparator, block average comparator, etc. Multi-input comparators with quadrilateral characteristics were developed in the 1960s. Static relays have an advantage in that they reduce the burden on current transformers (CTs) and potential transformers (PTs). The other advantages of these relays are that they can be operated fast, have a long life and require less maintenance because of absence of mechanical inertia and troubles caused by pitting of contacts.

After the advent of digital computers and the invention of the concept of the sampling-and-holding the voltage and current signals, it became possible to develop and instal digital (numerical) protection schemes.

Numerical impedance calculation methods are useful in digital techniques for implementation in transmission line protection. In the early 1980s, microprocessor-based power system protection schemes started replacing digital computers.

The concept of digital computing has stayed the same. But at present EHV and UHV transmission lines need to be protected by very fast acting reliable protecting schemes. For this purpose, digital distance relays have virtually replaced all previous electromechanical relays. The present trend in the field of relay development has seen the emergence of adaptive relaying schemes, and schemes based on artificial intelligence and wavelet algorithms.

5.3.1 Recent Developments

Traditionally, at lower voltage levels, power system protection relied on the measurement of power frequency components. Some new techniques were adopted for digital protection schemes. For ultra-high-speed protection, transient wave protection schemes, and other transient-based techniques have become popular and are in use nowadays. Because of their property of improved stability, adaptive protection schemes have also been widely adopted. Methods based on artificial intelligence (AI) and wavelet transforms have shown improved performance as well.

Thus there are many protection schemes based on different concepts. In every phase of development of power system protection, a compromise must be effected between economy and performance, dependability and security, complexity and simplicity, speed and accuracy, and credibility and conceivability. It is a well-established fact that the faster the fault is cleared, the smaller will be the disturbance that the fault will inflict on the power system network. The development of the relay technology with time is presented in Figure 5.2.

Nowadays fault detection is performed by numerical relays which, in many cases, consist of a digital signal processor (DSP) with additional measuring circuits and output circuits. One can thus expect progressive improvement in protection schemes with increasing speed and processing power of computers in the future.

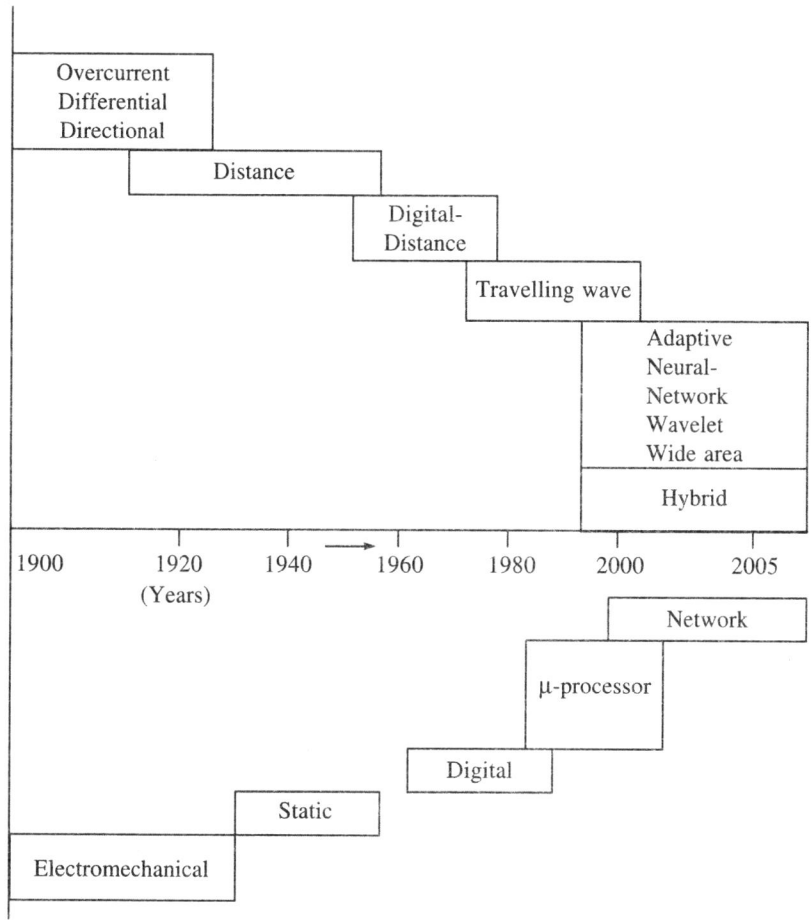

FIGURE 5.2 Development of protection relaying concepts and relays.

5.3.2 How do Protective Relays Operate?

For any location and type of protective relays, in the event of failure, there is some distinctive difference in one or more of the following quantities:

- Magnitude
- Frequency
- Duration
- Phase angle
- Rate of change
- Order of change
- Wave shape

Thus the relays, which may be of any type, simply recognise a particular difference and operate in response to the level of difference.

5.4

BASIC PHILOSOPHY OF PROTECTION SCHEMES

The three main aspects on the basis of which power system protection is analysed are discussed below.

(a) *Normal operation* means that there are no failures of equipment and/or no mistakes committed by personnel. This also means that there is no effect of any natural calamity on the power system.

(b) *Prevention of electric failure* means the provision of features in the design that are aimed at preventing failures. These features include adequate insulation, insulation co-ordination in accordance with the capabilities of lightning arresters, providing overhead ground wires and low tower footing resistance, and adoption of proper operation and maintenance procedures.

(c) *Mitigation of the effects of electric failure when it occurs* implies that certain protection provisions should be incorporated in the system. These provisions include limiting the magnitude of short-circuit current, provisions for promptly disconnecting the faulted sections, and features that investigate and send out an alert in the case of an incipient fault.

The principal function of protective relaying is to promptly remove any faulted section of the system from service. The fault conditions range from

(a) the balanced three-phase short circuit.

(b) to various unbalanced faults that cause the system to operate in any abnormal manner.

The secondary function of protective relaying is to:

(a) Provide information on location and type of faults. This information assists in expediting the repair.

(b) Distinguish between the normal and the faulted sections of the system properly.

5.4.1   Protection Zones

A power system network consists of different types of equipment and associated elements like generators, transformers, primary transmission, secondary transmission, primary distribution, secondary distribution, sub-stations, etc. Protective schemes depend on the type and nature of the equipment to be protected. It is thus wise to divide an electric power system into different protective zones, which is why power protection is classified as generator protection, transformer protection, transmission line protection, motor protection, busbar protection, etc. The planning for protective zones is done in such a way that the zones cover the entire power network collectively. Overlapping of the zones is allowed in order to avoid the unprotected (blind) areas. If an area remains unprotected, it means that any fault occurring in that area would not be cleared at all and such an area is called *blind spot*. For a fault occurring in the area of two overlapped zones, the circuit-breakers in both the directions will isolate the faulted area. The protection zones of a typical power system are shown in Figure 5.3. Thus, overlapping will trip all the circuit-breakers in both the zones even though a fault may have occurred on one side.

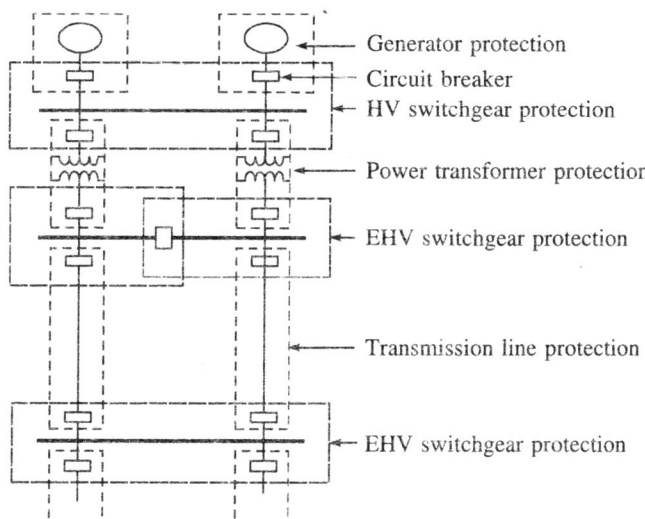

FIGURE 5.3   Protection zones of an electric power system.

5.4.2   Primary Protection

Primary protection or main protection is applied as the first line of defence. In Figure 5.3, it can be observed that the circuit-breakers are located in the connections to each power system element. So, disconnection of the faulted zone is possible only by tripping the circuit-breaker nearest the fault. In the case of overlapping of the two adjacent zones, more than one circuit-breaker may trip (including the really required one). And, this is unavoidable for ensuring a high degree of reliability. This is a preferred practice because blind spots cannot be left in the network. Since protective schemes are intended to protect the system in all eventualities, their reliability must approach the 100 per cent level and if not 100 per cent, it should definitely not be below 95 per cent. The desired degree of reliability can be achieved by designing the system properly and ensuring the proper installation and maintenance of the relays, circuit-breakers, trip mechanisms, transformers, wiring and other elements.

5.4.3   Back-up Protection

Back-up protection is applied as the second line of defence. Back-up relaying works if the primary protection fails to cause the tripping of the circuit-breakers. Back-up protection operates after a time delay to give the primary relay sufficient time to operate. An illustration of the back-up protection of transmission line XY is shown in Figure 5.4.

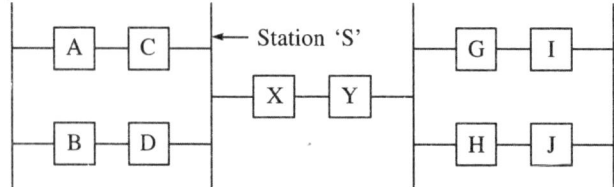

FIGURE 5.4   Illustration of the back-up protection of transmission line section XY.

If a fault occurs at station S, the back-up relays at locations A, B and Y provide back-up protection. In the same way, the back-up relays at A and Y provide back-up protection for faults occurring in line DB. When the back-up relaying functions, a larger part of the network is disconnected in comparison to the situation when the primary relaying operates correctly. This is because back-up relaying operates independently of those factors that might cause the primary relaying to fail. Back-up relaying must operate with a sufficient time delay in order to provide primary relaying sufficient time to function if it is able to.

A given set of back-up relays provides incidental back-up protection for faults in the circuit whose breaker is controlled by the back-up relays. Back-up relays are of the following three types:

1. *Relay back-up:* This is a local back-up scheme in which the main relays and their CTs, PTs, transducers, etc. are duplicated.

2. *Breaker back-up:* It is necessary to have a breaker back-up when a number of circuit-breakers are connected to a busbar. In this scheme, a time delay relay operates as part of the main relay and is connected in such a way as to trip all the other breakers on the bus if the proper breaker does not trip within the preset time.

3. *Remote back-up:* Back-up relays located in the neighbouring station provide back-up to the entire primary protective scheme, which includes the relay and other elements. This is mostly used as the cheapest back-up.

5.4.4   Quality Requirements of Protection

Selectivity (discrimination), reliability, sensitivity, stability and speed are the essential qualities that must be fulfilled by the protective scheme. These are discussed in detail below.

Selectivity or discrimination

This is the property of the system which enables it to discriminate conditions for which prompt operation is required from those for which no operation or time delay operation is required. For example, a relay should be able to discriminate or distinguish between a fault and an overload. The inrush of the magnetising current of the transformer may be comparable to the fault current. The relay should discriminate between these two and should not operate for the inrush current. Power swings in an interconnected system should also be ignored by the transmission line distance relays. Distance relays operate according to the *distance* between the relay's current transformer and the fault.

Reliability

Reliability implies the probability of failure. If the protection system fails to function properly, the interdependent mitigation features become largely ineffective. Thus it is imperative that the protective scheme be inherently reliable. The relaying scheme can be kept in proper condition through careful maintenance and accurate record-keeping of all the elements including the relay, circuit-breaker, PT, CT, transducers, battery and wiring, among other things. Simplicity and robustness of the relaying equipment also ensures their reliability under adverse system and environmental conditions. A typical value of the reliability of a protective scheme must not be less than 95 per cent.

We can therefore say that incorporating inherent design features, ensuring perfection in manufacturing practices, undertaking regular and thorough repair and maintenance of the protective equipment and appointing well-trained quality personnel to operate the system can all make the protective system reliable.

Sensitivity

The minimum level of the fault current (pick-up value) at which the relay operation takes place decides the sensitivity of the system. This fault setting is referred to the primary of the transducers. The sensitivity of the relay is expressed in the apparent power, i.e. the VA required to cause the relay operation. It implies that a 1 VA relay is more sensitive than a 2 VA relay. The sensitivity of the system depends on the sensitivity of the relays applied in the system.

A good relay should be sensitive enough to operate when the operating current just exceeds its pick-up value. The relay should not operate if the fault current is below its pick-up value.

Stability

Stability is that quality of the protective system which allows it to remain stable even if a large current flows in its protective zone due to a fault in the external zone. Even if the fault does not lie in its zone, the connected circuit-breaker is supposed to clear the fault.

Speed

The main goal of a protective system is to disconnect a faulted system element in the least possible time. The faster the speed of operation of the relaying scheme, the lesser will be the damage to the equipment and the network. Generally, the operating time of a protective relay constitutes one cycle on a 50-Hz base. Half-cycle relays are also used. The merits of fast fault clearing are that it:

- Helps avoid permanent damage to equipment and components of the apparatus.
- Reduces the chances of risks like fire hazards.
- Lowers the level of risk to the life of working personnel.
- Maintains the continuity of the power supply.
- Brings back the power system to the normal state sooner.
- Greatly improves the transient state stability limit of the power system.

5.5

CLASSIFICATION OF RELAYS BASED ON THEIR CONSTRUCTION

One way for classifying relays is according to their construction such as:

- Electromechanical relays
- Thermal relays
- Transductor relays
- Rectifier bridge relays
- Electronic relays
- Static relays
- Digital relays
- Microprocessor-based relays
- Network relays

These relays are described in detail in the following sections.

5.6

ELECTROMECHANICAL RELAYS

In electromechanical type of relays, when the actuating quantity exceeds a predetermined setting, a torque is developed which causes the moving element of the relay to travel and close a set of contacts in order to energise the trip coil of the circuit-breaker. The different types of electromagnetic relays as per their construction are:

- Attracted armature type
- Induction disc type
- Induction cup type
- Printed disc type
- Moving coil type
- Polarised moving iron type

Each of these relays is described in detail below.

5.6.1 Attracted Armature Type

This type of relay operates through an armature which is attracted to the poles of an electromagnet or through a plunger which is drawn into a solenoid. Either alternating or direct current can operate these relays. Direct current develops a constant torque but if alternating current symbolises the actuating quantities, the electromagnetic force exerted on the moving element, i.e. the armature or plunger, is proportional to the square of the flux in the air-gap. And if we neglect the effect of the saturation, the total actuating force may be expressed as

$$F = K_1 I^2 - K_2$$

where

  F = net force

  K_1 = a force conversion constant

  I = rms value of the actuating coil current

  K_2 = restraining force including friction

At the verge of picking up the relay, the net force is zero because the force generated by the actuating quantity is equal to the restraining force. And the operating characteristic is

$$K_1 I^2 = K_2$$

or

$$I = \sqrt{\frac{K_2}{K_1}} = \text{constant}$$

For higher fault current generally, pick-up current requirement is proportinately less as compared to the low fault current. This is described below.

When the magnitude of the actuating current I increases the predetermined setting, the net force causes the relay to operate by moving the moving element of the relay. With the picking up of the relay, the air-gap is shortened and a relatively smaller value of the coil current is now required to keep the relay picked up than that required to pick it up. This is considerably pronounced in dc relays. The reset is generally in the order of 90–95 per cent of the pick-up for ac relays and 60–90 per cent of the pick-up for dc relays. Adjusting the initial air-gap facilitates the adjustment of the pick-up. A higher pick-up calibration generally has a lower ratio of reset to pick-up. However, the reset value is of no consequence if the circuit-breaker operates at the instant at which the fault current passes through its zero. The plunger or armature type relays do not lend themselves to directional control, which is the reason why these relays are not used in directional protection relaying schemes.

The transients affect these relays because they operate very quickly, regardless of whether they are actuated by dc or ac quantities. This effect must be taken into consideration when the adjustment for any application is being determined. The operating time is less than 5 milli-seconds for modern relays, which are faster than the induction disc type or induction cup type relays, and are compact, robust and reliable. With one contact, the VA burden of this type of relay is about 0.08 W at pick-up. An attracted armature type relay is shown in Figure 5.5.

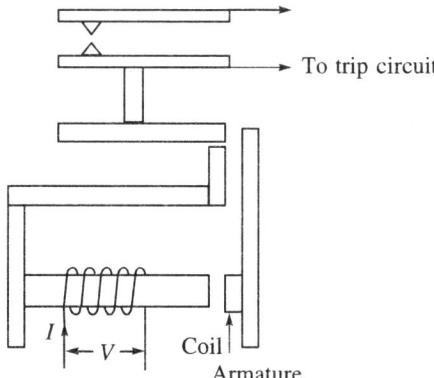

FIGURE 5.5 Attracted armature type relay.

5.6.2   Induction Disc Relay

The operation of this type of relay is based on the principle of electromagnetic induction. *Induction relays thus work only on ac quantities.* A disc is the moving element in the induction disc type relay and has the following two types of structures:

(a) Shaded pole structure

(b) Wattmetric structure

Shaded pole structure

In this, the disc, which is a moving element, is placed between the shaded and unshaded poles of the relay and mounted on a shaft. The disc is made of aluminium so as to have low inertia and, therefore, requires less deflecting torque for its movement. This relay also consists of a magnetic core carrying the exciting/operating coil. The operating coil is fed by the current proportional to the system current. The copper shading rings split the flux produced in the air-gap into two out-of-phase components. Its working principle is similar to that of a shaded pole type induction motor. The flow of current in the operating coil produces an electromagnetic torque, which, in turn, rotates the disc placed between the poles. The rotating disc closes the relay contacts and hence energises the trip coil of the circuit-breaker. In order to generate a restraining force, a spring or sometimes a weight is attached to the disc, which opposes the movement of the disc. The disc can thus rotate only if the operating torque overcomes the restraining torque. Figure 5.6 depicts a shaded pole type induction disc relay unit.

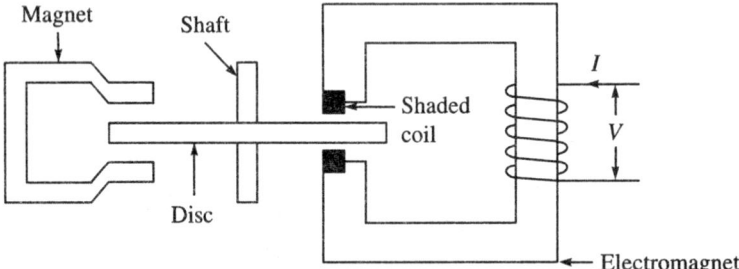

FIGURE 5.6   Shaded pole induction disc relay.

This type of structure is robust and reliable, and is used for overcurrent protection. This relay gives inverse time–current characteristics. It is also slower than an attracted armature type relay and an induction cup type relay. Its VA burden is in order of 2.5 VA, but it depends on the particular application. For the single actuating quantity, the torque is proportional to the square of the actuating current. The theory of torque production in an induction relay is commonly applied to the induction cup relay and induction disc relay; this theory is described in the following sections.

Wattmetric structure

This is the same structure, which is used for watt-hour meters. In this structure, there are two separate coils on two different magnetic circuits, each of which produces one of the two necessary fluxes for driving the disc (rotor). Figure 5.7 shows a wattmetric type of induction disc relay.

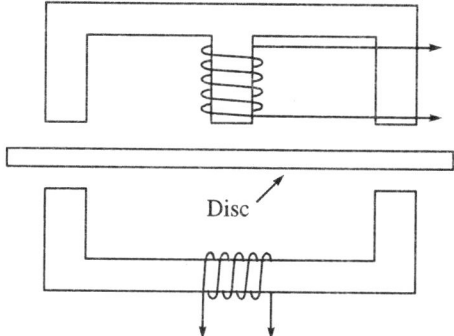

FIGURE 5.7   Wattmetric induction disc relay.

Torque production in an induction relay

When two fluxes, which are displaced in the space and time phase, intersect each other, they produce a torque. Let the flux Φ_1 be produced by the shaded pole and Φ_2 by the unshaded pole. The shaded pole flux lags the unshaded pole by an angle θ. So, we have

$$\Phi_1 = \Phi_m \sin \omega t$$

and
$$\Phi_2 = \Phi'_m \sin (\omega t + \theta)$$

The fluxes Φ_1 and Φ_2 induce voltages e_1 and e_2 in the disc due to the principle of induction. Voltages e_1 and e_2 circulate eddy currents in the disc of the relay. If the inductive effect of the disc is neglected, the eddy currents will be in phase with their respective voltages. The phasor diagram for this is shown in Figure 5.8.

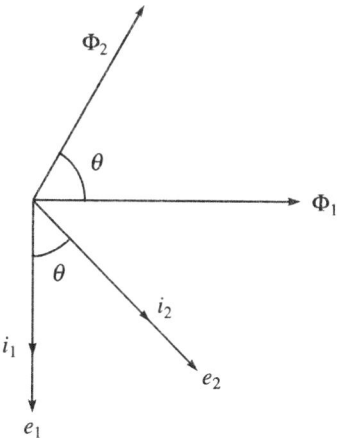

FIGURE 5.8   Phasor diagram for induction relay.

Now

$$e_1 \propto \frac{d\Phi_1}{dt} \propto \Phi_m \, \omega \cos \omega t$$

and

$$e_2 \propto \frac{d\Phi_2}{dt} \propto \Phi'_m \, \omega \cos (\omega t + \theta)$$

If the resistance is assumed to be constant, the eddy current will be directly proportional to the voltage, i.e.

$$i_1 \propto e_1 \propto \Phi_m \, \omega \cos \omega t$$

and

$$i_2 \propto e_2 \propto \Phi'_m \, \omega \cos (\omega t + \theta)$$

The resultant torque will be proportional to the difference $(\Phi_2 i_1 - \Phi_1 i_2)$ because flux Φ_1 interacts with eddy current i_2 and Φ_2 interacts with i_1. Φ_2 is leading Φ_1, hence the torque caused by Φ_2 and i_1 will be positive and that caused by Φ_1 and i_2 will be negative.

The resultant torque T shall thus be:

$$T \propto (\Phi_2 i_1 - \Phi_1 i_2)$$

$$\propto [\{\Phi'_m \sin (\omega t + \theta) \cdot \Phi_m \, \omega \cos \omega t\} - \{\Phi_m \sin \omega t \cdot \Phi'_m \, \omega \cos (\omega t + \theta)\}]$$

$$\propto \Phi_m \Phi'_m \sin \theta$$

When Φ_1 and Φ_2 are displaced by $90°$, the torque will be maximum. The rotation of the disc will take place from the unshaded pole to the shaded pole because Φ_2 leads Φ_1. The maximum torque is of constant magnitude, hence there is no possibility of vibration.

A permanent magnet of high retentivity steel is used for providing a damping torque. The motion of the disc is controlled by adjusting the position of the magnet. The ampere-turns requirement is designed to generate a minimum torque, which is required for the movement of the disc.

5.6.3 Induction Cup Relay

The construction of induction cup relay is similar to that of an induction motor, except that the rotor iron is stationary, leaving only the rotor conductor portion free to rotate. This structure produces a torque more efficiently than the shaded pole or the wattmetric structures because this construction keeps the magnetic leakage in the magnetic circuit to a minimum. An induction cup relay is more sensitive than an induction disc relay which makes it suitable for high-speed relaying schemes. Its operating time is about 0.01 second.

In this relay, the stationary iron core is placed between four or more electromagnets. The rotor is a hollow cylindrical cup that is free to rotate in the gap between the stationary iron core and the electromagnets. After getting energised, the electromagnets induce voltages in the rotor cup and then eddy currents. The eddy current caused by the flux of one pole interacts with the flux due to the other pole, and this interaction produces the working torque.

The reset-to-pickup ratio of induction cup relays varies from 95 to 100 per cent. Because of the friction and imperfect compensation of the control spring torque, the reset-to-pickup ratio of induction cup relays may not be 100 per cent. Figure 5.9 shows the construction of an induction cup relay.

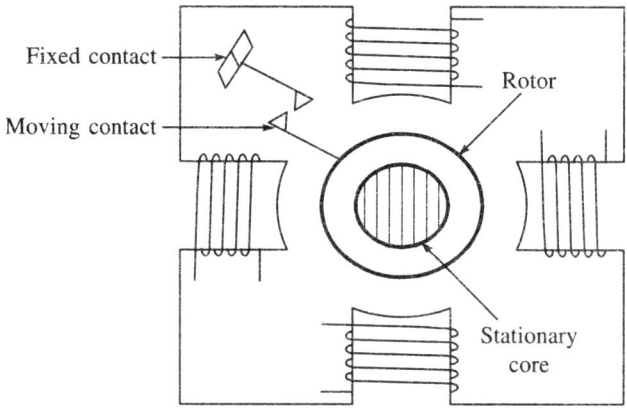

FIGURE 5.9   Induction cup relay.

Due to low weight of the rotor and efficient magnetic circuit, the VA burden in this relay is about three times less than that of induction disc type construction. This relay is widely used for distance and directional relaying schemes.

5.6.4   Printed Disc Relay

A printed disc relay works on the same principle as that used in dynamometer type instruments. A permanent magnet is used to produce a magnetic field in this relay. The output current of the CT is first fed to a rectifier and then to the printed disc. The construction of a printed disc relay is shown in Figure 5.10.

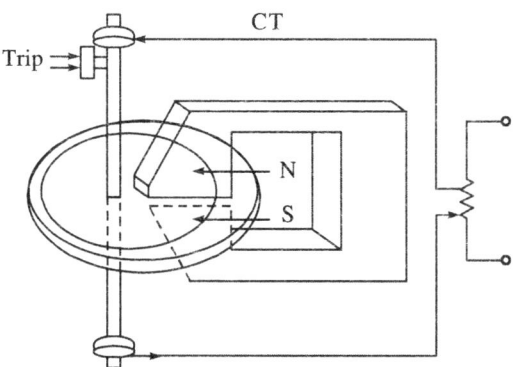

FIGURE 5.10   Printed disc relay.

A printed disc relay is usually 50 to 100 times more efficient in comparison to induction disc type relays. The basic principle of the torque production of this relay is that a current carrying conductor is placed in a magnetic field, resulting in the development of a torque.

The inclusion of a nonlinear network between the printed circuit of the disc and the rectified current input broadens the application of the printed disc relay.

5.6.5   Moving Coil Relay

This relay responds to a dc actuating quantity only and is also known as a permanent magnet moving coil relay or a polarised dc moving coil relay. This is a very sensitive relay and its sensitivity is in the order of 0.1 mW. Figure 5.11 shows the construction of a moving coil relay. Among the main components of this relay are a coil wound on a non-magnetic (aluminium) former, a permanent magnet, an iron core, a jewelled bearing, a spindle, and a phosphor bronze spiral spring.

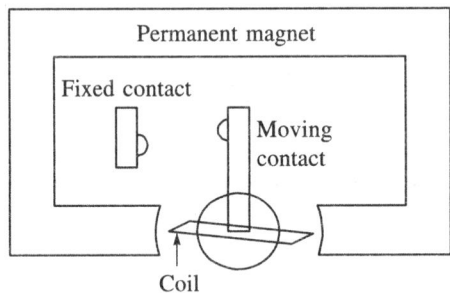

FIGURE 5.11   Moving coil relay.

An interaction between the field of the permanent magnet and of the coil produces an operating torque. The aluminium former is used to provide damping. Its operating time is about two cycles. The moving coil relay has inverse operating time–current characteristics and its operating torque is proportional to the current in the coil. It is used as a slave relay with rectifier bridge comparators.

5.6.6   Polarised Moving Iron Relay

A permanent magnet is used for polarisation, which increases the sensitivity of the relay. This relay is used only for direct current. Relay sensitivity can be enhanced in the range of 1.0 microwatt by using transistor amplifiers. It is also used as a slave relay with rectifier bridge comparators.

The permanent magnet produces a flux in addition to the main flux. It is more robust than the moving coil type relay because its current-carrying coil is stationary. The operating time of this relay is around 10 milliseconds. A polarised moving iron relay is shown in Figure 5.12.

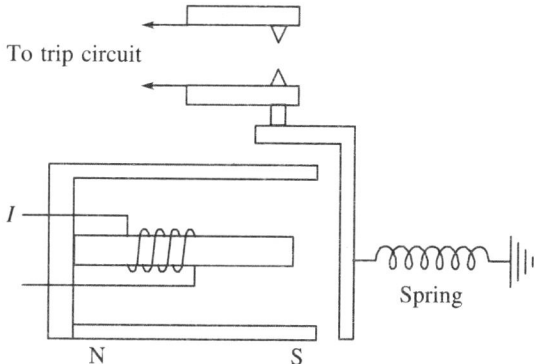

FIGURE 5.12 Polarised moving iron relay.

5.6.7 Demerits of Electromechanical Relays

Due to the fact that they have moving parts and contacts, the electromagnetic relays have some problematic drawbacks, which are listed below.

(a) The vibrations caused by external shocks affect the operation.

(b) The inertia of the moving parts sometimes results in defects in the contact racing. The relay coordination gets disturbed by any defect in the contact racing.

(c) Contact racing results in a high burden on the CTs and PTs.

(d) Contact pitting deteriorates the contacts and the relay starts mal-operating.

(e) These relays require a high operating time due to the inertia of the moving parts.

(f) The wear and tear of the moving parts necessitates frequent maintenance of these relays.

5.7
THERMAL RELAYS

Thermal relays are generally used for the protection of the apparatus. These relays work on the principle of indirect temperature measurement. When an abnormally high current flows through the circuit, the current raises the temperature due to i^2R heating. The main working element of the thermal relay is a bimetallic strip through which the fault current is made to pass. A bimetallic strip has two different metals of different thermal expansion coefficients. As the combined strip is heated up by the passage of the current, it gets deflected through a system of levers, which closes the relay contacts. The thermal relays are not as sensitive as the induction cup relays. They also have a very low drop-out rate because of the large thermal constant. Figure 5.13 shows the construction of a thermal relay.

The bimetallic strip is fixed at one end and heating by fault current causes uneven expansion of the strip at the other end. Uneven expansion of the strip causes it to bend upwards which deflects the cam. The movement of the cam causes the relay to operate. Thermal relays are used for the protection of transformers, motors, etc.

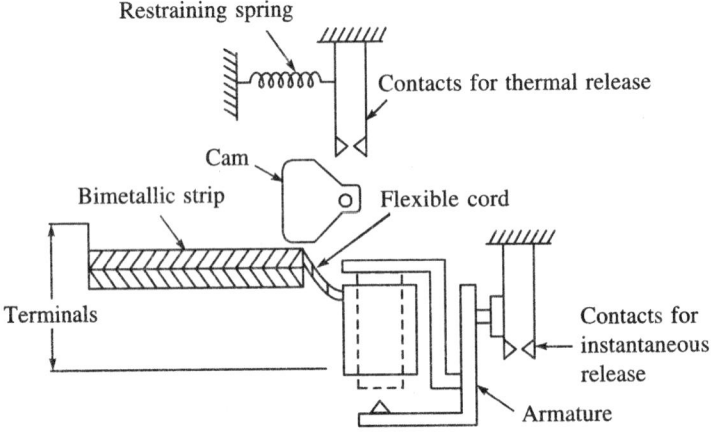

FIGURE 5.13 Thermal relay.

5.8
TRANSDUCTOR RELAYS

Transductor relays work on the principle of variation in the inductance. The unsaturated core of the coil offers a very high impedance whereas if the core is saturated, the coil is like a short-circuited circuit element. So, the nonlinear B–H curve of the magnetic core is used to operate this relay.

5.9
RECTIFIER BRIDGE RELAYS

In order to produce a working torque, electromagnetic action is utilised in rectifier bridge relays. After full-wave rectification, the signal is applied to the relay coils. These relays are used as both amplitude comparators and phase comparators.

5.10
ELECTRONIC RELAYS

These relays are made up of electronic components and do not have any moving part. They were developed during the 1940s. Because of the absence of the moving parts, these relays do not have all the drawbacks associated with electromagnetic relays.

5.11

CLASSIFICATION OF RELAYS BASED ON THEIR FUNCTIONS

Another way of classifying relays is according to their function such as:

- Overcurrent relays
- Undervoltage relays
- Underfrequency relays
- Directional relays
- Thermal relays
- Phase sequence relays, i.e. negative sequence and zero-sequence relays
- Differential relays and percentage-biased relays
- Distance relays, i.e. impedance relays, reactance relays, mho relays, offset and restricted relays
- Pilot relays, i.e. wire pilot relays, carrier channel pilot relays, fibre-optic-based relays, and microwave pilot relays
- Adaptive relays
- Neural-network based relays
- Wavelet-based relays
- Wide area-based relays
- Hybrid-based relays

These relays are used for the protection of system components such as busbar, transformer, generator, motor, and transmission line. These relays are described in detail in the following chapters.

5.12

CURRENT TRANSFORMERS (CTs)

Current transformers are required to step down the fault current (may be in the order of 100 A or more) to the low value of up to 5 A. The low value secondary current of the CT is fed to the measuring instruments for measurement purposes as well to the relays for protection purposes. A CT also isolates the relay circuit from the primary high voltage power circuit.

Current transformers used for protection schemes are designed to give the correct step-down ratio for several times the rated primary current. But the CTs used for instrumentation need to be accurate only over the normal working range of the currents.

The accuracy of a CT is expressed in terms of *a ratio error*, i.e.

$$\text{Percentage ratio error} = \frac{NI_S - I_P}{I_P} \times 100$$

where

N = nominal current ratio
   = ratio of the rated primary current and rated secondary current
I_S = secondary current
I_P = primary current

The CT ratio error depends on the exciting current of its core. The desirable CT accuracy (ratio error) is 2 to 3 per cent.

The *CT burden* is the load connected across the secondary of the CT. It is expressed in volt-amperes (VA).

Example 1 A three-phase transformer (33/6.6 KV), delta/star connected with current transformer on the low voltage side has ratio of 400/5. Calculate the ratio of the current transformer on the HV side.

Solution Current transformer on the LV side of the power transformer will have star connection. The line current on the star side of the power transformer

$$= 400 \times \frac{6.6}{33} = 80 \text{ A}$$

Delta connection has been provided in the CT on the star side of the power transformer. So, the current on the relay side of CT = 5 A.

And phase current in the CT secondary $= \dfrac{5}{\sqrt{3}}$ A

C.T. ratio on the HV side $= 80 : \dfrac{5}{\sqrt{3}}$

Answer: $80 : \dfrac{5}{\sqrt{3}}$

Note: Using the following MATLAB programming, we can solve the above problem.

```
clear
clc
%--------------Data Given-------------
%Transformer is 33/6.6 KV, Delta/Star connected
Vlv=6.6*10^3;
Vhv=33*10^3;
%% Vlv and Vhv are the LV and HV side voltage respectively
Nlv=400/5;
%Nlv=CT ratio on the LV side of the transformer
Ilv=400;
%Ilv=line current on the LV (star) side of the power
    transformer
Ictlv=5;
%Ictlv=the current on the relay (line current on the
    secondary) side of CT used on LV side
%--------------Solution----------------
%Current Transformer on LV side of the power transformer will
    have star connection %and delta connection on HVside of
    power transformer
Ihv=(sqrt(3)*Ilv*Vlv)/(sqrt(3)*Vhv);
%Ihv=line current on the HV (star) side of the power
    transformer
```

```
Icthv=5;
%Icthv=the current on the relay (line current on the
    secondary) side of CT used on HV side
Icthvp=Icthv/sqrt(3);
% Icthvp=the phase current on the secondary side of CT used on
    HV side
Nhv=Ihv/Icthvp;
%Nhv=CT ratio on the LV side of the transformer
%ANSWER: C.T. ratio on the HV side (Nhv)= 80:5/√3 OR
    27.7128:1.0000
```

5.13
POTENTIAL TRANSFORMERS (PTs)

Potential transformers are needed to step down the system voltage to a level suitable for measuring instruments and protective relays. The protective relay voltage rating is usually 110 V. The percentage ratio error for the PT is

$$\text{Percentage ratio error} = \frac{KV_S - V_P}{V_P} \times 100$$

where

K = nominal voltage ratio

= ratio of the rated primary voltage and rated secondary voltage

V_S = secondary voltage

V_P = primary voltage

The desirable PT ratio error is ±3%.

5.14
SUMMATION TRANSFORMERS

Summation transformers are sometimes required to derive single-phase quantities from three-phase fault current and voltage inputs. In a few cases and fault conditions, the current output of the summation transformer is found to be very small or negligibly zero. Special types of sequence filters are used to overcome this drawback.

5.15
PHASE–SEQUENCE CURRENT SEGREGATING NETWORK

A phase–sequence current segregating network is a phase–sequence filter circuit. The output of this circuit is just like the summation device, i.e. a summation transformer. The output current of this circuit can also be written as

$$I_{output} = K_0 I_0 + K_1 I_1 + K_2 I_2$$

where the constants K_0, K_1 and K_2 depend on the device on which the derivation of a single-phase quantity from a three-phase quantity is required.

5.16
ROUTINE CHECKS FOR SENSITIVITIES AND SECURE PERFORMANCE

The list of field checks required during installation and/or routine maintenance is given below.

Instrument transformer checks

(a) Validation of the expected secondary PT voltage and the CT current being received by the relay should be done by verifying the PT ratio and the CT ratio against those voltages and currents marked on the instrument nameplate.

(b) The operation of the PT neutral fuse shifts the PT neutral during ground faults. An open neutral does not allow the relay to operate for zero sequence current. Thus it is important to ascertain that there is no PT neutral fuse in the PT circuits.

(c) Grounding of the CTs and PTs should be carried out at only one place. Multiple grounds on PTs and CTs frequently cause significant zero sequence measurement errors.

(d) The phasing of the current and voltage circuits with proper phase sequence and phase rotation should be verified.

Standing voltage and current checks

A system unbalance or load unbalance can cause an excessive instrument error. Therefore the negative and zero sequence voltages and currents should be measured during normal load conditions and the proper phase balance of CTs and PTs should be ascertained.

Adjacent station comparison checks

Megawatt and Megavars ratings of the new installations should be validated by comparing those values against other proven meters in nearby stations.

Line construction checks

Various line phasings and configurations cause differences in the magnitude of the negative and zero sequence current flow for normal load conditions. In this situation, the determination of the relay settings should be desensitised to accommodate unbalancing caused by an unbalanced line.

Analysis of event reports

The information recorded in the valuable events reports data should be frequently analysed as such frequent analysis of data has been found to be very interesting, less expensive and rewarding.

Analysis of each microprocessor operation

Every single microprocessor relay operation should be analysed to ensure that the overall protection system is secure and reliable. It is imperative to get to the root of the every problem, question and uncertainty, and to carefully review even normal looking operations because they often have clues that can be used to avoid future trouble.

5.17
BASIC DEFINITIONS

In this subject few terms are used very frequently by the readers and service engineers. In order to have clear understanding of these terms, a glossary is given below in which, terms are defined to explain their technical meaning.

Protective gear or equipment: It includes protective relays, PTs, CTs and ancillary equipment to be used in a protective system.

Protective system: A protective system consists of protective gear equipment, which either secures isolation of the faulted element under pre-determined, usually abnormal conditions, or gives out an alarm signal, or performs both functions.

Protective scheme: A scheme, which may consist of several protective systems and is designed to protect one or more elements of a power system, is known as a protective scheme.

Relay: A relay is an automatic device by means of which an electrical circuit is indirectly controlled (opened or closed) and is governed by a change in the same circuit or another electrical circuit.

Protective relay: A protective relay is an automatic device that detects an abnormal condition in an electrical circuit and causes a circuit-breaker to isolate the faulted element of the system. In some cases, it may give out an alarm or visible indication to alert personnel on duty.

Protective zone: A power system network, which is divided into a number of zones from the protection point of view and has separate protective schemes for its protection. A protective relay has its own zone of protection.

Operating force/torque: A force/torque generated by the relay coils which causes the contacts of the relay to close is called the operating force/torque.

Restraining force/torque: A force/torque which opposes the operating force/torque is called the restraining force/torque.

Pickup: The minimum value of the actuating quantity (current, voltage, frequency, etc.) above which the relay begins to operate is called pickup.

Reset time: This is the time which elapses from the moment the actuating quantity falls below its reset value to the instant when the relay comes back to its normal (initial) position.

Drop-out: The threshold value of the actuating quantity (current, voltage, frequency, etc.) below which the relay is de-energised and returns to its normal position or state is called drop-out.

Operating time: It is the time which elapses from the instant at which the actuating quantity exceeds the relay pickup value to the instant at which the relay closes its contacts.

Setting: The value of the actuating quantity at which the relay is set to operate is called its setting.

Reach: The reach observes the maximum length of the line up to which the relay can protect the transmission line. This is usually used in connection with distance (or a component of the impedance) as seen by the relay, which is less than a pre-set value.

Under-reach: When a relay fails to operate even when the fault point is within its reach, (but it is at far end of the protected line), it is called under-reach.

Over-reach: When a relay operates a fault point, which is beyond its present reach (i.e. its protected length), it is called over-reach.

Seal-in relay: This is an auxiliary relay that is energised by the main relay's contacts. Its contacts are fixed in parallel to the main relay and are designed to relieve the contacts of the main relay from their current-carrying duty. It remains in the circuit until the circuit-breaker trips. Generally, the contacts of the seal-in relay are heavier than those of the main relay.

Reinforcing relay: This is also an auxiliary relay that is energised by the contacts of the main relay. Its contacts are in parallel with those of the main relay and it is also designed to relieve the main relay contacts from their current-carrying duty. The reinforcing relay is used to hold a signal from the initiating relay (main relay) for a longer period. As the contacts of the main relay are not robust, they are closed for a short time. But a seal-in relay is designed to remain in the circuit only till the circuit-breaker operates.

Back-up relay: A back-up relay operates after a slight delay, if the main relay fails to operate.

Primary protection: Primary protection acts as the first line of defence and is designed to clear the fault without any delay.

Back-up protection: The back-up protection acts as the second line of defence and is designed to clear the fault if the primary protection fails.

Flag indicator: Flag is a device, which gives a visual indication in the control room, as to whether a relay has operated or not.

Measuring relay: This relay performs measurement to detect an abnormal condition in the system to be protected. It is the main protective relay of the protective scheme, to which the energising quantities are applied.

Auxiliary relays: The main protective relays are assisted by the auxiliary relays. They repeat the operations of protective relays, control switches, etc. and relieve the protective relay of duties like tripping, time lag, sounding an alarm, etc. These may be instantaneous or may have a time delay.

Electromagnetic relays: These relays operate on the electromagnetic principle. Examples of electromagnetic relays include moving iron, moving coil, attracted armature, induction disc and induction cup type relays.

Static relays: These are solid-state relays and employ components like semiconductor diodes, transistors, thyristors, logic gates, ICs, etc. There are no moving parts in these relays.

Microprocessor-based relay: A microprocessor is used to perform the measurement of the electrical quantities or make comparisons, perform computations, and send tripping signals. Thus a microprocessor-based relaying scheme works on the application of a microprocessor in realising all sorts of relaying characteristics.

Overcurrent relay: A relay, which operates when the actuating current exceeds a certain predetermined value, is known as an overcurrent relay.

Undervoltage relay: This relay operates when the system voltage falls below a certain predetermined value.

Polarised relay: A relay, whose operation depends on the direction of current or voltage, is known as a polarised relay.

Directional or reverse power relay: A directional relay senses the direction of the flow of the power and detects whether the point of fault lies in the forward or reverse direction with respect to the relay location.

Time-lag relay: A time-lag relay operates after a certain predetermined time lag. The time lag may be due to the presence of a time-delay component or due to relay's inherent design feature. This relay makes the protection schemes suitable for time discrimination and is thus useful for control and alarm circuits.

Instantaneous relay: This relay works with no intentional time delay in its operation. It operates in a time span of 0.1 second or sometimes even less than this.

Inverse time relay: In this relay, the operating time is inversely proportional to the magnitude of the operating current.

Definite time relay: In this relay, the operating time is independent of the magnitude of the actuating current.

Inverse definite minimum time (IDMT) relay: This type of relay gives an inverse time characteristic at lower values of the operating current and a definite time characteristic at higher values of the operating current.

Induction relay: Induction relays, which include induction disc relays, induction cup relays, etc., operate on the principle of induction.

Moving coil relay: A moving coil relay, also called a permanent magnet dc moving coil relay, has a permanent magnet and a moving coil. The moving coil carries the actuating current.

Moving iron relay: This is a dc-polarised, moving iron type relay. It comprises an electromagnet, a permanent magnet and a moving armature in its construction.

Comparator: It is another name for the relay.

Phase comparator: This relay compares only the phase angle between the input quantities.

Amplitude comparator: This relay compares only the magnitudes of the input quantities.

Instantaneous comparator: This relay compares only the instantaneous values of the input signals.

Dual input comparator: A dual input comparator has two input signals.

Multi-input comparator: A multi-input comparator has more than two input signals.

Printed disc relay: This relay operates on the principle of a dynamometer. In this relay, there is a permanent magnet or an electromagnet and a printed disc. Direct current is fed to the printed circuit of the disc.

Thermal relay: The electrothermal effect of the actuating current is used for the operation of a thermal relay. It operates at a predetermined temperature in the protected component.

Distance relay: A distance relay measures impedance or a component of the impedance, i.e. resistance or reactance between the relay location and the fault. It protects transmission lines.

Impedance relay: An impedance relay measures the impedance between the relay location and the fault. It is a kind of a distance relay.

Modified impedance relay: It is an impedance relay with shifted characteristics. The voltage coil includes some current biasing.

Reactance relay: A reactance relay measures the reactance between the relay location and the fault. It is a kind of a distance relay.

Mho relay (admittance or angle admittance):   This directional relay is also applied in distance relaying schemes. It measures a particular component of the impedance to shift the mho characteristic on the R–X diagram. Its characteristic on the R–X diagram is that of a circle passing through the origin. This relay is also known as an admittance or angle admittance relay.

Conduction relay: The conduction relay is a mho relay whose diameter (passing through the origin) lies on the R-axis.

Offset mho characteristic: For an offset mho relay, the mho characteristic is shifted on the R–X diagram to include the origin.

Angle impedance relay (ohm relay):   The angle impedance relay, also known as ohm relay, has the characteristic of a straight line passing at an angle and cutting both the axes on the R–X diagram.

Quadrilateral relay: Quadrilateral relays are used for distance relaying and have a quadrilateral-like characteristic on the R–X diagram.

Elliptical relay: The elliptical relay, which is another type of distance relay, has an ellipse-like characteristic on the R–X diagram.

Frequency-sensitive relay: This relay operates at a pre-determined value of the system frequency. It may be an underfrequency relay or an overfrequency relay. An underfrequency relay operates if the system frequency falls below a certain value. An overfrequency relay operates if the system frequency exceeds a certain predetermined value of the frequency.

Differential relay: The differential relay operates in response to the difference of two actuating quantities.

Earth fault relay: Relays used for the protection of an element of a power system against earth faults are known as earth fault relays.

Phase fault relay: Relays used for the protection of an element of a power system against phase faults are known as phase fault relays.

Negative sequence relay: The actuating quantity of this relay is the negative phase sequence current. These relays are used to protect electrical machines against overheating due to unbalanced currents.

Zero sequence relay: The actuating quantity of this relay is the zero phase sequence current. This type of a relay is used for earth fault protection.

Starting relay or fault detector: This relay detects abnormal conditions and initiates the operation of other elements of the protective scheme.

Notching relay: The notching relay switches in response to a specific number of applied impulses.

Reliability: The reliability of a protective system is its ability to operate without failure. The reliability of a protective relay should be very high, a typical value being 95 per cent or more.

Sensitivity: Sensitivity is the measure of the degree of the effectiveness of a protective relay. A relay should be sensitive enough to operate when the magnitude of the actuating quantity exceeds its pickup value.

Selectivity or discrimination: The ability of a relay to discriminate between a faulty condition and normal conditions or between a fault within the protected section and outside the protected section is described as selectivity or discrimination.

Stability: Stability is the ability of the protective system to remain inoperative under all normal load conditions, and also in case of faults outside its zone. The relay should remain unaffected and stable in the event of external fault.

Blocking: Blocking implies the prevention of tripping of the relay, which may be due to the operation of an additional relay or due to its own characteristic.

Blinders: A blinder make the relay immune to certain faults.

Burden: The power consumed by the relay circuitry at the rated current is known as its burden.

Unit system of operation: This is the system which is able to detect and respond to a fault occurring only within its own zone of protection having absolute discrimination. Its zone of protection is well-defined. The unit system of protection includes differential protection of alternators, transformers or busbars, frame leakage protection, pilot wire and carrier current protection.

Non-unit system of protection: The non-unit system of protection does not have any absolute discrimination or selectivity. In this system, all the relays may respond to a given fault. For example, non-unit systems of protection symbolise distance protection and are time-graded, current-graded or both time- and current-graded.

Restricted earth fault protection: In restricted earth fault protection, the zone of protection is restricted only to the windings of the alternator or transformer. The scheme responds to the faults occurring within its zone of protection.

Residual current: The algebraic sum of all currents in a multi-phase system, denoted by I_{res}, is known as residual current. In a three-phase system, $I_{res} = I_A + I_B + I_C$

Transactor: This is an air-cored coil, which converts a current signal into voltage.

Transducer: Transducers produce the proper relay signals from the power system.

NC contacts: These are the normally closed contacts.

NO contacts: These are the normally open contacts.

Auxiliary contacts: Auxiliary contacts are contacts other than the main contacts in circuit-breakers.

Threshold operation: This operation is linked to the relay that is about to operate. It means that the fault has occurred at the balance point such as at the reach of the relay.

EXERCISES

1. Describe the nature, the causes and consequences of faults on a power system.
2. Explain the different types of faults according to their severity on the system.
3. Briefly discuss the basic philosophy of a protection scheme.
4. Briefly discuss the role of protective relays in a modern power system.
5. Describe the essential quantities of a protective relay.
6. Justify the statement: *The protective relay is a sensing device.*
7. Justify the statement: *The protective relay is a comparator.*
8. Discuss the zones of protection for a power system network.
9. What do you understand by primary and back-up protection? What is the role of back-up protection? What are the various methods of providing back-up protection?
10. What is meant by the pickup and reset values of an actuating quantity?
11. Describe the selectivity and stability of a protective relay.
12. Discuss the basic difference between the CT used for instrumentation and that used for protection.

13. Explain the concept of summation transformer and its use.

14. Describe how saturation affects the accuracy of CTs.

15. What is a linear coupler? Where is it used?

16. Discuss the different types of PTs with their areas of applications.

17. How is CT burden-specified?

18. Define the following terms:
    (a) Over-reach and under-reach
    (b) Balance point
    (c) Blinder
    (d) Flag indicator
    (e) Burden
    (f) Threshold operation
    (g) Seal-in-relay
    (h) Pick-up
    (i) Drop-out
    (j) Auxiliary contacts

19. Discuss the function of a circuit-breaker. List the different types of circuit-breakers you know. Record their applications in a tabular form.

20. Write short notes on:
    (a) Current transformers
    (b) CT burden
    (c) Potential transformers
    (d) Summation transformers
    (e) Phase-sequence filter

21. Explain the drawbacks of electromechanical relays.

Chapter 6

Protection Schemes

6.1

OVERCURRENT RELAYING

6.1.1 Instantaneous Overcurrent Relays

These relays work without an intentional time delay. Attracted armature type electromagnetic relays constructions are used as instantaneous overcurrent relays. In this relay, an armature (a piece of iron) moves into the field produced by a coil. An energising coil is wound over an iron core mounted on an iron frame. The armature is pivoted in order to make it free to move in line with the magnetic field. The movable contact moves with the movement of the armature and touches the field contact. There is no arrangement for time delay.

In the AC power supply, there is a reversal of the polarity and zero crossing in every half-cycle. Thus, the flux in the attracted armature also passes zero in every half-cycle as a result of which the armature begins to reset and release itself slightly from the pole. A low resistance copper shedding loop is used to surround one part of the split magnetic pole in order to eliminate the effect of passing zero in every half-cycle by the magnetised armature flux. The induced eddy currents in the loop cause a phase lag in the flux passing through the loop as compared to that in the other part of the pole. There is always some hold on the force because the flux does not come to zero on the entire pole face (the pole is split and one part of it is shaded).

During the design stage, due consideration should be given to the susceptibility of the relays towards any maloperation caused by transients and heavy mechanical vibrations.

These relays are useful for ensuring the feeder's protection, wherein current grading is feasible. These relays are also useful for protecting electrical equipment against short circuits, especially in the case of high fault current values.

The construction of the attracted armature-type electromagnetic relay is shown in Figure 6.1.

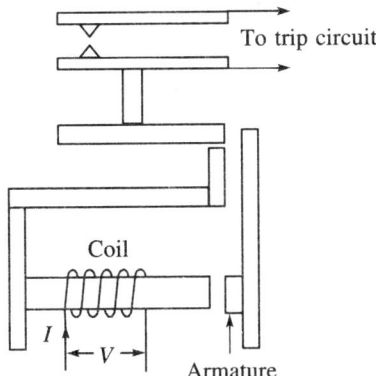

FIGURE 6.1 Attracted armature-type relay.

6.1.2 Time-delayed Overcurrent Relays

In order to achieve selectivity and back-up protection, it is essential to ensure a delay in the relay operation. A time-graded system needs to be applied in these schemes. Time-current grading implies that the operation time is inversely proportional to the fault current. If the fault current is higher, the relay should operate quickly to protect the system by minimising time delay and damage.

The general equation by which the time delay is realised is:

$$T = \frac{K}{[(\text{PSM})^n - 1]}$$

where

T = operation time of the relay

PSM = plug setting multiplier

= [fault current/pick-up current]

K and n = constants determine the operating characteristics of the relay

6.1.3 Definite Time Overcurrent Relays

The preset value of the fault current, which is utilised in these relays, is also termed as the "pick-up" value. This relay operates at the end of a definite set time. Any delay in the relay operation is caused by circuit components known as 'timing units'. These timing units could be any among magnetically operated devices with mechanical damping, motor-operated devices, thermal devices, electrical circuits containing reactance and non-linear impedances, electronic circuits or semiconductor circuits.

These relays are useful for protecting the radial feeder. These relays are also used to generate the overload alarm in generators and to ensure the protection of induction motors. Characteristics of these relay are shown in Figure 6.2.

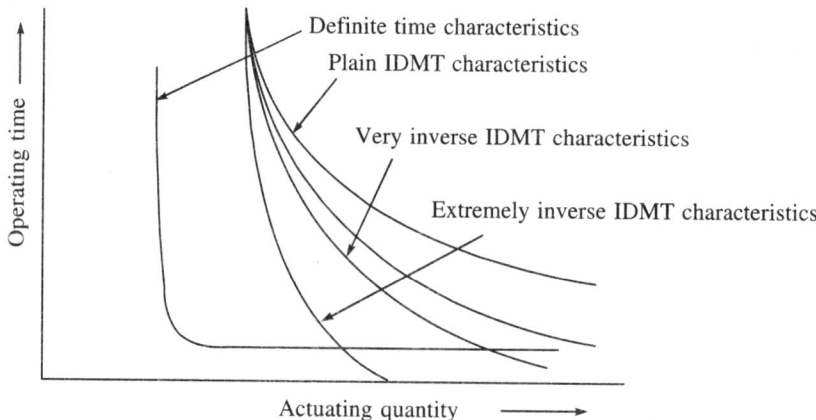

FIGURE 6.2   Characteristics of different types of overcurrent relays.

6.1.4   Inverse Time Overcurrent Relays

The operation time of the inverse time overcurrent relay is inversely proportional to the current passing through the relay coil. In this relay, the central limb of the E-shaped core is provided with a tapped current operating coil and a suitable lag coil is also provided on either one or both of the outer limbs.

The moving contact of the relay is attached with the aluminium disc whose shaft is restrained by a spiral spring. When the counter torque of the spiral spring is overcome by the operating electromagnetic torque, which is produced by the induction element on the disc, the disc starts rotating until the contacts are closed. An electromagnetic induction disc relay, which gives inverse time overcurrent characteristics, is shown in Figure 6.3.

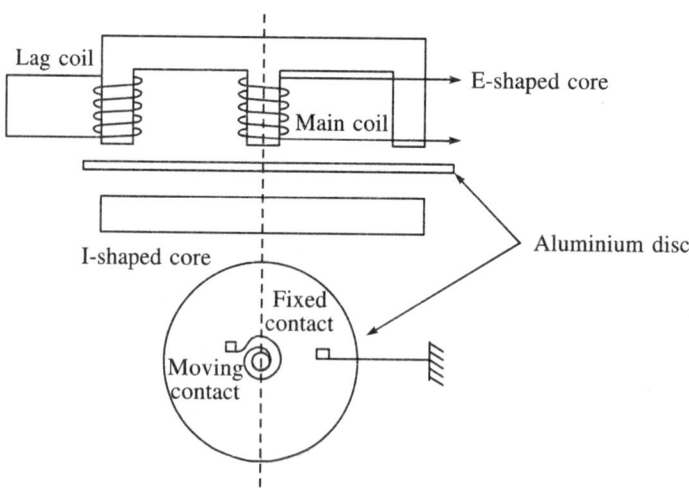

FIGURE 6.3   Electromagnetic induction disc relay, which gives inverse time overcurrent characteristics.

A working torque is produced by the interaction of two mutually displaced (both in space and phase) alternating magnetic fluxes. These two fluxes can be obtained by only a single energising quantity (current) by using a quadrature coil or a short-circuited lag coil. Splitting the pole and shading it by a low impedance-conducting loop can also help achieve the same purpose. The current setting can be controlled by tapping the operating coil because relays operate at a given ampere turn input to the coil.

The range of settings is provided by selecting the number of turns, which is in inverse proportion to each of the current values/current settings. Generally, relays have seven setting values, and cover a current range of 4:1. The taps are spring-loaded Plug Bridge which is a magnitude of current selection device. The taps are selected by inserting a single pin plug into one of seven positions. The disc speed is controlled by the braking system with the help of permanent magnet eddy-current damping. This is achieved by passing the disc between the poles of a horseshoe permanent magnet.

In order to facilitate discriminative time-graded overcurrent protection, disc type induction relays with inverse time-current characteristics are employed. Designing the disc in a proper shape (a slightly spiral shape) can help maintain constant disc speed throughout its movement. Round-shaped discs are generally not used here because the spring torque varies over the angle of travel as a result of which the effective driving torque decreases the speed of disc.

6.1.5 Inverse Definite Minimum Time Overcurrent Relays [IDMT-Type O/C Relay]

For the plug setting multiplier values below 10, an inverse time characteristic is obtained and IDMT characteristics are achieved at lower values of the fault current. IDMT relays are used to ensure distribution line protections. Some relays give inverse characteristics with better selectivity than in the case of plain IDMT relays.

In some situations, however, plain IDMT and inverse IDMT relays fail in selectivity. In these cases, extremely inverse IDMT relays are employed. These are suitable for protecting heavy machines against over heating. An extremely IDMT relay is also suitable for applications in load restoration.

6.1.6 Relay Coordination

Power system protection must possess, stable preventive operation for up to maximum short circuit current. To achieve the objective of no-nuisance tripping, power system protection engineers working on relay coordination must have fundamental knowledge of relay functions, switchgear, protection and associated controls, components of equipment installed in healthy zone or faulty zone, i.e. power transformers, cables, current transformers, potential transformers, circuit-breakers, transmission lines, etc.

Relay coordination is selective relay operation in the power protection system that shall discriminate operation of the specific relay to isolate the faulty section of power distribution from the healthy section of distribution system and to minimize the interruption of healthy section of the power distribution system.

Beyond a fault current level when current transformer gets saturated, the secondary current is not proportional to the primary current and may be much lower than the estimated value. Percentage impedance of the circuit plays very significant role in the relay coordination. It is thumb rule for the practice engineers to recognize the value of *percentage impedance* as the *percentage voltage drop* when the rated base MVA is flowing. For example, if the percentage impedance of the system up to a particular point is 15 per cent on 20 MVA base, it means 20 MVA is flowing and the voltage drop as measured at the point will be 15 per cent of the no-load voltage. Now, voltage drop per MVA is 15/20 such as 0.75 per cent. The no-load voltage of a transformer at normal tap remains generally higher than the system voltage to cope up requirement of maintaining the bus voltage under prescribed limit despite the impedance voltage drop due to rated loading.

A time coordination is necessary for IOC (inverse overcurrent) relay operation at any two successive relaying points. This is achieved by altering the TSM (time setting multiplier) ratio at two relaying points because operating time of an IOC relay is dependent on the ratio of the fault current to the current setting of the relay and not on the actual value of the current. TSM is described in the following topics. Even if the TSM setting is identical at the two relaying points, the time discrimination is necessary, if the CT ratio at two points differ. The reason behind this is because for a given fault current, the ratio of fault current to current setting varies which result in selection of CT ratio and plug setting.

Current transformer polarity must be checked properly, otherwise if it is reversed in the connections to the relay in any one phase, the currents will get imbalanced, i.e. the sum of three-phase current will not be zero in the secondary circuit despite balanced currents in the primary circuit. The secondary circuit unbalanced current due to reversal of polarity of one CT will now flow through the residual circuit operating the earth fault relay even under normal load conditions.

Current settings

Current settings are required to facilitate the relays to be made simple in construction, operating at lower current, cheaper and small in size. Fault current calculation and, therefore, analysis of the system is necessary before coordinating the relays. Students learn the various methods of fault current analysis in other subjects like power system analysis. Overcurrent relay, which is used with current transformer having 5 ampere, normally provides selection blocks for 50, 75, 100, 175 and 200 per cent of the normal rated current of the associated current transformer. It means that the relay windings are designed for 2.5 ampere, 3.75 ampere, 5.0 ampere, 6.25 ampere, 7.5 ampere, 8.75 ampere and 10 ampere, if CT secondary produces 5 ampere. Similarly, for use with CT rated at 1 ampere, normal full load relays are made with settings corresponding to from 0.5 ampere to 2.0 ampere. For some of the relays, time scale is also mentioned.

The time scale is simply a time setting index graduated from zero in fractions to unity, i.e. from 0 to 1.0. An accurate index serves the purpose of the movement of the moving contact of the relay up to desired distance of arc travel. This arc travel of the moving contact turns before the contacts close to initiate the tripping operation.

For reliable and accurate operation of the relay, one should know about:

(i) How to choose relay's plug settings for overcurrent?

(ii) How to choose time of operation, i.e. settings of the circular disc position?

As far as electrical power generating plants are concerned, the general system is radial feeders. Grading of current settings as well as time settings within the limits imposed by normal load conditions are generally recommended.

So, from the above description it is clear that operating time of any current operated relay is a function of current. Selective operation of the relays is required to protect the system within protected zone without affecting power supply to the healthy part of the system designed. The magnitude of the current at which relay should operate and the prescribed time within which it should trip are decided by calculating the plug setting of relays, which are provided with the settings on relays, installed or to be installed. These settings are PSM (plug setting multiplier) and TSM (time setting multiplier).

Relay operating time calculation for any fault current value needs calculation of the plug setting that is multiples of the setting achieved by dividing the fault current by the equivalent primary current of the relay setting. That is known as PSM.

Following steps are involved while calculating the PSM:

Step 1: Calculate the estimated fault current (short circuit current).

Step 2: Identify CT ratio.

Step 3: Adapt suitable percentage of relay setting.

Step 4: Calculate the equivalent primary current (CT primary current × percentage relay setting). This is actually relay setting current (RSC).

Step 5: Determine the equivalent plug setting multiplier (PSM) as

$$PSM = \frac{\text{Estimated fault current}}{\text{Relay setting current}}$$

Time settings

By altering the position of a stop against which the moving contacts of the overcurrent relay resets and thereby the distance between the fixed and the moving contacts is adjusted, the time setting can be done and indicated by a pointer on a calibrated scale. Generally, figures on the scale are actually multipliers for converting into operating times. If a relay is calibrated not to operate up to a particular percentage value say 30 per cent, it starts to operate with the current greater than 30 per cent the set current value. So, the time setting is observed by TSM (time setting multiplier). The TSM is the ratio of the operating time of the relay at a particular TSM setting to the operating time of the relay at a TSM of 1.0. TSM set at 1.0 is the actual time in seconds and indicated on the relay plate in single straight line. It is also indicated on graphical characteristic curve.

Following steps are involved while determining the TSM:

Step 1: Determine the time setting dial.

Step 2: Determine the operating time of the relay when set at 1.0 (means TSM is 1.0). T_m denotes it i.e. operating time of relay when set at 1.0.

Step 3:   Decide the actual time requirement of the relay operation (T).

Step 4:   Calculate the time setting multiplier (TSM) as

   T Actual time of the relay operation required

$$\text{TSM} = \frac{T}{T_m} = \frac{\text{Actual time of the relay operation required}}{\text{Operating time of the relay when set at } 1.0}$$

Note: The deciding parameters for determination of the grading margin, i.e. the time interval between the operations of two adjacent relays are:

*   The fault current interrupting time of the circuit-breaker
*   The overshoot time of the relay
*   Final margin on completion of operation

Example 1   Calculate the PSM suitable for a relay setting of 150%, if the fault current is 1500 A. The CT ratio is 150/5. Also determine the time of operation of the relay corresponding to the PSM. If the time setting dial is set at 0.2 and the time of operation of the relay when set at 1.0 is 1.56.

Solution   *PSM calculation:*
Given:

   Fault current = 1500 A

$$\text{CT ratio} = \frac{150}{5}$$

   % relay setting = 150%

So, the relay setting current, i.e. equivalent primary current

$$= \frac{\text{\% relay setting} \times \text{CT primary current}}{100}$$

$$= \frac{150 \times 150}{100} = 225 \text{ A}$$

Now,

$$\text{PSM} = \frac{\text{Estimated fault current}}{\text{Relay setting current}} = \frac{1500}{225} = 6.67$$

Here the PSM value, 6.67 indicates that the estimated fault current shall be 6.67 times the relay setting current

TSM calculation:
Given:

   Time setting dials that is TSM at PSM, 6.67 = 0.2

   Time of operation of relay when set at TSM, 1.0 that is T_m = 1.56

So, The time of operation of relay when PSM is 6.67

$$= 0.2 \times 1.56$$

$$= 0.312 \text{ s}$$

Answer: 6.67; 0.312 s

Note: Above problem can be solved by using the following MATLAB programming.

```
clear
clc
%--------- data given----------
If=1500;
n=150/5;
%If and n are the fault current and CT ratio respectively
T=0.2;
% T is the time setting dial
Tm=1.56;
% time of operation of relay when time setting dial is set at
    1.0
RS=150;
% RS is the relay setting
% -----------solution-------
np=150;
ns=5;
% np is the primary and secondary CT current respectively
Irs=RS*np/100;
% Irs is the relay setting current
PSM=If/Irs
% PSM is the plug setting multiplier at TSM 0.2
Tp=T*Tm
% Tp is the time of operation of relay when PSM is 6.67
```

Example 2 A relay having rating 5 A is connected to a supply through a current transformer of current ratio 400/5. If a fault current of 3360 A flows in the circuit, determine the time of operation of the relay having setting of 120% and TSM = 0.5. Let the operating time for PSM = 7 is 3 s.

Solution Given:

Fault current such as primary current = 3360 A

TSM = 0.5

Pick-up value of the relay = 5 A

Relay setting = 120%

So,

$$\text{The operating current of the relay} = 5 \times \frac{120}{100} = 6 \text{ A}$$

The plug setting multiplier (PSM) of the relay

$$= \frac{\text{Secondary current (= fault current/CT ratio)}}{\text{Relay setting current, i.e. operating current of the relay}}$$

$$= \frac{3360/(400/5)}{6} = 7$$

Since, the operating time of the relay for PSM = 7 is 3 s

The actual operating time = 3 × TSM

$$= 3 \times 0.5 = 1.5 \text{ s}$$

Answer: 1.5 s

Note: Above problem can be solved by using the following MATLAB programming.

```
clear
clc
% -------------data given--------
If=3360;
n=400/5;
%If and n are the fault current and CT ratio respectively
TSM=0.5;
T=3;
% T is the time of operation of relay when, PSM is 7
Ip=5;
% rated current of relay
RS=120;
% relay setting
% -----------solution-------------
Io=Ip*RS/100;
%Io is the operating current of the relay
Isf=If*(1/n);
% Isf is the secondary fault current
PSM=Isf/Io;
Ta=T*TSM
% Ta is the actual operating time
```

Example 3   IDMT characteristics of the relay are shown in Figure 6.4. Design the time–current grading of the system as given below. TSM at the relay Z is 0.1.

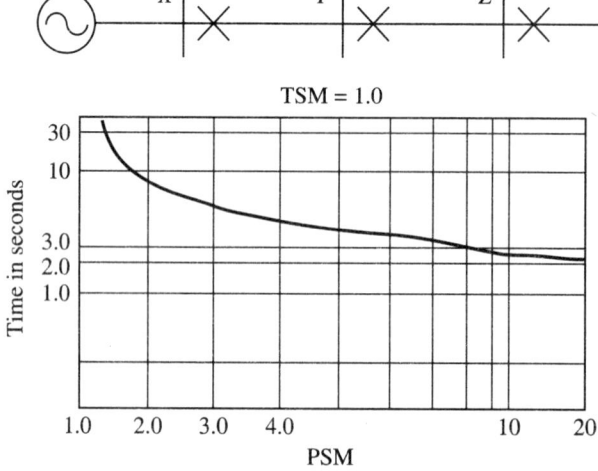

FIGURE 6.4   IDMT curve on log-log graph paper.

Relay point	CT ratio	Fault current	Current setting
X	400/5	5000 A	125%
Y	200/5	4000 A	125%
Z	200/5	2000 A	100%

Solution   Let us take up the relay point Z first.

The current in the secondary of the relay corresponding to the maximum fault current flow of 2000 A

$$\frac{2000}{200/5} = 50 \text{ A}$$

PSM at 100% current setting $= \dfrac{50}{5} = 10$

From the relay characteristics [Figure (6.4)]; the operating time corresponding to the PSM, $10 = 2.8$ s.

For the TSM $= 0.1$, the operating time of the relay at Z

$$= 2.8 \times 0.1 = 0.28 \text{ s}$$

Discrimination between the relays at Z and at Y is required to be achieved because fault may take place just before Z or just after Z without change in the amount of the current.

If 0.5 s is considered as the discrimination time (time of operation of relay at Z plus operation time of the circuit-breaker and plus the travel time of the relay at Y), the operating time of the relay at Y, if the fault occurs at Z

$$= 0.28 + 0.50$$

$$= 0.78 \text{ s}$$

Now, if the fault occurs near the relay at Z, the secondary current in the relay at location Y

$$= \frac{2000}{200/5} = 50 \text{ A}$$

Now, the relay operating current, if the relay setting is 125%

$$= 5 \times \frac{125}{100} = 6.25 \text{ A}$$

So, $\text{PSM} = \dfrac{50}{6.25} = 8$

And the operating time corresponding to PSM, 8 is 3 s from the curve and the operating time of the relay at Y when graded w.r.t. relay at Z is 0.78 s.

So, the TSM of the relay at $Y = \dfrac{0.78}{3} = 0.26$

Now, considering if the fault occurs near Y

$$\text{PSM} = \frac{4000}{6.25 \times (200/5)} = 16$$

And operating time corresponding to this PSM = 2.5 s

So, the actual operating time of the relay at $Y = 2.5 \times 0.26$

$$= 0.65 \text{ s}$$

Here, it should be noted that the CT ratio of the relay at X is higher as compared to the relay at Y, so the current discrimination is inherent.

The percentage setting of the relay at $X = 125\%$

So, the PSM of the relay at X when fault occurs near Y

$$= \frac{4000}{6.25 \times (400/5)} = 8$$

The operating time corresponding to PSM, 8 = 3.0 s

And hence the operating time of the relay at X when graded w.r.t. $Y = 0.65 + 0.50 = 1.15$ s

$$\text{TSM} = \frac{1.15}{3} = 0.38$$

Again, if the fault occurs near X, the PSM will be

$$= \frac{5000}{6.25 \times (400/5)} = 10$$

The operating time corresponding to PSM, 10 = 2.8 s

So, the actual operating time of the relay at $X = 2.8 \times 0.38$

$$= 1.064 \text{ s}$$

> **Answer:** Operating times of the relays at X, Y and Z respectively are 1.064 s, 0.65 s, 0.28 s. Operating time of the relay at X when graded w.r.t. Y is 1.15 s. Operating time of the relay at Y when graded w.r.t. Z is 0.78 s.

Note: Above problem can be solved by using the following MATLAB programming.

```
clear
clc
%---------data given---------
Ifz=2000;
Ify=4000;
Ifx=5000;
%Ifz, Ify & Ifz are fault current flow in the line when fault
   occures near point Z, Y & X respectively
nz=200/5;
ny=200/5;
nx=400/5;
% nz, ny, nx are the CT ratio, CT used at point Z, Y & X
RSz=100;
RSy=125;
RSx=125;
% RSz,RSy & RSx are the relay setting of the relays at point Z,
   Y & X
```

```
ns=5;
% ns is the secondary CT current
%-------------------Solution--------------------------%
%Let us take up the relay point Z first, i.e. Fault occurs near
    Z%
Ifsz=Ifz/nz;
%Ifsz is the secondary fault current in the CT at location Z, if
    fault occurs near the relay at Z.
Irsz=RSz*ns/100;
%Irsz is the relay (at point z) operating current, if the relay
    setting is 100%:
PSMz=Ifsz/Irsz;
% from the relay characteristics [Figure 6.4]; operating time
    (T10) corresponding
% to PSM, 10: = 2.8 s
T10=2.8;
TSM=0.1;
% For the TSM = 0.1, the operating time (Toz) of the relay at Z
    is
Toz=T10*TSM
%Toz is the operating times of the relays at Z
%-------for the relay at point y, but fault occurs near Z------
Td=0.5;
%Td is the discrimination time
Ty=Td+Toz
%Ty is the operating time of the relay at Y when graded w.r.t. Z
Isy=Ifz/ny;
%Isy is the secondary fault current in the CT at location Y, if
    fault occurs near the relay at Z.
Irsy=RSy*ns/100;
%Irsy is the relay (at point y) operating current, if the relay
    setting is 125%:
PSMy=Isy/Irsy;
% from the relay characteristics [Figure 6.4]; operating time
    (T8) corresponding
% to PSM, 8: = 3 s
T8=3;
TSMy=Ty/T8;
%TMS of the relay at Y
%---------------Fault occurs near Y-------------------
Ifsy=Ify/ny;
%Ifsy is the secondary current in the relay at location Y, if
    fault occurs near the relay at Y.
PSMy1=Ifsy/Irsy;
```

```
T16=2.5;
Toy=T16*TMSy
%Toy is the operating times of the relays at y
%--------------for the relay at point x, but fault occurs
    near y-------
Isx=Ify/nx;
%Isx is the secondary current in the relay at location x, if
    fault occurs near the relay at y.
Irsx=RSx*ns/100;
%Irsx is the relay(at point x) operating current, if the relay
    setting is 125%:
PSMx=Isx/Irsx;
Tx=Toy+Td
%Tx is the operating time of the relay at X when graded w.r.t. Y
TSMx1=Tx/T8;
%TSM of the relay at Y
%---------------Fault occurs near X-----------
Ifsx=Ifx/nx;
%Ifsx is the secondary current in the relay at location x, if
    fault occurs near the relay at x.
PSMx1=Ifsx/Irsx;
Tox=T10*TSMx1
% Tox is the operating times of the relays at x
%--------------------------ANSWER--------------------
% operating times of the relays at X, Y and Z respectively are
    1.0733 s, 0.6500 s and 0.2800 s
% operating time of the relay at X when graded w.r.t. Y is
    1.1500 s
% operating time of the relay at Y when graded w.r.t. Z is
    0.7800 s
```

Circuit-breaker operation time

If the relays are set for operation in sequence, successive relays' operation time such as the time interval between the time settings should be finalized first. This time interval depends on interrupting time of the circuit-breakers. It is mandatory requirement that the fault current must be completely interrupted before the discriminating relay ceases to be energized.

The time delay of the relay installed nearest to the most fault prone area should have shortest time delay. The shortest time delay setting is dependent on the minimum possible value of the time multiplier setting. Theoretically, one can say that it must be zero but it is about 0.1 second in order to allow a safe contact gap. For the others, time delay may be multiplier of 0.1 second. The desirable job for an overcurrent inverse time relay is to clear the fault in minimum time. But, if there are a number of circuit-breakers operating in series, proper discrimination and close adjustment of time settings and current settings are necessary. This is necessary to achieve desirable stability with rapid clearance of the fault. *So, the minimum difference in operating times between the relays should be sufficient to cover the time for the circuit-breaker to trip and its arc to be cleared plus overswing preventing time due to inertia. In addition to this a safety margin is also required.*

For overcurrent protection, practical discrimination of 0.4 to 0.5 seconds time interval between successive relays is considered (assuming the relays are operating on the definite minimum part of their characteristics).

For better accuracy of the relay tripping, the two most important features, the actual fault current and the CT ratio must be preferably checked. Because practical values of relay setting depend on the actual faults currents, which may in fact be obtained on the CT ratios at each point.

As far as determination of the circuit-breaker rupturing capacity (please refer chapter 'High Voltage Circuit-breaker') is concerned, circuit resistance is usually neglected during fault KVA calculation. But while calculating the fault KVA for relay setting, circuit resistance as well as reactance must be considered otherwise fault current calculated would be in excess of the true value. After deciding the current setting keeping into mind the provision of adequate protection against continuous overload, the operating time for the corresponding maximum fault current multiplier then is determined from the relay characteristic curve. After this, the time setting multiplier is set to obtain the desired grading interval relative to the other relays. One should not ignore or forget the cross-check for faults at various points in the different sections of the electrical power network.

Earth fault relay settings

Overcurrent relays are used for earth fault protection. These overcurrent relays may be of instantaneous type or have inverse time characteristics but the relays having both the features are better to prefer. While using the overcurrent relay for earth fault protection, it should be ensured that no setting lower than 50 per cent of the current transformer rating is desired. During calculating the anticipated earth fault such as primary currents, the earth-limiting resistor (allowance for neutral point) must be taken into account. The impedance of the shunting effect due to residual current may have considerable effect on the earth fault relay current. The impedance consisting of the CT and overcurrent relay coils vary due to the load variation and the relay current settings accordingly. Earth fault relays are generally provided with a range of settings of 20 to 80 per cent or 10 to 40 per cent expressed as a percentage of the rated secondary current of the CT. However, during deciding the settings for earth fault relays, following points should be considered carefully:

(i) If the ammeters are inserted in the secondary connections, that should be of low impedance (as low as possible) otherwise they may introduce errors.

(ii) To lessen the shunting effect, low overcurrent setting (as low as possible) should be used.

(iii) An injection current approaching maximum fault conditions should be used so that the CT and relay iron cores are in a magnetic state.

(iv) When overcurrent relays are used in combination with three current transformers the earth fault or leakage current relay should be connected on the residual-current principal.

6.2

DIFFERENTIAL RELAYING

6.2.1   Circulating Current Differential Relays

In this relaying scheme, the pick-up current is equal to the difference of the currents coming to the operating coil, and its working principle can be explained by Kirchhoffs Current law. This scheme entails a comparison of currents entering and leaving the equipment. As shown in Figure 6.5, if the currents entering and leaving the equipment are not equal, the difference of the currents flows in the third branch, which is an indication of the fault. The differential current proportional to the fault current passes through the operating coil of the relay and the relay operates if the differential current exceeds a preset value.

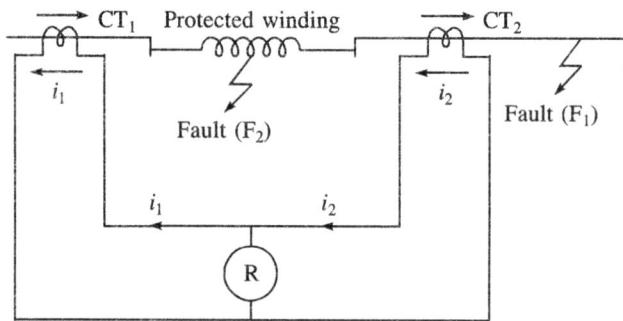

FIGURE 6.5   Circuit for circulating current differential relay.

If we observe the relay circuit shown in Figure 6.5, in case of an external fault, say at F_1, the current entering and leaving the protected winding of the equipment will be the same in magnitude and phase relation. It means that the instantaneous values of the secondary of the CT, i_1 and i_2, are the same at all the instants. Thus the difference $(i_1 - i_2)$ will be zero and the relay does not operate. But if the fault occurs within the protected zone, say at F_2, the unequal fault currents, say i_1 and i_2, will be distributed in both the directions and the differential or spill current $(i_1 - i_2)$ will flow through the operating coil of the relay. As this spill current exceeds the pick-up value, relays operate and isolate the protected equipment from the system.

In this scheme, care should be taken to correctly connect the polarity of the CTs, otherwise the third branch will get a current $(i_1 + i_2)$ in the place of current $(i_1 - i_2)$.

Circulating current differential protection is applied for the differential protection of generators, restricted earth fault protection of the transformer, busbar differential protection and for protection of motors.

6.2.2   Opposed Voltage Differential Relays

This scheme is illustrated in Figure 6.6. In this scheme, the voltage induced in the secondary winding of CT_1 is proportional to the current i_1 and the voltage induced in the secondary winding of CT_2 is proportional to the current i_2. Thus, the current proportional to the difference

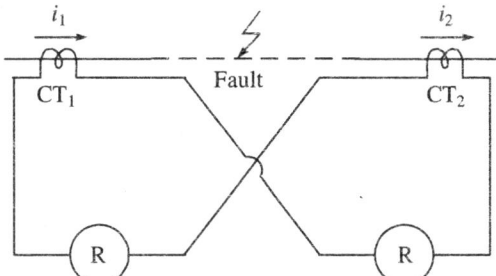

R signifies the instantaneous overcurrent relay

FIGURE 6.6   Opposed voltage differential relay.

of the currents $(i_1 - i_2)$ flows through the operating coil of the relay. Since the CTs are identical on both sides, in case of external faults and normal operating conditions, the secondary currents of the CTs oppose each other and the induced voltages are balanced by giving the difference between the voltages across the operating coil zero.

During an internal fault, however, a spill current proportional to $(i_1 - i_2)$, which may be proportional to $(i_1 + i_2)$ if there is a fault at both the ends, flows through the operating coil of the relay. If this current exceeds the pick-up current, the relay operates and isolates the protected equipment from the system.

In this scheme, no current flows through the secondaries of the CTs under normal conditions, which facilitates its application in the pilot wire unit protection of small feeders. Since the current in the secondaries of the CTs is zero under normal conditions, all the primary ampere-turns are used for exciting the transformer.

6.2.3   Biased (Percentage) Differential Relays

The circulating current differential protection scheme ensures absolute discrimination because both the CTs employed are exactly identical (theoretically) and have the same saturation characteristics. But in actual practice, the two CTs are never seen to have absolutely identical characteristics, even if they are manufactured by the same manufacturer.

This differences in the characteristics of the CTs leads to the flow of spill current even if there is no internal fault. If this spill current exceeds the preset value of the relay, it may result in an undesirable relay operation.

The biased or percentage differential relay has two coils, one of which is the biased coil, that operates as the restraining coil, while the other is the operating coil. The relay operates when the operating torque produced by the operating coil exceeds the restraining torque. This occurs if there is a heavy external fault. Thus biasing the differential relay is needed to prevent this type of an unwanted relay operation. The biasing is done by connecting the operating coil at the mid-point of the restraining coil as shown in Figure 6.7. The setting of the biasing setting is done according to the ratio of the minimum current flowing through the operating coil to the average restraining current.

$$\% \text{ Bias} = \frac{(i_1 - i_2)}{(i_1 + i_2)/2} \times 100$$

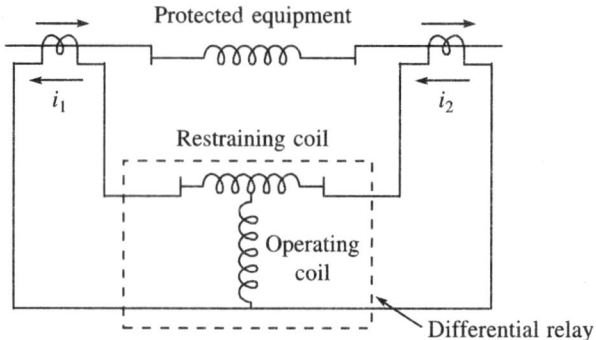

FIGURE 6.7 Biased (percentage) differential relay.

The condition for relay operation is that the pick-up ratio:

$$\frac{(i_1 - i_2)}{(i_1 + i_2)/2} > \text{Preset biased current}$$

and

$$(i_1 - i_2) > \text{basic set current}$$

This scheme is used to ensure the differential protection of alternators and large transformers.

6.3
DIRECTIONAL RELAYING

6.3.1 Basic Theory

Directional relays are actuated from two different independent (current-current and voltage-current) sources. The polarising quantity of the directional relay produces one of the fluxes and is taken as the reference point against which the phase angle of the other quantity is compared. Thus, the phase angle of one quantity is kept fixed while the other quantity undergoes wide changes in the phase angle.

Angle θ of the common relay equation for the working force; $(F \propto \Phi_1\Phi_2 \sin \theta)$ changes because of the independent actuating sources. Various relays such as Watthourmeter structure, induction cup structure, double induction loop structure and single induction loop structure are used in directional relaying schemes. The two actuating quantities may be current-current and voltage-current sources.

In a *current-current directional relaying* scheme, the actuating current quantities emanate from two separate current transformers. Assuming that there is no saturation, after substituting the fluxes with the currents, say I_1 and I_2, we get the following expression for the working torque:

$$T = K_1 I_1 I_2 \sin \theta - K_2$$

Here:

θ = phase angle between the rotor fluxes produced by the currents I_1 and I_2
I_1 and I_2 = the rms values of the actuating currents
K_1 and K_2 = constants

For the symmetrical structures of the relays, θ may be taken as the phase angle between I_1 and I_2 of the above equation. The phasor diagram for the current-current directional relay is shown in Figure 6.8. It is obvious that the maximum torque will appear when the coil currents are 90° out of phase; but in terms of the currents supplied from the actuating sources, the maximum torque will appear at some angle other than 90°. The condenser or resistor shunting may therefore be done with one of the actuating coils. Now, the current I_1 will be the resultant current supplied by the source to the coil and the resistor in parallel.

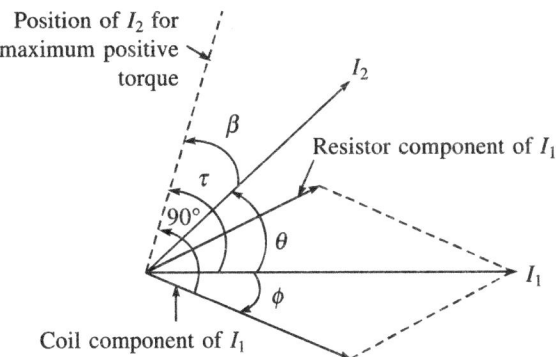

Position of I_2 for maximum positive torque

β

τ

90°

θ

ϕ

I_2

Resistor component of I_1

I_1

Coil component of I_1

FIGURE 6.8 Phasor diagram of current-current directional relay.

The working torque equation shall thus be:

$$T = K_1 I_1 I_2 \sin(\theta - \Phi) - K_2$$

Here, if I_2 leads I_1, angle θ is considered as positive and if the coil component of current I_1 lags behind the resultant current I_1, angle Φ will be considered negative. For example if, $\theta = 60°$ and $\Phi = -25°$, the torque will be:

$$T = K_1 I_1 I_2 \sin 85° - K_2$$

Angle τ is known as the *angle of maximum torque*. It is the value of θ at which the maximum positive torque generation takes place. This is the reason why τ is used for all relay design purposes in place of Φ. This is applicable for all the relays, irrespective of whether they are symmetrical or not. Hence, the general torque equation for the directional relay can be written as:

$$T = K_1 I_1 I_2 \cos(\theta - \tau) - K_2$$

or

$$T = K_1 I_1 I_2 \cos\beta - K_2$$

where $\beta = \tau - \theta$.

In a *voltage-current directional relaying scheme*, one actuating quantity is fed from the potential transformer and the other from the current transformer output. Thus, in terms of the voltage and current actuating quantities, the working torque equation will be written as:

$$T = K_1 VI \cos(\theta - \tau) - K_2$$

Here:

θ = the angle between V and I

τ = the angle of maximum torque

V = voltage (rms value) applied to the voltage coil circuit

I = the current-coil current (rms value)

The θ and τ are positive regardless of the nature of the relationship between V and I. The phasor diagram for the maximum torque in a voltage-current directional relay is shown in Figure 6.9.

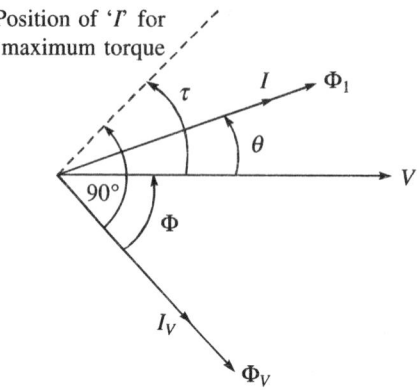

FIGURE 6.9   Phasor diagram of voltage-current directional relay.

If there is no series impedance with the voltage coil, angle τ is generally 20° to 30° leading while angle Φ is obviously 60 to 70° for most of the voltage coils. The insertion of series resistance and capacitance in the voltage-coil circuit can help change the angle between voltage V and I_V to almost any lagging or leading value. Figure 6.10 shows the operating characteristics of a voltage-current directional relay.

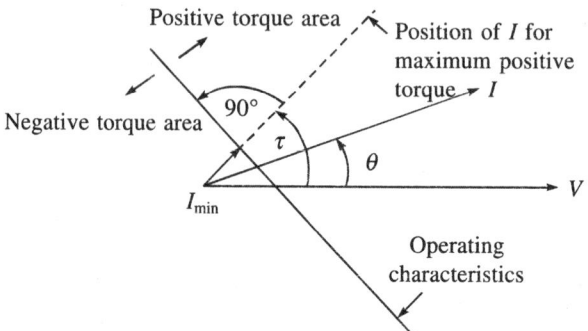

FIGURE 6.10   Operating characteristics of a voltage-current directional relay.

The voltage is the reference point for the voltage-current directional relay here because it is the polarising quantity and its value should be assumed to be constant. The operating characteristic is in the form of a straight line and for different magnitudes of the reference voltage, the operating characteristics will be depicted by another straight line parallel to the line shown in Figure 6.10. The parallel lines representing the operating characteristics will be expressed by:

$$VI_{min} = \text{constant}$$

I_{min} is the minimum pick-up current. However, for practical reasons, the current must be slightly higher to cause pick-up. There should thus be several numbers representing the operating characteristics, with each one of them signifying the possible values of the reference voltage.

6.3.2 Directional Overcurrent Relays

These relays are combinations of directional and overcurrent relay units in the same enclosing case. They are used to ensure phase or fault protection, and may be installed in any combination of directional relays with inverse time overcurrent relays and instantaneous overcurrent relays. One of these combinations, i.e. a balanced-beam current balance relay, is shown in Figure 6.11. In this case, of the two types of actuating torques produced, one is known as 'overcurrent' while the other is known as a 'directional' type torque. Due to the directional feature in the relay, the overcurrent unit remains inoperative until the directional unit is operated (closed). This is facilitated by connecting the directional unit contacts in series with the shading-coil circuit. Directional unit contacts can also be connected with one of the two flux-producing circuits of the overcurrent unit. The contacts of the overcurrent unit alone are present in the trip circuit.

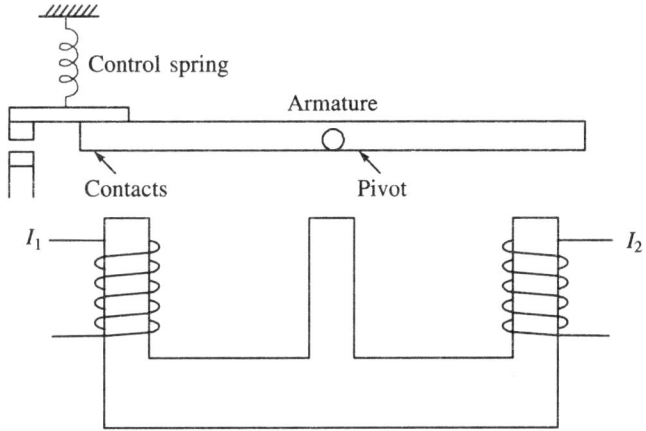

FIGURE 6.11 Balanced-beam current balance relay.

If the negative torque effect of the control spring is neglected, the operating torque equation will become:

$$T = (K_1 I_1^2 - K_2 I_2^2)$$

At the verge of the relay operation, the restraining torque becomes equal to the operating torque making the value of the net torque zero. Therefore:

$$K_1 I_1^2 = K_2 I_2^2$$

Hence, the operating characteristics are realised by the following equation:

$$\frac{I_1}{I_2} = \sqrt{\frac{K_2}{K_1}} = \text{constant}$$

The operating characteristics of a current balance relay are shown in Figure 6.12. This characteristic also includes the controlling spring effect, which requires a certain minimum

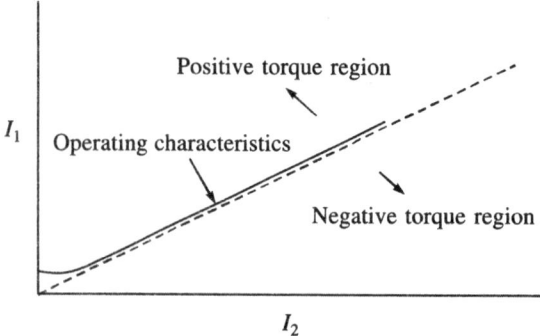

FIGURE 6.12 Operating characteristics of a current balance relay.

current I_1 to pick up when I_2 is zero. As the value of the current increases, the spring effect becomes less and less effective. The relay picks up for the ratios of I_1 and I_2, which are represented by points above the operating characteristics.

When the ratio of one of the currents to the other does exceed an effective value, the current balance relay inherently offers selective protection when a short circuit takes place in any of the circuits.

It is preferable to install separate directional and overcurrent units because they are easy to set separately. However, this separate installation of relays increases the sizes of the directional and overcurrent units, thereby imposing a greater burden on the current transformer.

6.3.3 Directional Overcurrent Setting

The three types of settings which are required for the proper operation of the directional overcurrent relays for phase faults are instantaneous setting, pick-up tap setting and time dial setting.

Instantaneous setting needs to be done in case of the maximum worst possible fault current, which is 1.2 to 1.3 times rated operating current. This factor (1.2 to 1.3) selection is necessary to prevent overreaching of the operation range, which could result in an unexpected interruption of power supply to the consumers.

Pick-up tap setting needs to be done in keeping with the facility offered to the user of selecting the taps within a fixed range. The lower limit of the range is decided after the inclusion of the overloading effect and the power swing. The lower limit can be fixed by multiplying the maximum load current through the relay by a factor of 1.2 to 1.3. The upper limit is decided by multiplying the smaller value of the line-to-line fault current by a factor of 0.5 to 0.6. This factor also helps clear the fault in the back-up region. For the all lines in the normal remote bus fault current, the factor remains in the range of 0.1 to 0.2. Time dial setting is done according to the need of operation of each overcurrent relay.

6.4

DISTANCE RELAYING

Overcurrent relays, differential relays or directional relays operate if the amount of current or power exceeds a preset value. But the distance relays are governed by the ratio of the applied voltage to the current. These relays include impedance relays, reactance relays and mho relays. The distance relaying scheme entails the use of a dual input comparator, and as mentioned above, it works on the basis of a working torque, which is produced by the current coil and voltage coils. For example, an impedance relay operates when the ratio V/I is less than a predetermined value. The current coil of the relay receives the line current while the voltage coil of the relay is supplied with the bus voltage. The reach of the distance relay cannot be shifted by the switching operation, or a change in the generation capacity or the type of fault. Phase distance relays, which are used for the phase fault involving two or more phases, are fed with delta voltage and delta current (line voltage and line current). For line to ground fault, the relay is supplied with star voltage and star current, i.e. the phase voltage and phase current.

6.4.1   Application of Impedance Relay

Figure 6.13 shows a simple application of the impedance relay in the transmission line. The protected zone impedance is Z_L if the fault occurs say at point F_1 inside the protected zone. The relay compares the impedance of the part of the line between fault point F_1 and the place where the relay is installed ($Z_F = V/I$). The relay will operate because $Z_F < Z_L$. But the relay will not operate if the fault occurs outside the protected zone because in that situation $Z_F > Z_L$.

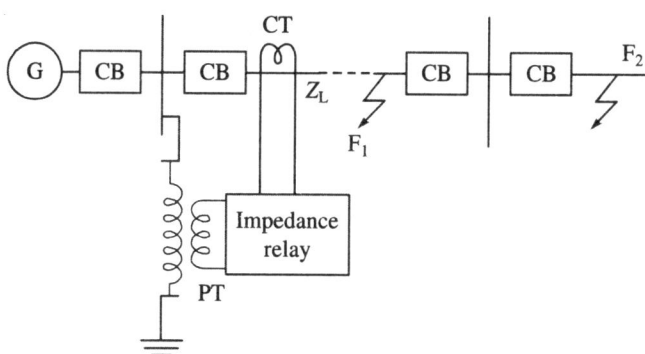

FIGURE 6.13   Application of impedance relay in the transmission line.

Distance relaying works both as the primary as well as the backup protection. Distance relays are of two types: the *definite distance relay* operates instantaneously for a fault occurring up to a predetermined distance from the relay and the *time distance relay* is used when the operation of time is proportional to the distance of the fault point from the relay. The time distance relay operates faster in case of a fault occurring closer to the relay than that occurring farther away from the relay fault closer to the relay operates earlier than the fault farther away from the relay. If the relay closer to the fault point fails, the next relay works as backup protection.

Generally, a particular location is divided into three zones of protection and hence three units of impedance relays are required for the protection of the specified location. The stepped time-distance characteristics of impedance relays for three zones of protection are shown in Figure 6.14.

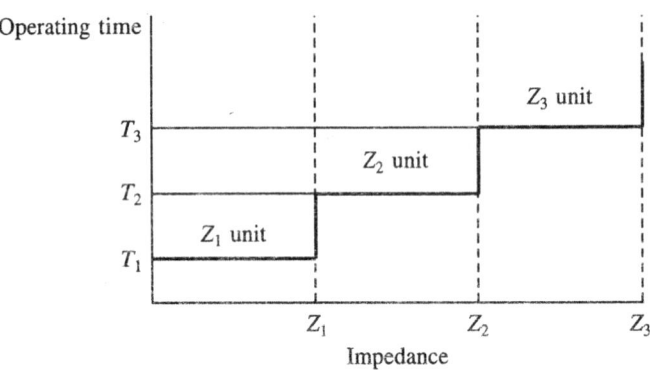

FIGURE 6.14   Operating time/impedance sketch for an impedance distance relay.

The protected zone of the first unit is known as the 'first zone of protection' and it is preferable to adjust its limit to protect only up to 80 to 90 per cent of the protected line. This is done to avoid an undesirable overreach. It operates instantaneously, within one to two cycles.

The second unit protects the rest of the protected line, which is beyond the reach of the first unit. The second unit's adjustment covers the operation under the arcing and transients at the end of the line, which also cover the underreach tendency. Thus, in order to take into account the underreaching tendency, the second zone is generally set to reach up to 50 per cent of the shortest adjoining line section. The second zone of the protective unit operates after a fixed time delay with an operating time 0.2 to 0.5 seconds.

The third zone of protection acts as back-up protection for adjoining line. The setting of the third zone of protection covers the first line such as the protected line in addition to the second line and 25 per cent of the third line. The time delay for the third unit is usually 0.4 to 1.0 second. Impedance relays are used for the protection of short length transmission lines.

6.4.2   Application of Reactance Relay

There is actually an overcurrent element, which develops a positive torque and a restraint current voltage directional element in the reactance relaying scheme. The directional unit is arranged in such a way as to develop the maximum negative torque when its current lags behind its voltage by 90°. Induction cup type or double induction loop type structures are used to make reactance relays.

The torque equation for the reactance relay is:

$$T = K_1 I^2 - K_2 VI \sin \theta - K_3$$

Here, θ is positive if I lags behind $V(-K_3)$ is the spring effect. The net torque is zero at the balance point and hence:

$$K_1I^2 = K_2VI \sin \theta + K_3$$

By dividing the equation by I^2 we get:

$$K_1 = K_2 (V/I) \sin \theta + K_3/I^2$$

$$(V/I) \sin \theta = Z \sin \theta = X = (K_1/K_2) - (K_3/K_2I^2)$$

By neglecting the effect of the control spring, we get:

$$X = K_1/K_2 = \text{constant} = K$$

The relay operates when $X < K$.

This means that the operating characteristic of a reactance relay is such that all the impedance radius phasors whose heads lie on this characteristic have a constant X-component. It is obvious that this is a straight line. The resistance component of the impedance does not have any effect on the operation of the relay. The operating characteristic of a reactance relay is shown in Figure 6.15. Like the impedance relay, three units of reactance relays are also used for the protection of a section of the line. Reactance relays are used for the protection of medium length transmission lines. A high-speed unit, which is able to protect 80–90 per cent of the line section, is used as unit-I. It operates instantaneously, within one to two cycles. Unit-II is used to protect the rest of the line section. The reach of the second unit extends to up to 50 per cent of the adjacent line section. The second zone of the protective unit operates after a fixed time delay with an operating time of 0.2 to 0.5 seconds. Unit-III is used as back-up protection for the adjacent line section. The time delay for the third unit is usually 0.4 to 1.0 second. The stepped time-distance characteristics of reactance relays for three zones of protection are similar to those of impedance relaying shown in Figure 6.14.

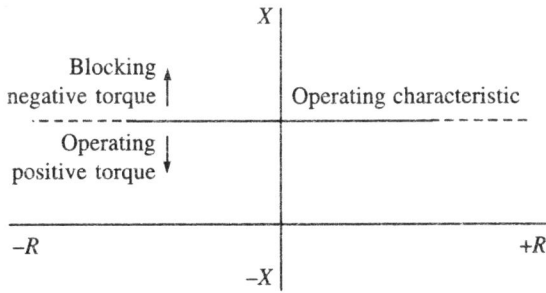

FIGURE 6.15 Operating characteristic of a reactance relay.

The schematic diagram of the induction cup relay used for reactance is shown in Figure 6.16. The current flowing in the upper and lower poles of the coil produces the polarising flux. The operating current also produces a flux in the right hand side pole. The secondary winding of the right hand pole is closed through a phase shifting circuit. This is the reason why the flux in the right hand side pole remains out of phase with the fluxes of the upper and lower poles.

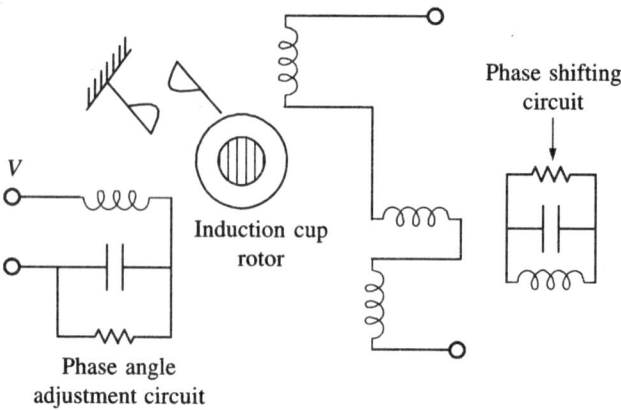

FIGURE 6.16   Reactance relay schematic diagram.

When the polarising fluxes interact with the flux in the right hand side pole, an operating torque K_1I^2 is produced. The restraining torque is produced due to the interaction of the polarising flux with the flux produced in the winding on the left hand side pole. This restraining torque is proportional to $VI\cos(90 - \Phi)$. The phase angle adjustment circuit is connected in series with the voltage coil.

6.4.3 Application of Mho Relay

Mho relays are generally four-pole induction cup type electromagnetic relays. The voltage coils energise the upper and lower poles, which produces a polarising flux. The operating quantity of the current energises the left pole.

The voltage energises the right hand pole. The impedance setting is done with the help of adjusters which may be referred to as Adjuster-A and Adjuster-B. One of these adjusters is available on the primary winding of the internal CT taps while the other is available on the secondary winding of the internal CT. The switching in resistors in series with the restraint winding controls the second and third zone reach settings. Figure 6.17 shows the induction cup type Mho relay.

The torque characteristic of an Mho relay is expressed by the following equation:

$$T = K_1VI\cos(\Phi - \alpha) - K_2V^2 - K_3$$

The flux produced by the operating current I interacts with the polarising flux to give an operating torque $[K_1VI\cos(\Phi - \alpha)]$. The phase shifting circuit fixed at the left pole carries the variable resistor, which helps in adjusting the angle α.

The interaction of the fluxes produced by the right hand pole takes place with the polarising fluxes, which produce a restraining torque $[K_2V^2]$. $(-K_3)$ is the spring effect and the condition for the relay to operate is given by the following equation:

$$K_1VI\cos(\Phi - \alpha) > K_2V^2$$

or

$$(I/V)\cos(\Phi - \alpha) > (K_2/K_1)$$

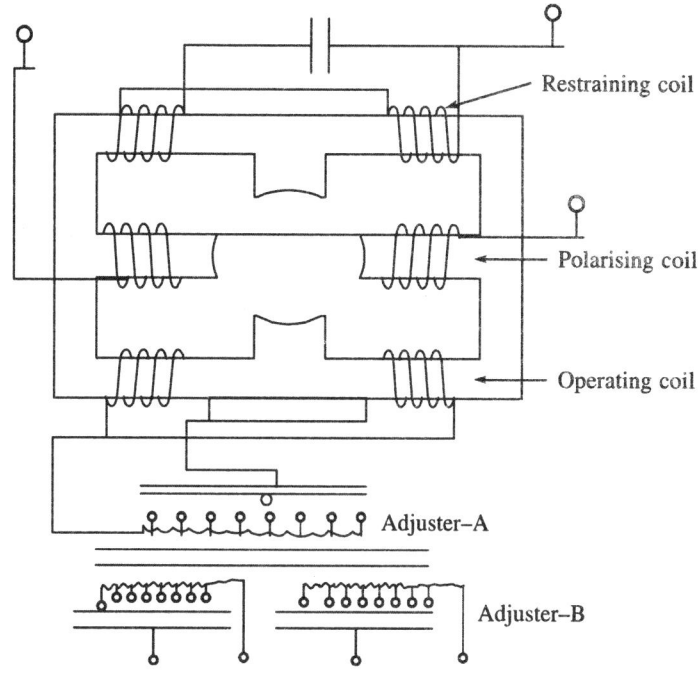

FIGURE 6.17 An Mho relay.

or
$$Y \cos (\Phi - \alpha) > (K_2/K_1)$$

or
$$1/[Y \cos (\Phi - \alpha)] < K$$

Here:

Y is the admittance and K is the ratio of K_2 and K_1.

The above expression can be written in terms of the impedance as follows:

$$Z/[\cos (\Phi - \alpha)] < K$$

Like the impedance relay, three units of the Mho relays are also used for the protection of a section of the line. Mho relays are used for the protection of long transmission lines. High-speed units which are able to protect 80–90% of the line section constitute unit-I. It operates instantaneously, within one to two cycles. Unit-II is used to protect the rest of the line section. The reach of the second unit extends to up to 50 per cent of the adjacent line section. The second zone of the protective unit operates after a fixed time delay with an operating time of 0.2 to 0.5 seconds. Unit-III is used as back-up protection for the adjacent line section. The time delay for the third unit is usually 0.4 to 1.0 second. The stepped time-distance characteristics of Mho relays for three zones of protection are similar to those for impedance relaying shown in Figure 6.14.

6.5

TRANSLAY RELAYING

Translay relaying is a modified form of the voltage balance differential relay system, which works on more or less the same principle used in the voltage balance system. In this system, the secondary CT voltages at the ends of the feeders are compared. Since the CTs used in the translay relaying scheme have to supply power only to a relay coil, they have a normal design without air gaps. Translay relay as applied to a single-phase system is shown in Figure 6.18.

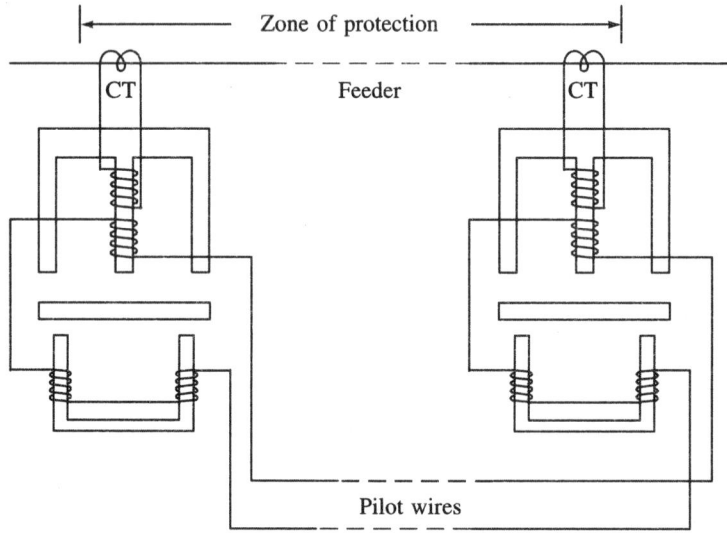

FIGURE 6.18 Translay relay applied to a single-phase system.

The upper shunt magnet acts as a quadrature transformer and produces the pilot terminal voltage. The terminal voltage varies with the primary current. Current does not flow in the pilot wires if the currents at both ends are equal because in that case the voltages induced would also be equal. In other words, it can also be said that under healthy conditions, the current supplied at the two ends of the protected feeder is the same and the primary windings of the relays carry the same current, as a result of which the relaying scheme does not operate. If fault occurs on the protected feeder, the current leaving the feeder is different from the current entering it, Hence unequal voltages are induced in the secondary winding of the relays and the current circulates between the two windings, which creates the working torque to move the disc. The resulting operating torque depends upon the position and nature of the faults in the protected zone and at least one element of either relay will operate under any fault condition.

EXERCISES

1. With the help of a neat diagram, explain the construction and working of a non-directional overcurrent relaying scheme.

2. Describe the constructional details of differential overcurrent relaying.

3. Describe the principle of the operation of percentage differential relaying with a neat schematic diagram and relevant characteristics.

4. How is definite minimum time achieved in an IDMT relay? Describe the function of a shedding ring in the IDMT relay.

5. What do you mean by the term 'Relay coordination'? Describe.

6. Describe the importance and use of PSM and TSM for the overcurrent relays.

7. Explain the working principle of the distance relay.

8. Describe the Translay scheme of protection.

9. Explain the working principle of the directional relaying scheme.

10. Give the connection diagram of an Mho relay and describe its working.

11. Explain a reactance-relaying scheme.

12. Show a comparison of the impedance relaying, reactance relaying and Mho relaying schemes.

Chapter 7

Microprocessor-based Digital Relaying

7.1
INTRODUCTION

Electromechanical relays have been employed as power system protection devices from the very beginning, after the application of fuses as protective devices. However, the use of electromechanical relays is a time-consuming process because of the need for maintenance in these relays and their complicated design. Microprocessor-based relays have many advantages over electromechanical relays in terms of cost, engineering design and maintenance.

The advent of Large Scale Integrated Technology (LSIT) has resulted in the development of sophisticated and efficient microprocessors. These technologically advanced microprocessors have helped popularise microprocessor-based digital relaying schemes, which have virtually replaced the conventional relaying schemes in the developed countries. Developing countries are also systematically phasing out conventional relaying schemes in favour of microprocessor-based digital relaying schemes.

Advantages

Following are the advantages that microprocessor-based digital relaying schemes have over conventional relaying schemes:

(i) These relays are compact and flexible, and the overall scheme takes up less panel space.
(ii) Their programmable characteristics make them extensively applicable.
(iii) They impose a very small burden on CTs and PTs, which enables relays to be set up for higher sensitivity.
(iv) They are capable of processing and displaying the signals very efficiently, accurately and without unnecessary time delay.

(v) Their multi-functional characteristics enable microprocessors to perform the relaying functions of several systems.

(vi) They are more reliable and sensitive than other relays.

(vii) They help to significantly reduce the total number of components of the relaying scheme. Generally it is said that if the cost of the electromechanical relaying scheme is 1 p.u., the cost of the microprocessor-based relaying scheme would be 0.35 p.u., and if the cost of installation is also included, then it would be 0.5 p.u.

(viii) The design and wiring of these relays is simpler and cheaper than those of others.

(ix) The cost of installation and the need for frequent maintenance and testing are greatly reduced in the case of these relays.

(x) The use of these relays can facilitate an improvement in the understanding of the power system network and its reliability. Microprocessor based-relaying scheme can report the events taking place in the system, verify the faults, analyse the voltage dip and identify the fault location.

(xi) Microprocessor-based relaying has the capability to co-ordinate easily with other devices of the network.

(xii) This relaying scheme provides data throughout the utility.

(xiii) A second ground on the feeder outside the sub-station establishes a return path through the feeder neutral which is seen by the CTs on the grounded feeder as an inflow to the bus. In this situation, the fault current is accounted as double and the current imbalance trips the bus differential relays. Thus the fast bus trip eliminates inadvertent ground paths in the bus differential zones.

(xiv) The microprocessor-based relaying scheme facilitates monitoring of the trip coil and triggers off an alarm in case of loss of voltage across the relay or the failure of the self-test for the relay.

(xv) Microprocessor-based relaying also has the capability to enhance protective functions through customisation of flexible overcurrent settings, changeover scheme initiation, breaker failure protection, underfrequency loadshedding, etc.

(xvi) A microprocessor can facilitate communication by connecting a feeder relay's output contact to a bus relay's input. When a feeder relay senses fault which occurs near to it, descriminates it and communicates to the bus relay which prevents the bus relay's definite time element from picking up the bus lockout relay.

In the microprocessor-based digital relaying scheme, three-phase alternating quantities received from the power system through CTs and PTs are sampled simultaneously or sequentially at uniform time intervals (4 to 32 samples per cycle). These analog signals are then converted into the digital form through an A/D (Analog to Digital converter). After conversion into digital form, signals are transferred to the digital processor. Digital signals are in the form of coded square pulses that represent discrete data. These signals are in binary form. Digital processors are set with pre-specified values. In co-ordination with the control and data outputs, the digital processor compares the dynamic inputs with the pre-specified values and decides to generate an alarm signal and, if required, a trip signal, to the output device.

A block diagram of a microprocessor-based digital relaying scheme is shown in Figure 7.1.

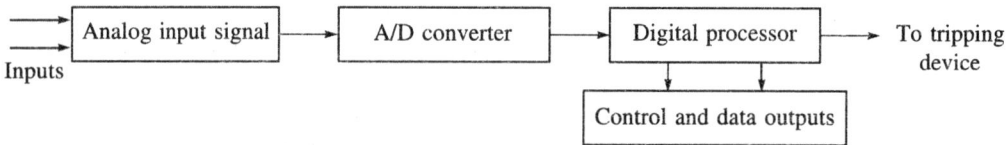

FIGURE 7.1 Block diagram of a microprocessor-based digital relaying scheme.

Microprocessors should be kept in a clean environment and be properly shielded from external influences such as temperature variations and contaminants. The system earthing from which they receive their control voltage should also be done properly.

7.2
DIGITAL LOGIC COMMUNICATION

A digital relationship is the key to relay-to-relay communication. In terms of microprocessor-based relay logic, the status of a logic point is represented by a logical or binary value, i.e. 1 or 0. The *Relay Logic Status* has the following three states:

(i) The state of a relay element includes the 'picked up' or 'dropped out' states.
(ii) The state of an output contact includes the 'closed up' or 'open' states.
(iii) The states of controlled inputs are 'asserted' or 'de-asserted'.

Example

The shared logic information communication among the relays ensures proper co-ordination at the remote stations. Sharing of the relay logic status between the schemes enables the directional distance or overcurrent relays at both ends of a transmission line to trip with little or no intentional time delay for clearing of faults at any point in the protective zones.

The shared logic information communication is also the basis for the formation of the following schemes:

(i) Permissive tripping scheme
(ii) Inter-tripping (direct or transfer tripping) scheme
(iii) Block tripping scheme

Figure 7.2 depicts the relay's share logic status in the pilot logic communication scheme.

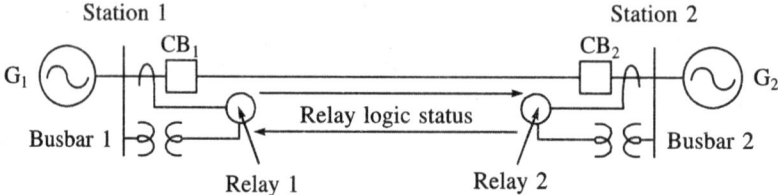

FIGURE 7.2 Relay share logic status in pilot logic communication.

The relay logic status elements are shown in Table 7.1.

TABLE 7.1 Relay Logic Status Element

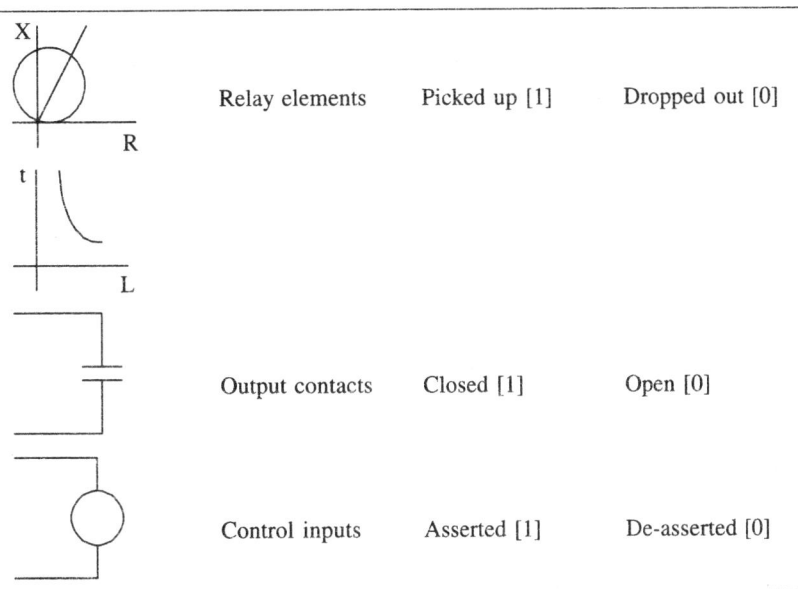

Relay elements	Picked up [1]	Dropped out [0]
Output contacts	Closed [1]	Open [0]
Control inputs	Asserted [1]	De-asserted [0]

Basic schemes are generally required to share a single logic point whereas complex schemes necessitate the sharing of multiple logic points. Apart for the three schemes mentioned above (permissive tripping scheme, inter-tripping scheme and block tripping scheme), the other schemes used in the relays include:

(i) Remedial action schemes
(ii) Status monitoring schemes
(iii) Remote control schemes

7.3
DIRECT RELAY-TO-RELAY DIGITAL LOGIC COMMUNICATION

The inherent digital logic processing capability is built in with the communication capability of microprocessor-based relaying. This capability is used for developing the sharing schemes of relay logic status between relays. Each microprocessor-based relay has a communication port that is capable of receiving and sending digital messages. In addition to this capability, the microprocessor-based relay also processes digital data representing the status of the relay measuring elements, control inputs and control outputs. A combination of these two capabilities of the advanced microprocessor thus facilitates direct relay-to-relay digital logic communication.

It is possible to send the status of eight programmable internal relay elements (encoded in a digital message) from one relay to the other through a serial communication port (for example through a patented EIA-232 serial communication port).

Each relay is wired through the communication channel, and this direct relay-to-relay digital logic communication technique creates eight additional virtual outputs on each relay corresponding to eight virtual control inputs on the other relay.

The internal relay elements RMB1 to RMB8 are the eight virtual inputs. These virtual inputs follow or mirror the respective status of the virtual outputs TMB1 to TMB8 in the sending relay as shown in Figure 7.3(c).

Direct relay-to-relay digital logic communication is described in Figures 7.3(a), (b) and (c).

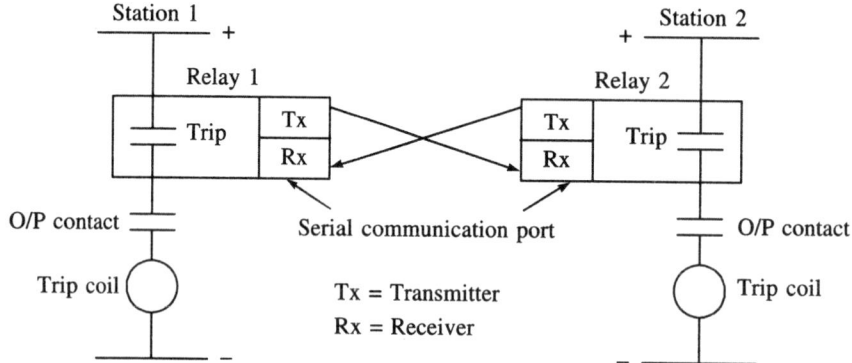

FIGURE 7.3(a)   Hardware connection of direct relay-to-relay digital logic communication.

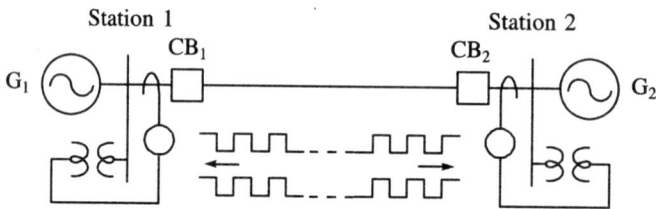

FIGURE 7.3(b)   Single line diagram of direct relay-to-relay digital logic communication circuit showing the signal flow.

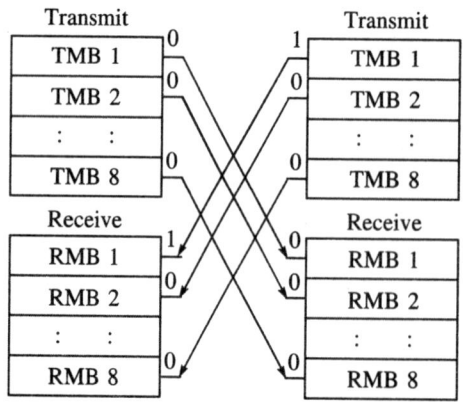

FIGURE 7.3(c)   Direct relay-to-relay digital logic bits transformation.

Any change in the TMB1 status of Relay-2 at station 2 from a logical 0 to a logical 1 causes the RMB1 status of Relay-1 at station 1 to change from logical 0 to 1. This creates a virtual connection between the two relays as the Receive Mirrored Bits (RMBs) of one relay follow the status of the respective Transmit Mirrored Bits (TMBs), sent from the other relay. Each RMB is assigned a control input function including permissive trip, block trip, etc., in the programming.

Each TMB is programmed to an output contact, with a logic equation that represents the status of an internal relay element, control input, output contact or any combination of these elements.

The direct relay-to-relay digital logic transformation technique has the following advantages:

(i) It produces the equivalent of eight traditional relay communication channels between relay terminals, which increases the functionality of the system and makes it cost-effective.

(ii) This technique also eliminates the need for expensive traditional communication equipment.

7.4
DIGITAL MESSAGE SECURITY

In the traditional schemes, the communication equipment performs the necessary signal integrity checks before handing over the message to the relay system. In the direct relay-to-relay digital logic transformation technique, as discussed above, the relay assumes the responsibility for ensuring digital message security.

A digital message sent from one relay to another generally consists of two bytes. Each byte contains eight data bits. The status of each bit is represented as a logical 0 or 1 in the digital message. Each byte of the message carries a part of the eight-relay logic status bits representing the programmable Transit Mirrored Bits (TMBs). The relative position of the status of each bit in each message frame is shown in Figure 7.4(a).

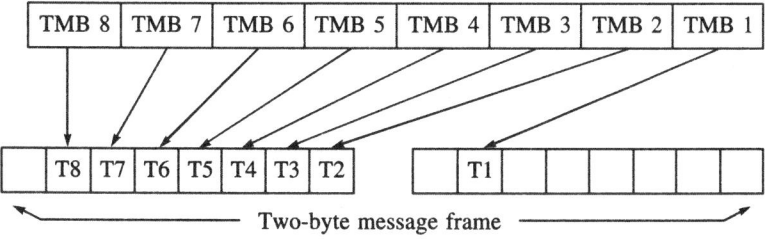

FIGURE 7.4(a)   Relay logic status bits in digital frame.

Multiple security measures are required to ensure that the eight-relay logic status elements are communicated from one relay to the other in the correct order. In this scheme, each byte of the two-byte message has a one-bit byte flag system to identify the correct byte sequence. The second byte of the message includes a six-bit CRC (Cyclic Redundancy Check) table calculated from the status of the eight-relay logic status bits. This is shown in Figure 7.4(b).

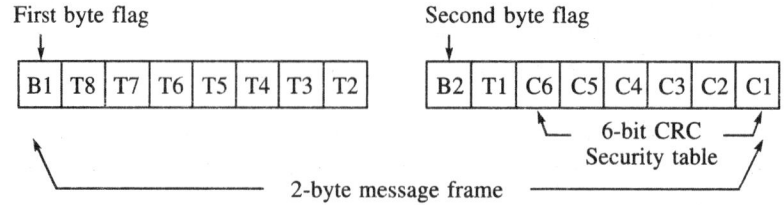

FIGURE 7.4(b)   Byte Flag and CRC Security Bits in Message Frame.

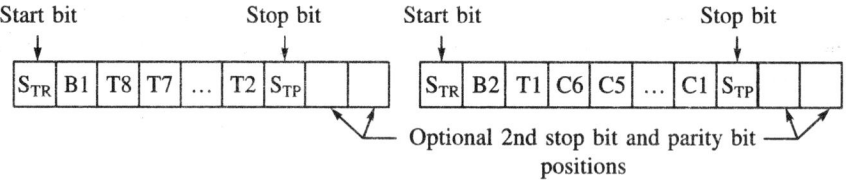

FIGURE 7.4(c)   Asynchronous message framing bits.

While ensuring that the byte flags are in the correct order, the relay checks each message. It calculates a CRC (Cyclic Redundancy Check) value from the received message bits and checks that this value matches the received CRC. The relay also checks message framing to ensure proper start, stop, and parity bits. It then performs a timing test to ensure that a message is received for each message sent. The enormous message is rejected, if any one of the checks fails. This is shown in Figure 7.4(c). Each byte of the asynchronous message is preceded by a start bit and followed by up to three bits, which can include one or two stop bits and a parity bit.

7.5
RELAY INTERFACE WITH UTILITY

In addition to the protection functions, microprocessor-based relaying schemes provide metering data, status information, targets and fault location. The data is accessed through relay communications ports, local displays or other human-machine interfaces (HMI) by the various employees of the concerned organisations, who use this data for their specific purposes. For example, planning engineers need to analyse load demand data collected from the relay while operators need to know targets and fault locations. Protection engineers need to analyse event reports in order to explain a fault on a particular line. A relay-centric view is shown in Figure 7.5.

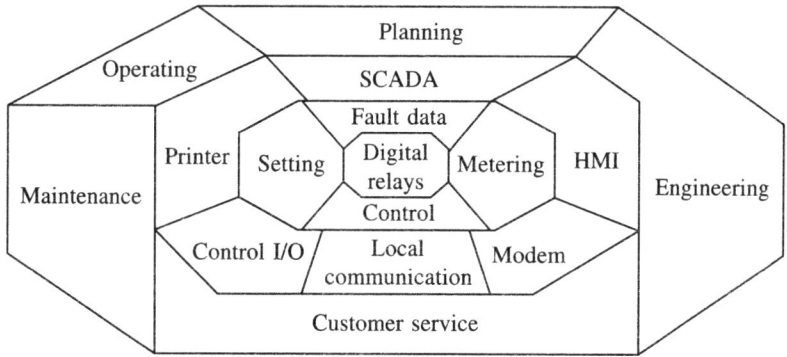

FIGURE 7.5 Relay interface with utility.

7.6
MICROPROCESSOR-BASED OVERCURRENT RELAY

Microprocessor-based relays are extensively used in the power industry. They operate when the circuit current exceeds the predetermined value. The microprocessor uses a multiplexer for sensing the fault currents in a number of circuits and sub-circuits. It should be noted that, the microprocessor accepts signals *in the voltage forms*, which is why in this scheme, the CT fault current derived is first converted into proportionate voltage signals, and then fed to the rectifier, multiplexer, analog to digital converter and the microprocessor. A block diagram of the scheme is shown in Figure 7.6.

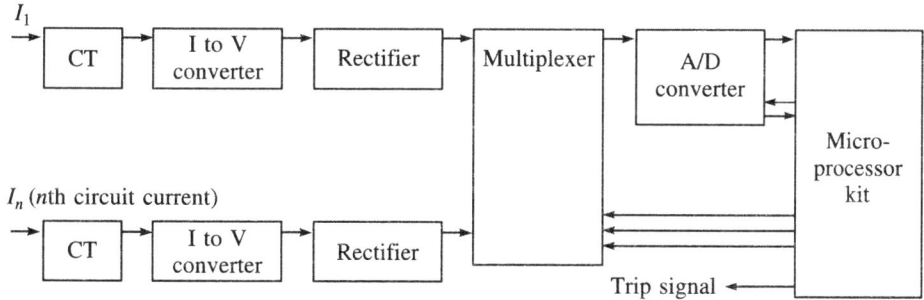

FIGURE 7.6 Block diagram of a microprocessor-based O/C relay.

The output of the rectified voltage signals is fed into the multiplexer. The microprocessor (microcomputer) then sends command for switching ON the desired channel of the multiplexer in order to obtain the rectified voltage in the particular circuit.

After this, since the microprocessor needs digital signals, the output of the multiplexer is fed into the analog to digital converter (ADC). Again, the microprocessor sends signal to the ADC for starting the conversion and reads the end of the conversion signal to examine whether the conversion is over and compares the signal with the pre-determined pick-up value. A programme flowchart for an overcurrent relay is shown in Figure 7.7.

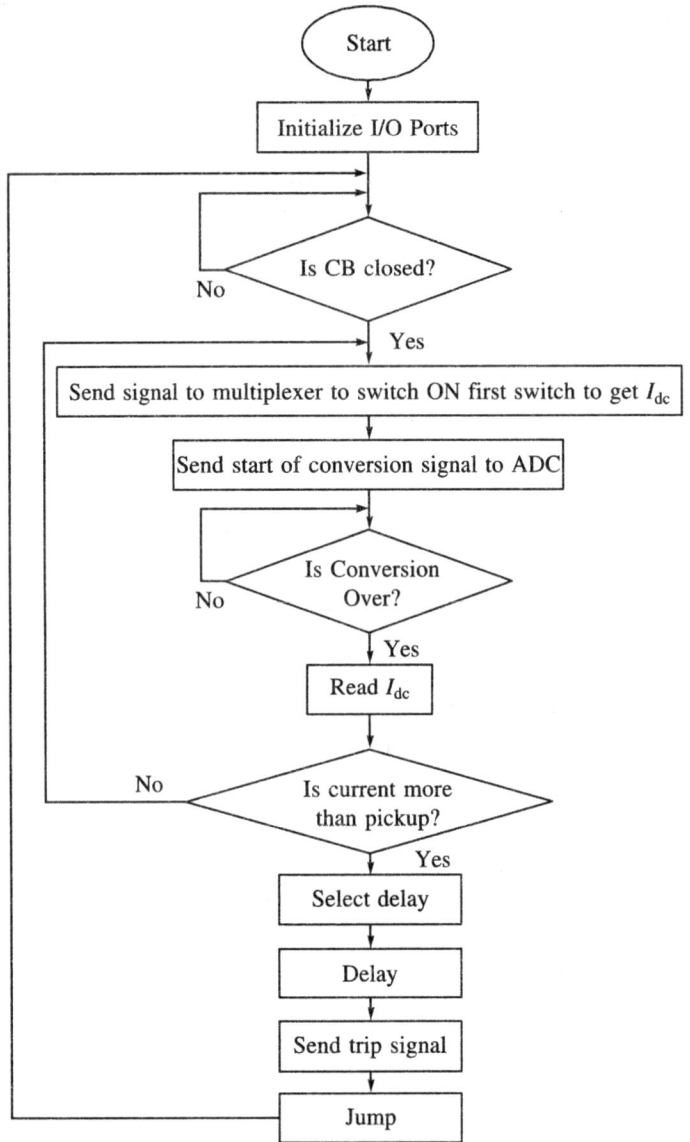

FIGURE 7.7   Flow chart of a microprocessor-based O/C relay.

7.7
MICROPROCESSOR-BASED IMPEDANCE RELAY

The ratio of the voltage (V) and current (I) is the impedance of the transmission/distribution line between the relay location and the fault point. Thus the characteristics of the impedance relay can be realised by comparing the voltage and current at the relay location. In the microprocessor-based impedance relaying, for the comparison of the voltage and current, DC voltage (V_{dc}) and DC current (I_{dc}) are used at the relay location to facilitate a comparison of the voltage and the current. AC fault voltage and current are rectified at the relay location in order to make them suitable for feeding in the microprocessor. The condition that needs to be satisfied for the operation of the relay is depicted in the following equation:

$$K_1 V_{dc} < K_2 I_{dc}$$

or

$$\frac{V_{dc}}{I_{dc}} < \frac{K_2}{K_1}$$

or

$$\frac{V}{I} < K$$

or

$$Z < K$$

Here, K_1, K_2 and K are constants. The calculation of the values of K for different zones of protection is done and stored in the memory in the form of data in order to obtain the desired characteristics. A block diagram of the scheme is shown in Figure 7.8.

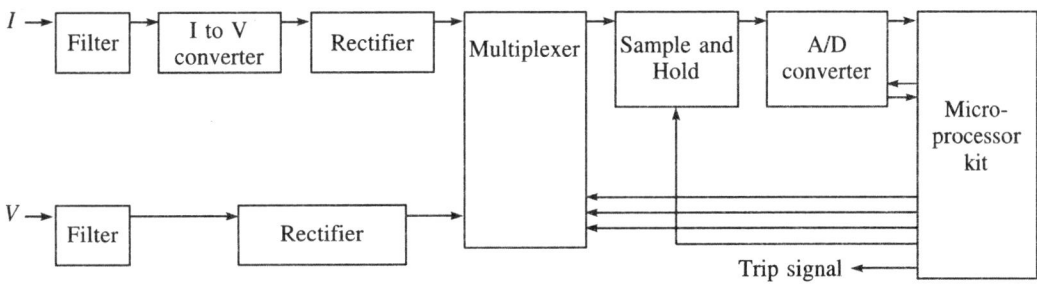

FIGURE 7.8   Block diagram of a microprocessor-based impedance relay.

With the help of PT and CT, the fault voltage and current are stepped down to the desired value and the output of the current signal is converted into the proportional voltage signal by using a current-to-voltage converter. After that a precision rectifier rectifies the signals.

The rectified signals are then fed into the two different channels of the multiplexer, which are switched ON sequentially by proper commands from the microcomputer. The output of the multiplexer is fed into the analog-to-digital converter through a sample and hold and a data acquisition system.

The microcomputer reads the DC voltage and the current, (V_{dc} and I_{dc}), calculates the value of Z as seen by the relay and then compares it with the values of Z_1, Z_2 and Z_3, which are the values of the impedances for zones I, II, and III, respectively, of the protection system. A programme flowchart for the impedance relay is shown in Figure 7.9.

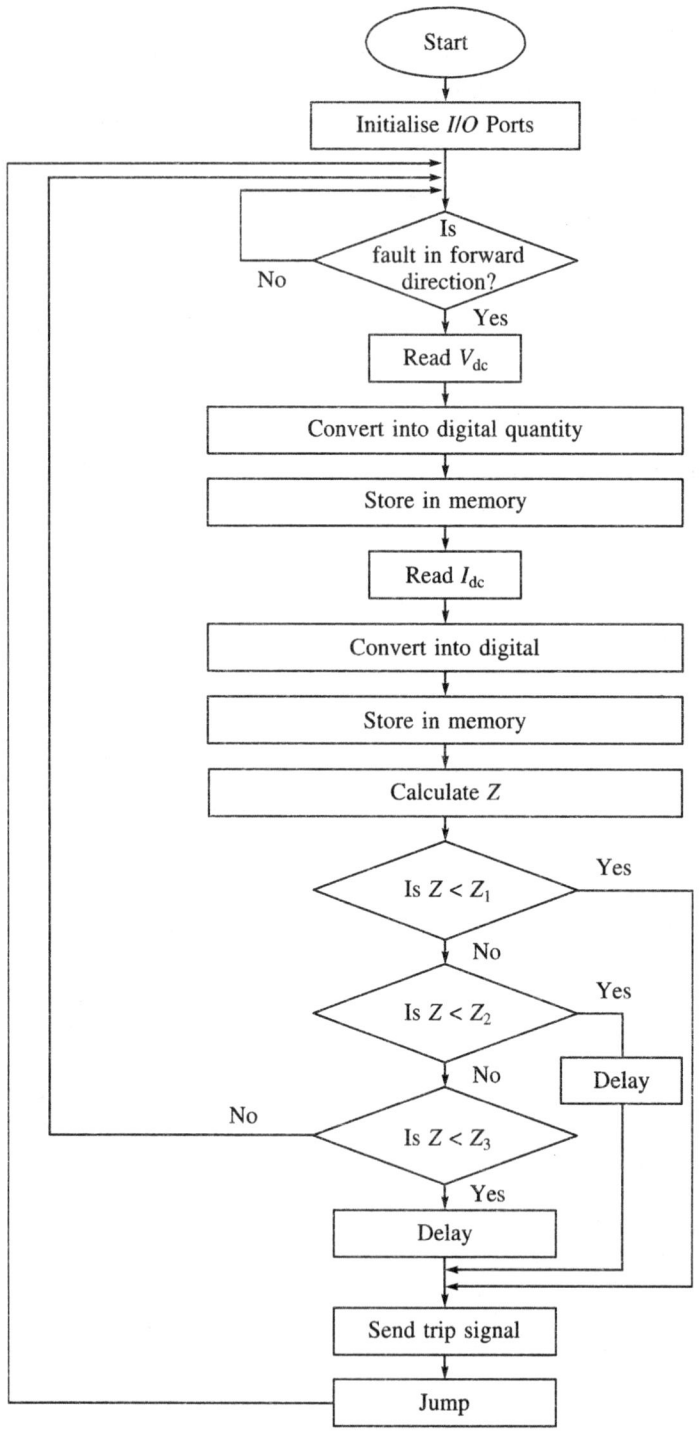

FIGURE 7.9   Flow chart of a microprocessor-based impedance relay.

7.8
MICROPROCESSOR-BASED REACTANCE RELAY

A comparison of the instantaneous values of the voltage at the moment when the fault current crosses its zero is undertaken in this relay. Thus the characteristic of the reactance relay is realised by comparing the instantaneous values of the voltage at the moment of the current zero against the rectified current.

The block diagram of a microprocessor-based reactance relay interface for the realisation of the relay characteristic is shown in Figure 7.10. The output of the zero crossing detectors is read and examined by the microprocessor. The microcomputer determines whether the fault current has crossed its zero point. If the fault current crosses its zero value, the microcomputer sends the command signal to the multiplexer to switch ON the channel and gets the instantaneous value of the voltage through the analog-to-digital converter. After that, the microcomputer sends the command signal to the multiplexer to switch ON the channel and gets the rectified value of the fault current. Thereafter, the microcomputer calculates the reactance, X. The instantaneous value of the voltage at the moment of current zero is $V_m \sin \Phi$.

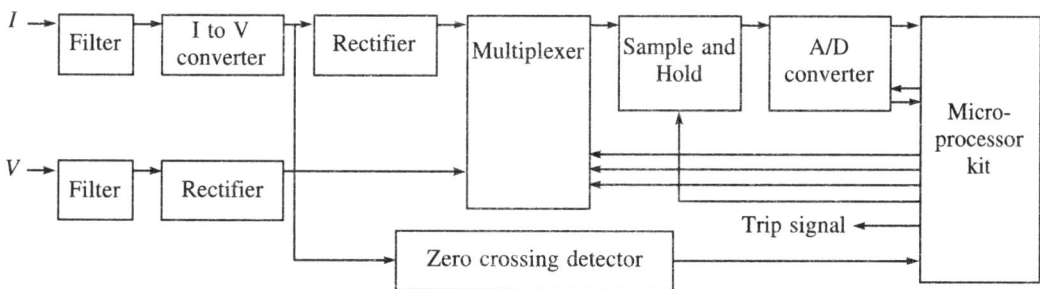

FIGURE 7.10   Block diagram of a microprocessor-based reactance relay.

The condition to be satisfied for the relay operation is depicted in the following equation:

$$V_m \sin \Phi < K_1 I_{dc}$$

or

$$\frac{V_m \sin \Phi}{I_{dc}} < K_1$$

Since, V_m and I_{dc} are proportional to the rms values of V and I, respectively, the above equation may be written as:

$$\frac{V \sin \Phi}{I} < K$$

or

$$Z \sin \Phi < K$$

or

$$X < K$$

The microcomputer calculates the X value as seen by the relay and compares it with X_1, the reactance of the fist zone of protection. A tripping signal is sent instantaneously to the circuit-breaker if the compared measured value of X is less than the predetermined value of the first zone reactance X_1. In case X_1 is less than X but X is less than X_2, the tripping signal will be sent after a preset delay. In case X is more than X_2 but is within the directional unit protection zone, the tripping signal is sent after a greater predetermined delay. The programme flowchart of a microprocessor-based reactance relay is shown in Figure 7.11.

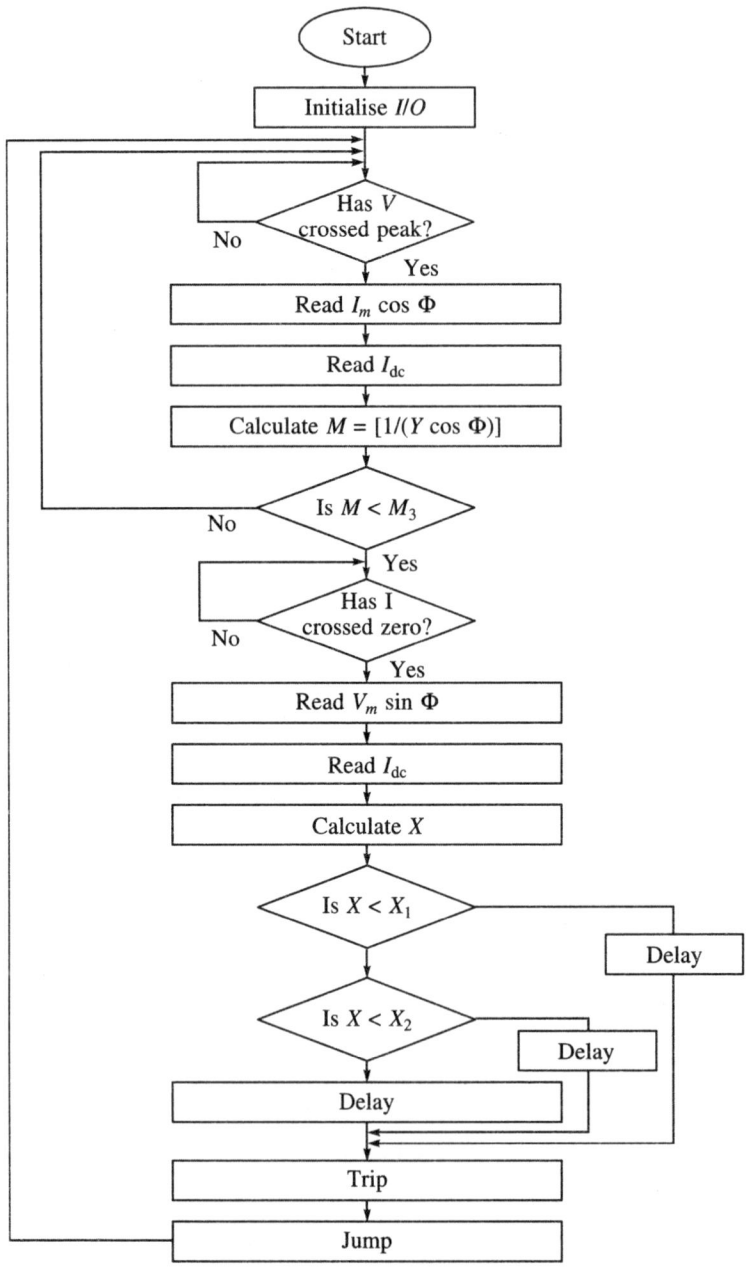

FIGURE 7.11   Flow chart of a microprocessor-based reactance relay.

7.9
MICROPROCESSOR-BASED MHO RELAY

An Mho relay is inherently directional and detects a fault in the forward direction. The characteristic of this relay on the *R-X* (impedance) diagram is a circle passing through the origin. *Its characteristic occupies the least areas on the R-X diagram, which is why it is the least affected by surges that remain on the line for a comparatively longer duration. This is the reason why the Mho relay is most ideally suited for long line protection.* The Mho relay's tripping area can be further reduced by using blinders.

The characteristic of the Mho relay is realised by comparing the instantaneous values of the current at the moment of the voltage peak against the rectified current. $I_m \cos \Phi$ is the instantaneous value of the current at the moment of the voltage peak.

The following equation depicts the condition which needs to be satisfied for the operation of the relay:

$$I_m \cos \Phi > K_1 V_{dc}$$

or

$$\frac{I_m \cos \Phi}{V_{dc}} < K_1$$

Since, I_m and V_{dc} are proportional to the rms values of I and V, respectively, the above equation may be written as:

$$\frac{I \cos \Phi}{V} < K_2$$

or

$$Y \cos \Phi > K_2$$

or

$$\frac{1}{Y \cos \Phi} < \frac{1}{K_2}$$

or

$$M < K$$

where, K_1, K_2 and K are constants.

With the introduction of angle θ while feeding the voltage and current signals to the relay in the design, the above equation is redefined as follows:

$$\left[\frac{1}{Y \cos (\Phi - \theta)} \right] < K$$

Changing the angle θ can result in the shifting of the Mho relay characteristic towards the R-axis in order to increase its tolerance to the arc resistance. The value of θ should preferably be less than 75° to facilitate reasonable arc resistance tolerance, especially when a fault occurs near the bus.

Figure 7.12 shows the inclusion of angle θ and arc resistance tolerance. The schematic block diagram shown in Figure 7.13 depicts the realisation of the Mho relay's characteristics.

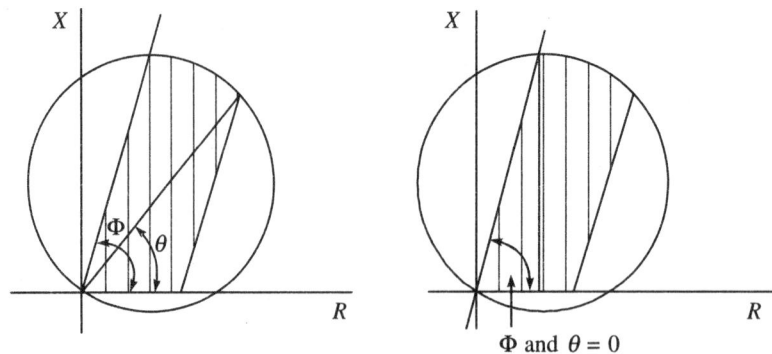

FIGURE 7.12   Mho relay's characteristics with different values of angle θ.

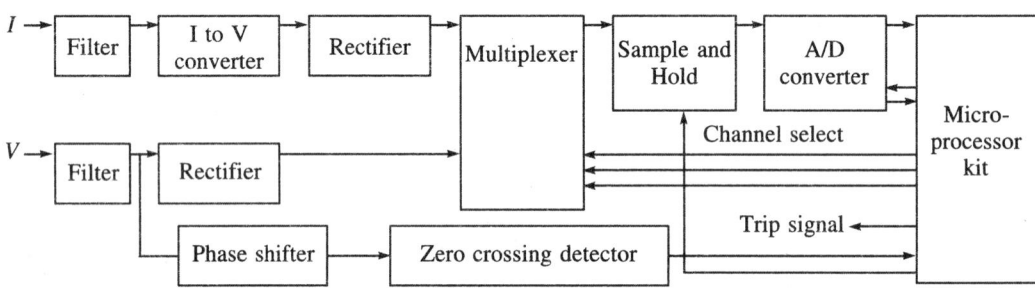

FIGURE 7.13   Block diagram of a microprocessor-based Mho relay.

A phase shifter is required to shift the phase angle by 90°. Then the output of the phase shifter is fed to the zero crossing detectors in order to obtain the pulse required for the microcomputer.

The microcomputer reads the output of the zero crossing detectors and checks whether the voltage has crossed its peak value or not. When a voltage pulse at the instant of its peak is achieved, a command signal is sent to the multiplexer to switch ON the channel in order to obtain the instantaneous value of the current at the moment of peak voltage. This instantaneous value of the current is now fed to the analog-to-digital converter. The digital output of the converter is now stored in the memory. The rectified voltage V_{dc} through the multiplexer and the analog-to-digital converter is also stored in the memory of the microcomputer. Figure 7.14 shows the programme flowchart for an Mho relay.

7.10
SCADA INTERFACING AND METERING

For the operation of the system, real time voltage, current, power (watt) and reactive power (VAR) data are required.

Most operating and dispatch centres need these data and fault locations. The advent of fault locating digital relays has made it possible to access this information.

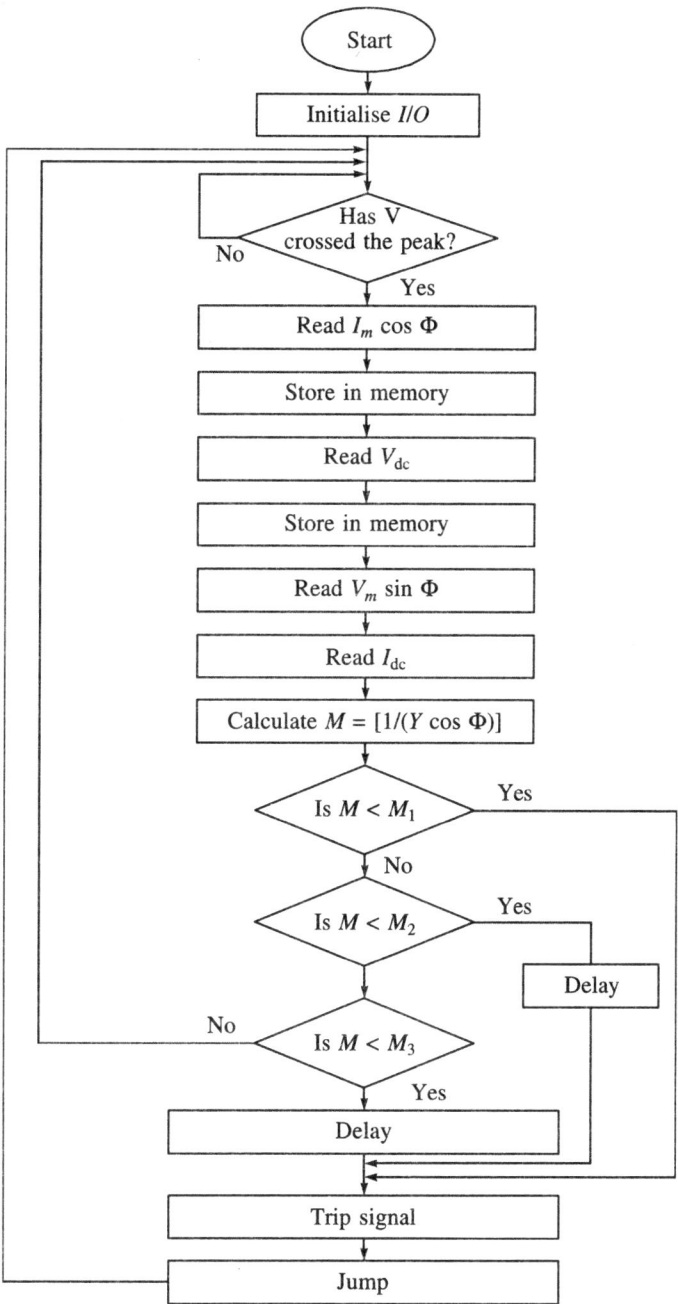

FIGURE 7.14   Programme flow chart for an Mho relay.

Many modern relay trip units operate on digital principles which allow the direct acquisition of digital data, thereby permitting a direct interface of the digital relays and relay trip units. In order to produce better results at lower cost in terms of security, accuracy and efficiency, it is advisable to maintain the digital relay data in digital form. The other requirements which are fulfilled by the digital data pertain to relay targets, relay elements, event history, breaker interruption data, relay settings and relay self-test status. Figure 7.15 demonstrates the communication interface between the digital relays and a multi-port digital relay trip unit.

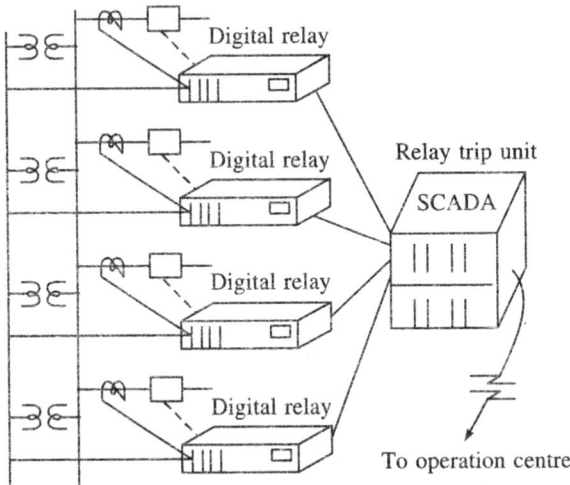

FIGURE 7.15 Communication interface between digital relays and multi-port relay trip unit.

Fault location on distribution feeders is achieved with limited accuracy. But the fault locating data is useful for identifying the sections of the faults, and for deputing and dispatching the fault repairing team.

An ideal distribution line is an overhead line which may be assumed to have the same conductor size throughout its length. The line is not fed the unwanted signals through grounded transformer banks, capacitors, generators or other sources. Figure 7.16 shows how engineers, operators and customers can access the relay remotely via modem communication. SCADA is discussed in greater detail in Chapter 13 (Network Relays).

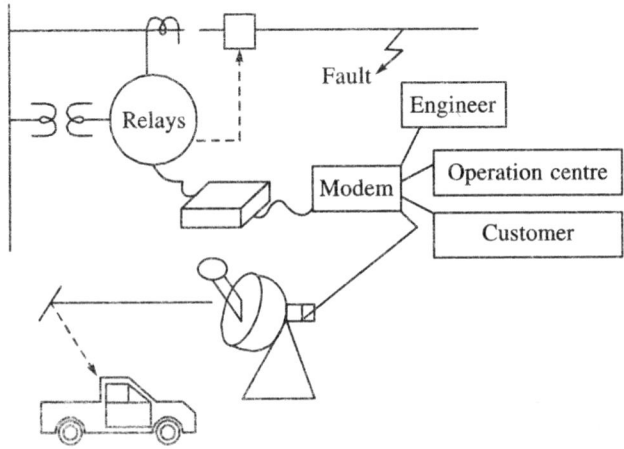

FIGURE 7.16 Remotely accessing a relay for fault and targeting data.

7.11
RELAY TESTING

Relay testing is two types:

1. Installation testing
2. Routine testing

Installation tests are performed to ensure that the relay settings are correct, which, in turn, would ensure that the scheme is designed correctly for the intended application. In a scheme designed with electromechanical relays, it is imperative for each discrete relay to be tested and calibrated properly. The steps followed for the testing are as follows:

(i)   Each relay should be connected to the test equipment.

(ii) Different setting arrangements need to be made.

(iii) The relay should be tested as per the established test routine.

The calibration process is, however, very time-consuming.

The next step of testing is to test the "trip-check" to ensure proper wiring and trip circuits. Since a number of discrete devices are used in the scheme, the trip-checks sometimes become very time-consuming. This is especially so in case of an incorrect design or wiring error, when locating the error becomes very problematic and may require many hours of troubleshooting. As far as the microprocessor-based relaying scheme is concerned, it operates by using computer programming, which makes it simple to test and verify.

The installation tests for the microprocessor-based relaying scheme should be designed in such a way as to ensure that the relay settings have been entered correctly. The relay pick-up should also be checked at the critical points. Using a software command to close the relay outputs is simpler than connecting the voltage and current tests sources to the relay to perform fault simulations. Once the relay has been fully tested, the software that defines the operating characteristics of the relay would also have been verified. Therefore, it is not necessary to subject each relay to any more detailed tests, if the relays are of the same type and the same version.

Routine tests are performed to ensure that the relays are functioning within the established specifications. These tests may be carried out at a gap of one to three years and are performed on electromechanical relays which are more or less similar to those performed during the installation process. Routine tests are conducted to confirm that all the internal components are operating within the specified tolerances, and that all the contacts and external circuits are functioning properly. If the relay is functioning properly, it implies that the relay algorithm will also operate correctly. The routine maintenance in a microprocessor-based relaying scheme consists of verifying that the data acquisition system along with the inputs and outputs is working as per the established specifications. If the relay self-check status shows that the relay is measuring the analog currents and voltages properly, it means that the relay will function properly. The other checks are required to verify the output contacts and logic inputs.

7.12

APPLICATIONS OF MICROPROCESSOR-BASED RELAYS

Microprocessor-based relaying schemes also have many other features which are not available in electromechanical relays. These include fault location, advance metering, control capabilities and event reporting. In addition to these, microprocessor-based relaying schemes can also be used for any application in which electromechanical relays are used. Following are the applications of microprocessor-based relaying schemes:

(i) Microprocessor-based relaying schemes are more cost-effective than electromechanical relays. This makes microprocessor-based relaying schemes perfect replacements for old protection schemes. The ability to locate faults is a unique feature offered by these schemes. The fault locating information not only helps reduce the patrol time and expenditure otherwise incurred on the location of faults but also facilitates the evaluation of problems in the power network.

(ii) The cost-effectiveness of microprocessor-based relaying schemes helps justify their usage during the construction of new sub-stations or power transmission lines.

(iii) In most cases, microprocessor-based relaying schemes incorporate the entire logic required to operate a particular communications-aided scheme. Since external auxiliary relays are not required for the operation of the scheme, this also helps reduce design and material costs.

(iv) A multiple setting group can be provided within a single device in some microprocessor-based relaying schemes. The relays facilitate a number of independent settings for applications such as a bus-tie breaker. This means that there are a number of independent relays within a single relay.

(v) In case of relays which offer multiple setting groups, the settings may be changed at the source stations without sending out a technician or operator.

(vi) The installation of microprocessor-based relaying schemes also ensures a wide application of the system as these relays offer a large variety of protection elements and schemes and find applications at many voltage levels.

(vii) Most microprocessor-based relaying schemes record the system conditions in the event of protective element operation. These records are useful for analysing the performance of power systems. They also offer the facility of reviewing and analysing event data, which is a valuable maintenance tool. Further, this analysis of event data is more accurate and useful than that obtained through simulated tests because the relay performance in the case of microprocessor-based schemes is recorded in the actual system.

Thus, in view of their various advantages and the unique features and innovations they offer, microprocessor-based relaying schemes are now being widely used in many utilities.

EXERCISES

1. Explain the merits of microprocessor-based relaying schemes.

2. Describe the applications of microprocessor-based relaying schemes.

3. Showing a hardware connection, describe the direct relay-to-relay digital logic communication system.

4. Explain how an overcurrent relay is realised by using microprocessors.

5. Explain how an impedance relay is realised by using microprocessors.

6. Describe the realisation of a reactance relay through the use of microprocessors.

7. Explain how an Mho relay is realised by using microprocessors.

8. Write a brief note on SCADA Interfacing and Metering.

9. Write a note on Digital Message Security.

10. Describe the requirement and importance of microprocessor-based relay testing.

11. Write a note on the interfacing of microprocessor relays with utilities.

Chapter 8

Static Relays

8.1
INTRODUCTION

The action of the relay which involves sensing and determination of the fault, and isolation of the faulty section of the power system allows the healthy section to provide its services without interrupting the power supply. Quick fault clearing can enable the power network to transmit a larger amount of power to the users without loss of synchronism and instability. In fact, the evolution of static relays came about because of their characteristics of rapidly clearing faults and maximising the system's power transfer capability, accuracy and reliability.

Initially, when development work started on static relays in 1940, thermionic valves were used. The relays developed then were known as electronic relays. After the invention of transistors and other electronic components during the decade 1950–60, these components were incorporated in static relays. These newly developed solid-state relays have all the advantages of electronic relays in addition to the advantages listed below.

Advantages of static relays

(i) Static relays impose a low burden on CTs and PTs, which, in turn, reduces the VA (Volt Ampere) rating requirement and both the size and cost of current transformers and potential transformers.

(ii) A lower VA rating of current transformers implies that voltage levels have to be kept low for the same current rating. The CT operation is required in the lower part of linear characteristics, which helps prevent the CT core saturation, and consequently facilitates an improvement in the CT and PT accuracy.

(iii) Unlike electromagnetic relays, static relays have no moving parts. This helps eliminate problems like contact erosion, contact bounce, vibration and shocks, dryness in the contact, spring restraint and arcing, thereby increasing the efficiency of static relays.

(iv) Due to the absence of friction, static relays have a high effective torque, which helps improve the relay response.

(v) The absence of moving parts in static relays also increases the precision level of their characteristics.

(vi) The amplification facility imparts greater sensitivity to static relays.

(vii) The compact size of the static relays is made possible by the availability of small-sized and effective ICs, which facilitates a saving in the panel space.

(viii) The absence of moving parts in these relays also reduces the incidence of wear and tear, thus bringing down the need for maintenance.

(ix) Greater flexibility can be achieved in these relays because of their ability of filtering harmonics, differentiation, integration, and sensing of the negative phase sequence component of the current and voltage.

(x) Static relays have a low resetting time for facilitating the rapid automatic reclosing of circuit breakers. This resetting time helps achieve proper selectivity.

(xi) Static relays also provide a high drop-off to pick-up ratio, low transient overreach and low overshoot.

Limitations of static relays

(i) Voltage transients generally occur due to incidences of lightning and circuit-breaker operations, among other things. When these voltage surges override the fault signals or normal signals, they cause a malfunctioning of the static relays. Severe surges or voltage spikes may damage electronic components.

(ii) Solid-state devices like static relays are adversely affected by changes in temperature.

(iii) Static relays are made up of several discrete components like resistors, capacitors, transistors, etc. The overall reliability of the static relays depends on the reliability of these components. Thus, in order to achieve high reliability in the relays, each of their components should be selected with proper care and rating. Soldering and connections should also be undertaken with great care.

(iv) Since static relays need battery or auxiliary power supply for functioning, this has to be provided.

8.2
STATIC RELAY COMPONENTS

A list of the electronic devices used in static relaying schemes is given below. It is presumed that students have already studied all these components in previous courses like basic electronics, analog electronics, integrated electronics, digital electronics and power electronics. These devices are:

1. Diodes, zeners
2. Transistors
3. Thyristors
4. Rectifiers
5. Voltage rectifier circuits
6. Smoothing circuits
7. Transistor amplifiers
8. Filter circuits
9. Logic circuits (OR gate, AND gate, NOR gate etc.)
10. Multi-vibrators (monostable, bistable and astable)
11. Differentiating circuits
12. Integrating circuits
13. Operational amplifiers
14. Level detectors
15. Time delay circuits
16. Output circuits
17. DC auxiliary supply
18. Surge absorbers/surge suppressors
19. Comparators.

8.3
COMPARATORS

A relay's main principle is based on the comparison of two operating parameters (one of which is the operating quantity, while the other is the restraining quantity). Quantities to be compared may be either in amplitude or in phase. The amplitude and phase relations are the functions of the system conditions. The comparator, which is the heart of the relay, decides its working characteristics. A comparator may thus be defined as a device that makes these comparisons. There are two types of comparators. These are as follows:

  (i) Amplitude comparators
 (ii) Phase comparators

8.3.1   Amplitude Comparators

Amplitude comparators compare the magnitudes of the two quantities, i.e. the operating quantity and the restraining quantity, regardless of the value of the phase angle. When the magnitude of the operating quantity exceeds that of the restraining quantity, the relay operates and sends the trip signal to the circuit-breaker.

There are three types of amplitude comparators. These are as follows:

  (i) Integrating amplitude comparator
 (ii) Instantaneous amplitude comparator
(iii) Sampling amplitude comparator

Integrating amplitude comparator

This type of relay is mainly of two types—the circulating current type and the voltage opposed type.

The *circulating current type* of relays are assembled in two ways—firstly, through rectifier bridges with a slave relay, and secondly, through rectifier bridges with static output devices. A rectifier bridge with a slave relay is shown in Figure 8.1. In this relay, the input signals are the operating and restraining currents. These are $S_1 = Ki_1$ and $S_2 = Ki_2$, respectively. The relay operates when $S_1 > S_2$. Thus, two full wave rectifiers are needed, one for the operating quantity and the other for the restraining quantity. The outputs of these relays are applied to the slave relay, which is a DC polarised relay.

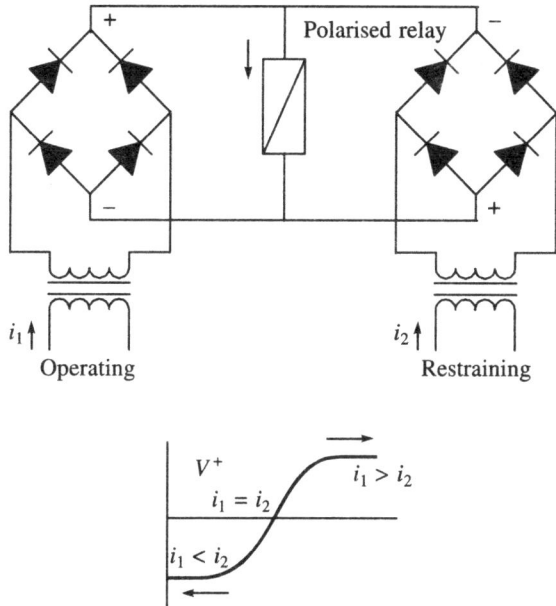

FIGURE 8.1   Integrating type rectifier bridge comparator with slave relay.

Rectifier bridges with static output devices without a polarised relay may also constitute the integrating type amplitude comparator. This circuit, which consists of an averaging circuit of the difference of the rectified currents ($i_1 - i_2$), along with the polarity detector, is shown in Figure 8.2. This relay operates if the average value of the output is positive.

The *opposed voltage comparator* works on the principle of averaging the difference of the rectified voltages. The block diagram of this comparator is shown in Figure 8.3. The bridge of this relay is not very sensitive at low inputs and the comparator has no limiting action on the output devices for the voltage and current.

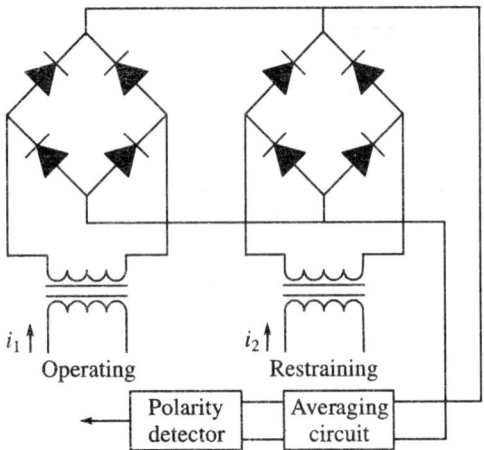

FIGURE 8.2 Integrating type rectifier bridge comparator with static o/p device.

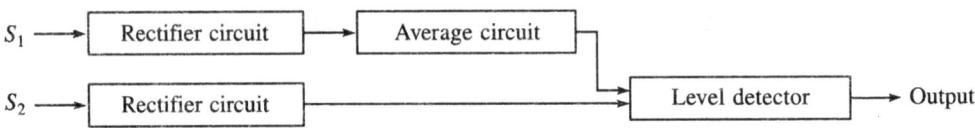

FIGURE 8.3 Block diagram of opposed voltage comparator.

Instantaneous amplitude comparator

The block diagram of this relay comparator is shown in Figure 8.4. Basically, this is similar to the opposed voltage comparator. Instantaneous amplitude comparators are of two types: the averaging type and the phase splitting type.

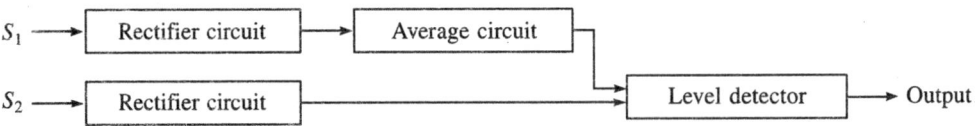

FIGURE 8.4 Block diagram of instantaneous comparator.

In order to provide a fixed restraint level, the restraining signal is rectified and then smoothened properly in the averaging type instantaneous amplitude comparator. But the operating signal is not smoothened while it is rectified fully. In order to achieve the operating condition, the peak of the operating signal must exceed the restraining level. In this relay, smoothening is generally done with the help of the capacitor, which causes a delay in operation. This is the reason why phase splitting type comparators are preferred over instantaneous amplitude comparators. In the phase splitting type comparator, phase splitting is done before rectification. A continuous output signal is achieved by smoothening both the operating and restraining inputs. The operating and restraining signals are smoothened before the comparison is made.

Sampling amplitude comparator

As its name indicates, in this comparator, the signals, before being compared, are sampled either at the same instant or at different instants. In case of the sampling of one signal, comparison of the signal proportional to its average rectified values with sampled signal is made. A block diagram of the sampling amplitude comparator for the reactance relay is shown in Figure 8.5. In this relay, the voltage signal is sampled and then compared with the average value of the current passing through the zero value.

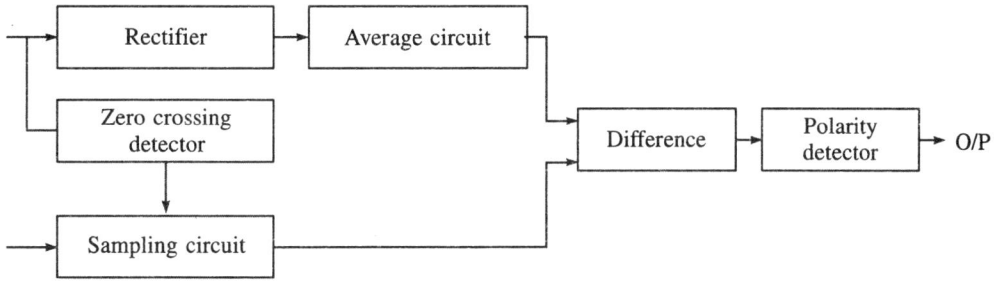

FIGURE 8.5   Block diagram of sampling amplitude comparator for reactance relay.

Let the instantaneous value of the voltage at the current passing through zero be $V \sin \Phi$, where Φ is the power factor angle. The operation of the reactance relay takes place, if reactance (x) as seen by the relay, is less than K, which is the design reactance of the system.

8.3.2   Phase Comparators

This comparator mainly facilitates a comparison of the phase relation between two input signals, say S_1 and S_2. The output required for operating the relay is obtained when the phase relationship between S_1 and S_2 varies within certain limits. For instance, if θ is an angle by which S_1 lags behind S_2, the condition for the relay operation will be given by the following equation:

$$-\alpha_1 \leq \theta \leq \alpha_2$$

It is a sine comparator if $\alpha_1 = 0°$ and $\alpha_2 = 180°$. And it is a cosine comparator if both α_1 and α_2 are 90°. The two types of popular phase comparators are:

  (i) Coincidence type phase comparators, and
 (ii) Vector product type phase comparators.

  Coincidence type comparators are made as per the following techniques:

  (i) Block spike phase comparison,
 (ii) Phase splitting technique,
(iii) Integrating phase comparison, and
 (iv) Rectifier bridge phase comparison.

Block spike phase comparator

This is a coincidence type phase comparator. It works on the principle of measuring the period of coincidence. Suppose α is the phase difference between two signals and the period of coincidence of the signals is, say $\psi = (180° - \alpha)$. This implies that for the desired operation α at less than +90°, the coincidence period will be greater than 90°. Now, for this condition an operating angle will be:

$$-90° \leq \alpha \leq 90°$$

In block spike phase comparison, one input is converted into a spike and the other into a square wave at the instant of passing the signal either at zero value or peak value. The phase comparator output is depicted in Figure 8.6.

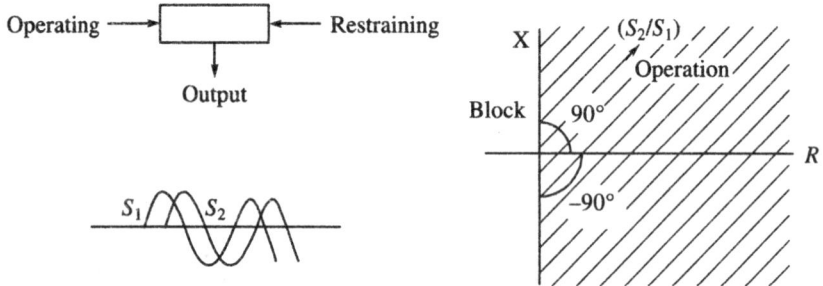

FIGURE 8.6   Illustration of phase comparator output when operating angle is within –90° to + 90°.

In order to obtain the output of the coincidence of the two signals, the spike and the square wave, the two signals are fed into an AND get. It is necessary to shield the relay against magnetic and electric circuit in order to prevent any unwanted spike due to any external interference or switching operation.

Phase splitting comparator

In the phase splitting comparator, both the input signals are split into two components each. Split phase components are shifted by an angle of +45° with reference to the original signal. Now, the four components are fed into an AND gate as shown in Figure 8.7. The AND gate gives the output at the instant when all the four signals are simultaneously positive.

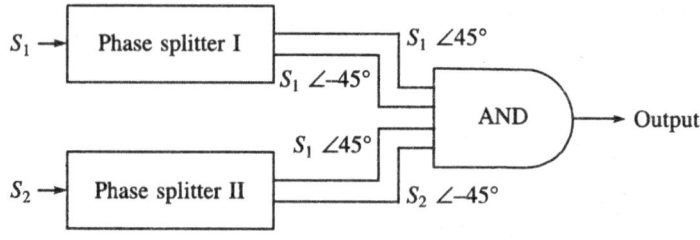

FIGURE 8.7   Block diagram of phase splitting comparator.

The main advantage of this comparator is that it is not affected by an unwanted spike due to any external interference or switching operation. Since the phase shifting process takes some time, there is some time delay in this comparator, which makes it slower than the block spike comparator.

The output of the AND gate is obtained for $-90° \leq \alpha \leq 90°$.

Integrating phase comparator

In the integrating phase comparator, the measurement of the period of coincidence is done after integerating the signals, S_1 and S_2. This integration of the signals is done by feeding them into the AND gate, which is also known as the coincidence detector. Both the sinusoidal signals first need to be converted into square waves and then fed into the AND gate. The output of the AND gate is fed into the integrator circuit whose integrated output signal is detected by the level detector. The level detector is generally a thyristor circuit. If the integrated signal level goes beyond the predetermined value, there is a corresponding time of coincidence. The relay sends the trip signal to the circuit breaker. Figure 8.8 shows the block diagram of an integrating phase comparator.

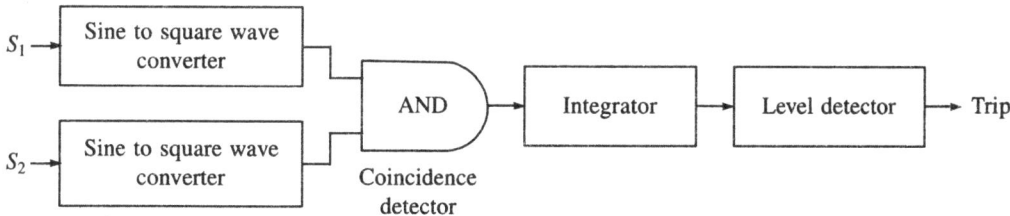

FIGURE 8.8   Block diagram of integrating phase comparator.

Rectifier bridge phase comparator

This comparator is widely used in distance relaying schemes. Precision rectifiers are preferred for better realisation of the characteristics. One rectifier bridge phase comparator is shown in Figure 8.9. If the reverse current is less than the forward current, the current can flow in both the directions. A diode can function as a gate as long as it is kept open by a forward current. The operating current is kept at half of the gating current. In this circuit, i_1 is the operating signal while i_2 is the gating signal. In the first half cycle, the gating current $i_2/2$ flows through the diodes D_1 and D_2. This current opens the diodes and current i_1 flows through D_1 in the forward direction and through D_2 in the reverse direction. The voltage drop across the terminal PQ due to the current flow i_1 will be of positive polarity while the voltage drop due to the gating current will be zero because the current here flows in the reverse direction.

The gating signal flows through D_3 and D_4 in the next half cycle. Now the current i_1 flows in the reverse direction which is why the opposite polarity output appears across PQ. This output is then fed to the polarity detector, RC charging and the level detector. Obviously the output will be positive during a positive coincidence period and negative during an anti-coincidence period.

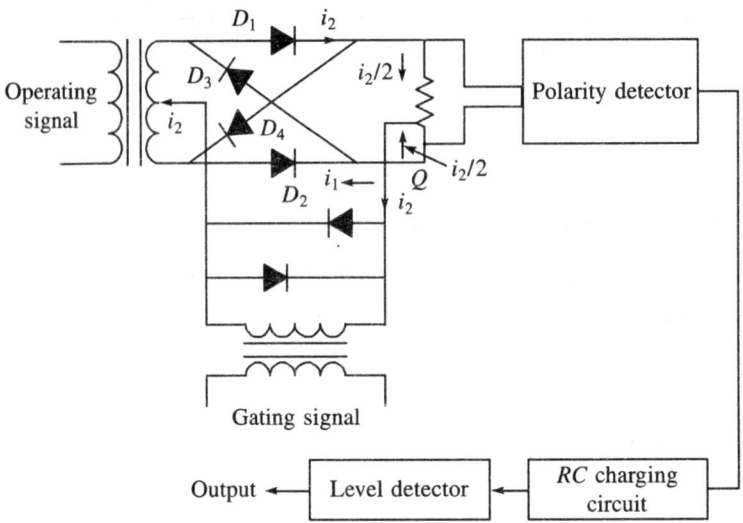

FIGURE 8.9   Rectifier bridge phase comparator.

The vector product type phase comparators are:

(i) Hall effect type phase comparator, and

(ii) Magneto-resistivity type phase comparator.

Hall effect phase comparator

This comparator falls under the category of vector product phase comparators. These comparators operate on the principle of the Hall effect, the effect invented by E.H. Hall. *According to this effect, if a current passes through X-direction of a Hall element placed in a magnetic field of the Y-direction, Hall voltage is induced in the Z-direction across the edge of the element.* Indium arsenide (InAs) is a very good Hall element. Another Hall element is indium antimonide (InSb).

Hall effect comparators are, however, not generally preferred because they have certain demerits. These are listed below.

(i) They cause errors due to rising temperatures.

(ii) They have a low output.

(iii) The cost of these comparators is high.

Magneto-resistivity phase comparator

These relays are better than Hall effect relays and work on the principle of the Gauss effect. As per this effect, the resistivity of a semiconductor varies, if it is subjected to a magnetic field. In this device, a voltage signal, say V_1, is applied to a semiconductor disc for producing a magnetic field and another voltage signal, say V_2, sends a current through the disc at a right angle to the magnetic field. Suppose Φ is the angle between the two voltages V_1 and V_2. The current flowing through the disc will be in proportion to $V_1V_2 \cos \Phi$. This concept is used for phase comparison in this comparator.

8.4
STATIC OVERCURRENT RELAY

The overcurrent relay is normally used to individually protect the major parts of the machinery against abnormally high currents. These are also used as back-up protection for feeders.

A static overcurrent relay is set to a reference current value (the preset value of the current), which is actually the pick-up current within the reach of the relay. Static overcurrent relays are principally single-input non-directional comparators. The fault current is stepped down through the CT and passed through a triggering circuit. If this fault current exceeds its reference as per the base voltage, a voltage pulse is generated. This voltage pulse is then applied to the gate of the SCR for triggering off the circuit and generating a tripping signal. A block diagram of the overcurrent relay is shown in Figure 8.10. At the first step, the voltage, which is generally proportional to the secondary current of the CT, is rectified by a rectifier. This rectified output is fed into a logarithmic circuit in order to obtain its logarithmic value.

In the next step, an integrator integrates the output of the amplified (to a value of I^n) anti-

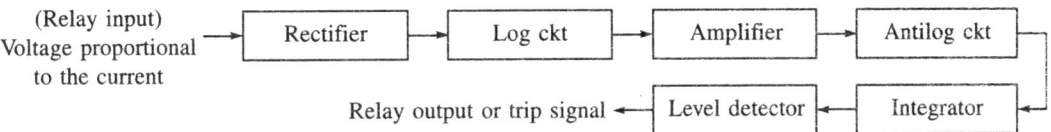

FIGURE 8.10   Block diagram of the static overcurrent relay.

logarithmic signal. The output of the integrator is then compared by the level director. When the voltage of the integrator output crosses the preset value (reference value) of the voltage, the level detector gives the output trip signal.

Demerits of static overcurrent relays

(i) Since they are single-input comparators, they are less reliable.
(ii) These relays are non-directional and need to be supplemented with directional units for fulfilling the requirements of the latter.
(iii) With change in the type of fault, reach of the static relay changes which in turn changes the descrimination of the fault level and leads to wrong tripping.

Due to the above-mentioned demerits, overcurrent relays are used to protect only distribution systems, feeders and transmission lines not exceeding 33 kV.

8.5
STATIC DISTANCE RELAYS

Distance relays are two-input comparators. Two input signals are fed into the comparators from the supply end wherein there is a possibility of the flow of fault current. The operation of these comparators depends upon the change in the relationships in the two input quantities. Distance relays are widely used for transmission line protection.

The threshold characteristic of a two-input comparator used for distance relaying is a circle on whose X-axis lies the real co-ordinate, which represents the ratio of the two quantities, and on its Y-axis lies the imaginary co-ordinate which represents the ratio of the two quantities. This circle diagram, which represents the threshold characteristics, is applicable to a linear system only. The following two types of relationships are used for the determination of the threshold characteristics of these relays:

(i) Complex Z-plane, also known as complex β-plane.

(ii) Complex Y-plane, also known as complex α-plane.

8.5.1 Impedance Relay

The impedance relay measures the impedance of the line up to the point of the fault as shown in Figure 8.11. For a fault in the line at any point F:

$$V = IZ$$

where Z is the impedance of the line up to the point of fault F and the fault is assumed to be a direct short circuit with zero fault impedance.

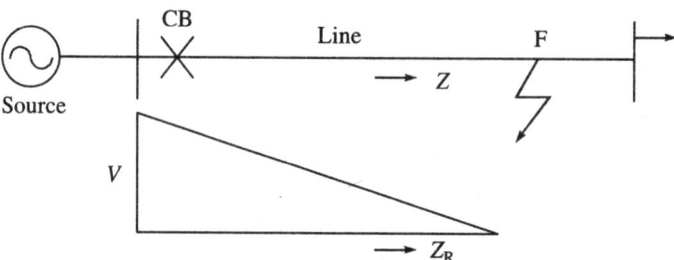

FIGURE 8.11   Representation of fault in a line.

The impedance relay works on the same principle as that of the amplitude comparator. Suppose:

$Z =$ Impedance up to the fault point

$Z_R =$ Relay setting impedance, a replica impedance proportional to the set impedance (impedance of the line up to the reach of the relay).

The job of the relay is to continuously monitor the line current I through the CT and bus voltage V through the PT. The line current passes through a transactor of equivalent impedance Z_R. Thus the realisation of a signal IZ_R is done by passing the current I through the transactor, whose impedance is Z_R.

Thus, when the relay works as an amplitude comparator, the operating signal to the relay can be denoted as follows:

$$S_1 = IZ_R$$

The restraint signal can be denoted as:

$$S_2 = V = IZ$$

By dividing the above two equations by common current I, the final signals achieved are denoted in the following equations:

$$S_1' = Z_R$$

$$S_2' = Z$$

The computation of the impedance Z up to the fault point and then the comparison with the relay setting impedance Z_R is carried out by the relay. The tripping condition can be denoted as follows:

$$Z < Z_R$$

If the fault occurs at the reach of the relay such as at the threshold, it can be denoted as follows:

$$Z = Z_R$$

8.5.2 Directional Relay

If the signals are available for the amplitude comparator, the signals for the phase comparators are achieved by finding the sum or difference of the signals as amplitude comparator. The directional relay is a dual of the plane impedance relay. It means that the signal emanated as an amplitude comparator for the plane impedance relay becomes the signal emanated as a phase comparator in the directional relay and vice versa. This means that if a plane impedance relay is an amplitude comparator, the directional relay is a phase comparator.

Thus, for the *amplitude comparator*:

$$S_1 = [IZ_R + V(=I_Z)]$$

$$= I[Z_R + Z]$$

or

$$S_1 \propto [Z_R + Z]$$

and

$$S_2 = [IZ_R - V(=IZ)]$$

$$= I[Z_R - Z]$$

or

$$S_2 \propto [Z_R - Z]$$

And, for the *phase comparator*:

$$S_3 = IZ_R$$

or

$$S_3 \propto Z_R$$

and

$$S_4 = V(=IZ) \propto Z$$

and α is the phase angle between the signals S_3 and S_4.

The tripping condition is achieved if the phase angle α is denoted as follows:

$$-90° < \alpha < 90°$$

And at the threshold, $\alpha = \pm 90°$.

The characteristics of the relay, represented by a straight line passing through the origin and tripping direction, is shown in Figure 8.12.

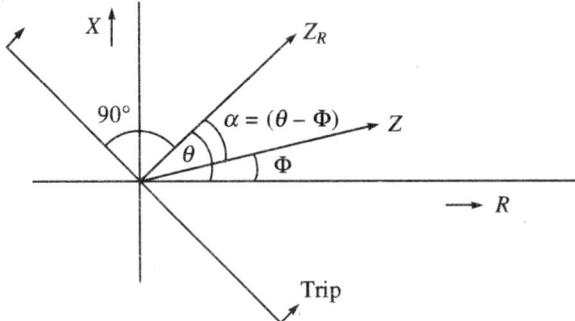

FIGURE 8.12   Directional relay characteristics.

[The signals considered above are in the Z-plane. The signals in the Y-plane can be similarly obtained by dividing the signals in the Z-plane by Z_R or by multiplying them by Y_R {where $Y_R = (1/Z_R)$}].

8.6
STATIC POLYPHASE RELAYS

A single polyphase relay replaces at least three single-phase relays which are individually used for the protection of the phases. A polyphase (three-phase) relay consists of three modules, one each for the R, Y and B phases. Phase modules R, Y and B are basically single-phase comparators. Proper input signals need to be fed to the comparators through suitable measuring circuits so that the polyphase relay responds to all types of faults. Figure 8.13 shows the block diagram of the relay. The outputs of the phase modules R, Y and B are fed into the OR gate, which gives the final trip signal.

Since there will be different values of the short circuit line currents I_R, I_Y and I_B, the fault measuring schemes do not give the correct reach for phase-to-phase faults and phase-to-ground faults owing to different fault line currents.

The measuring circuits give correct measurements only if the zero sequence current is taken into account. In order to compensate for this, fault point voltages need to be provided for each phase module.

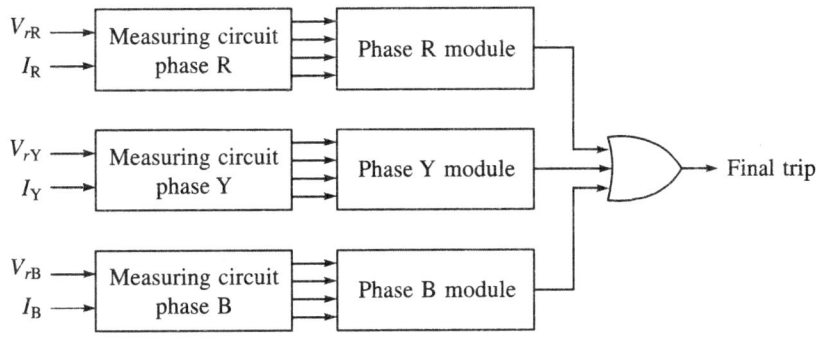

FIGURE 8.13 Block diagram for polyphase (three-phase) distance relay.

Let the compensated fault point voltages V_x, V_y and V_z be provided with line-to-ground voltages V_{rR}, V_{rY} and V_{rB}. The derivation of the compensated fault point voltages is described below.

Let Z_L be the line impedance and

V_{fR} = Phase R line to neutral fault point voltage considering the zero fault impedance

$$= V_{rR} - I_R Z_L$$

By putting the sequence components of I_R and the sequence impedances of Z_L in the above equation, we get:

$$V_{fR} = V_{rR} - I_{R1} Z_{L1} - I_{R2} Z_{L2} - I_{R0} Z_{L0}$$

$$= V_{rR} - (I_R + K I_{R0}) Z_{L1}$$

Now, the three compensated voltages will be as denoted in the following equations:

$$V_x = V_{rR} - (I_R + K I_{R0}) Z_R = V_{x1} + V_{x2} + V_{x0}$$

$$V_y = V_{rY} - (I_Y + K I_{R0}) Z_R = \alpha^2 V_{x1} + \alpha V_{x2} + V_{x0}$$

$$V_z = V_{rB} - (I_B + K I_{R0}) Z_R = \alpha V_{x1} + \alpha^2 V_{x2} + V_{x0}$$

Here, α is an operator and $Z_{L1} = Z_R$ is the replica impedance.

$$V_{x1} = V_{rR1} - I_{R1} Z_R$$

$$V_{x2} = V_{rR2} - I_{R2} Z_R$$

$$V_{x0} = V_{rR0} - I_{R0} Z_R$$

In these relays, V_{rR} and $-V_x$ are the signals sent to the phase R module. Similarly, V_{rY} and $-V_Y$ are the signals sent to the phase Y module and V_{rB} and $-V_Z$ are the signals sent to the phase B module. The characteristics of polyphase relays constitute a combination of the characteristics each of the phase modules. And each module's characteristic is similar to the mho relay's characteristic.

Mathematically:

$$S_1 = V_{rR}$$

and

$$S_2 = -V_x = (I_R + KI_{R0})Z_R - V_{rR}$$

$$\left(\frac{-V_x}{V_{rR}}\right) = \frac{[(I_R + KI_{R0})Z_R - V_{rR}]}{V_{rR}}$$

$$= \frac{Z_R}{V_{rR}/(I_R + KI_{R0})} - 1$$

$$= \frac{Z_R}{Z_L} - 1$$

$V_{fR} = 0$ as the R phase is short-circuited.

Thus, if

$$\frac{-V_x}{V_{rR}} = \alpha = \frac{[Z_R - Z_L]}{Z_L}$$

for tripping, the condition is $(-90° < \alpha < 90°)$.

EXERCISES

1. What is a static relay? Make a list of the components used in the static relaying scheme.

2. What do you mean by the threshold characteristic of a static relay?

3. Write the advantages and limitations of static relays.

4. Describe different types of amplitude comparators.

5. Describe different types of phase comparators.

6. What is meant by duality of the impedance relay? Derive its signals both as an amplitude comparator as well as a phase comparator.

7. Giving a block diagram, explain the static overcurrent protection scheme.

8. Write a note on static distance relaying.

9. Write a brief note on static directional relaying scheme.

10. Describe the advantages of poly-phase relays. Discuss with the help of a neat diagram the theory and principle of the operation of a poly-phase relay.

Chapter 9

Travelling Wave Relays

9.1
INTRODUCTION

The power frequency components of the power system were the main parameters being utilised for measurement in power system protection. The high frequency signals were filtered out and not used for the purpose of protection in the conventional methods in order to prevent interference. But high frequency components contain a wide range of information about the:

(i) Fault type
(ii) Location
(iii) Direction, and
(iv) Time duration of the fault.

It is thus clear that high-frequency transient signals produced by the fault contain more information about the fault than power frequency signals.

At present, power engineers are mainly concerned with power system stability because of the interconnected networks in the deregulated power supply system. There is therefore a need for ultra high speed (u.h.s.) clearing of the fault, which improves the transient stability of the power system. The development of u.h.s. circuit breakers like the SF_6 circuit breaker has facilitated the ultra high-speed isolation of the fault through utilisation of the concept of travelling wave phenomena.

Transients or travelling wave signals generated by the fault provide first-hand information about a possible disturbance on the line. It is thus prudent to use travelling wave signals to detect faults, as this is undoubtedly the quickest possible scheme for fault detection. The protection scheme based on fault-initiated travelling waves measures the distance to the fault by using the time taken for a wave to travel from the relaying point to the fault and back.

A fault on the high voltage line, which occurs at a non-zero voltage, causes the rapid discharging of pre-fault charges on the line. This discharging of the charges generates surges on the transmission line. These surges are propagated on the line in the form of waves, which travel with the speed of light in both directions on the line from the fault point and are reflected along the line. In fact, the reflections take place whenever the waves arrive at the impedance discontinuities, for example at line terminals, junctions, or the fault itself. The repetition of the reflections of these waves causes a decay in high frequency voltage and current transients. Each wave carries a composite form of frequencies of a wide range from a few kilocycles to several megacycles. These waves have a steep rising front but a comparatively slower decaying tail. The characteristic impedance of the line depends upon the line parameters.

The transients become prominent in EHV transmission lines because of the large length and domination of the capacitive reactance. Thus, the utilisation of the transient signals known as travelling waves can help identify faults in the power transmission lines.

9.2
SUPERPOSITION IN A FAULTED NETWORK

In Figure 9.1(a), the L-G fault occurs at point F and the voltage and currents signals are measured at the relay location R. The fault inception in the line causes the post-fault voltage and current v_R and i_R, respectively, at the location R. Now, let the steady state pre-fault voltage and current be v_{RP} and i_{RP}, respectively, as shown in Figure 9.1(b).

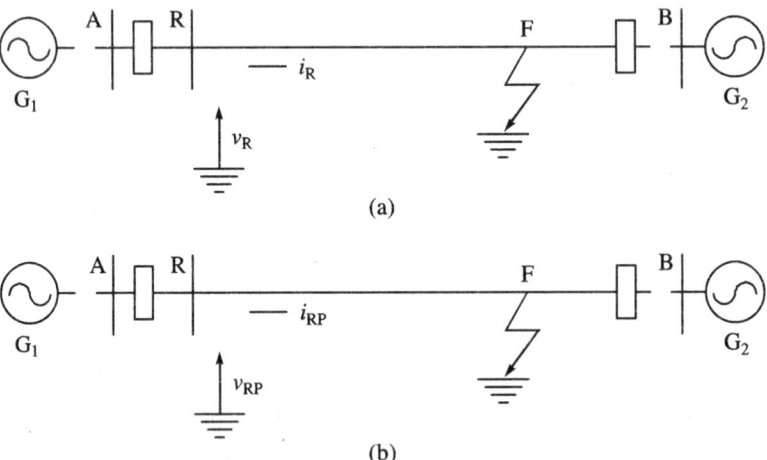

FIGURE 9.1   Representation of network before and after fault.

There will be deviations in the pre-fault voltage and the current values after the fault. Post-fault voltage and current differ in terms of the values of deviations. Let, $\Delta v_R(t)$ and $\Delta i_R(t)$ be the fault-generated voltage and current deviations from the steady state pre-fault values at the relay location, respectively.

Thus, by the superposition of the pre-fault signals and the deviation signals, we obtain the following equations:

$$v_R(t) = v_{RP}(t) + \Delta v_R(t) \tag{9.1}$$

$$i_R(t) = i_{RP}(t) + \Delta i_R(t) \tag{9.2}$$

The superimposed values shown in equations (9.1) and (9.2) are generally used in travelling wave relaying algorithms in order to facilitate the evaluation of travelling waves.

The voltage deviation (Δv_R) depends upon the following factors:

(i) Amplitude of the voltage at the fault location at the instant of the fault
(ii) Impedance of the line and sources
(iii) Fault resistance

If the fault occurs at the instant of the maximum voltage, the voltage deviation (Δv_R) also becomes maximum. If, on the other hand, the fault occurs at the instant of the voltage zero, the deviation is zero.

The magnitude of the current deviation (Δi_R) depends upon the fault inception angle and contains an exponentially decaying component known as the *DC offset*. The DC offset value is the maximum for faults occurring when the voltage is at its zero value, and this takes place when the fault occurs at its maximum voltage. The loop resistance of the fault reduces the magnitudes of both the deviations, Δv_R and Δi_R.

9.3
TRAVELLING WAVE PROPAGATION

Electrical disturbances on lines end up propagating voltage and current transients in the form of travelling waves. The superimposed voltage and current waves travelling along the line depend on the line characteristics. *Wave reflection takes place whenever the waves arrive at the impedance discontinuities, for example at line terminals, junctions, or the faults itself.*

The Bewley Lattice diagram is generally used while illustrating the travelling wave phenomena. The Bewley Lattice diagram of a single-phase power system is shown in Figure 9.2.

Long transmission line is represented by distributed parameters. Assume that a fault occurs at a position that is D_f km away from the relay. The travelling waves generated by the occurrence of a fault are propagated along the line. As the backward travelling wave V_1 arrives at the source G_1 (Generator number one) behind the relay, reflection occurs. The reflected wave V_{r1} returns along the line towards the fault point. At that point, if the fault resistance is not zero, a part of the wave is reflected while the other part is transmitted. The reflected wave V_{r2} returns to busbar-1 after some time.

If the time interval, say t_0, between the arrival of the reflected wave V_{r1} and the backward wave V_{r2} is identified, the distance D_f, can be estimated from t as per the following equation:

$$D_f = \frac{vt_0}{2} \; ; \text{ (where } v \text{ is the speed of the wave)}$$

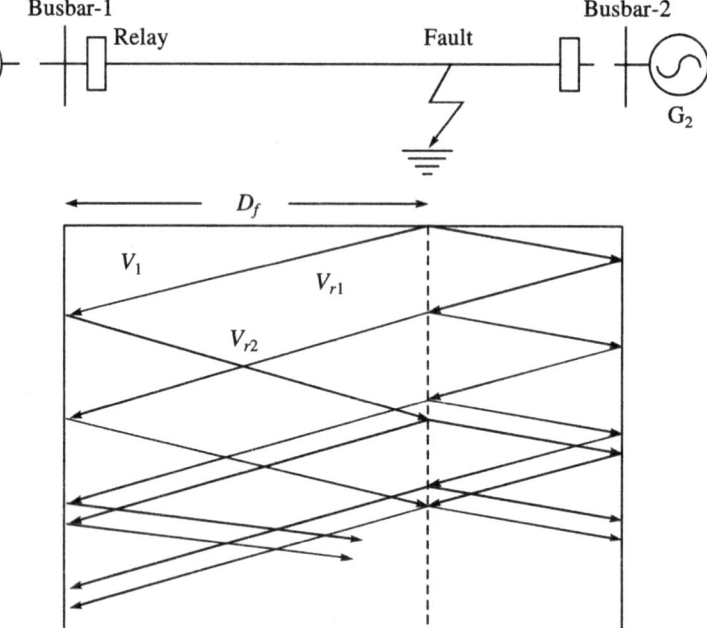

FIGURE 9.2   Principle of travelling wave propagation.

Identification of the signal V_{r2} thus constitutes the key problem of the travelling wave protection.

The magnitude of the reflected and transmitted wave fronts depends on the fault impedance of the transmission line. The reflected wave components travel back towards the R region (Refer to Figure 9.2). The wave fronts, which travel towards busbar B from the fault location F, are reflected from busbar B and reach F again. A part of these wave fronts is transmitted towards R while the remaining part is reflected back towards busbar B. The wave fronts are propagated and reflected continuously in this manner until they are damped out.

The amplitude and shape of the voltage and current signals on a transmission line may change in the lossy line. The general solution of the wave equation may be given as follows:

$$v(x,\ t) = F_1(x - at) - F_2(x + at) \tag{9.3}$$

$$i(x,\ t) = [F_1(x - at) + F_2(x + at)]/Z_0 \tag{9.4}$$

where

a = Propagation velocity of the surge in the line

Z_0 = Surge impedance of the line

In terms of the incremental components, the voltage and current deviations occurring due to a fault can be expressed as follows:

$$\Delta v(x,\ t) = f_1(x - at) - f_2(x + at) \tag{9.5}$$

$$\Delta i(x,\ t) = [f_1(x - at) + f_2(x + at)]/Z_0 \tag{9.6}$$

where f_1 and f_2 are the backward and forward travelling waves calculated from the incremental components of the voltage and current signals. By solving the equations (9.5) and (9.6) above, we get the following:

$$\text{Backward signal, } S_1(t) = \Delta v(x, t) + Z_0 \Delta i(x, t) = 2f_1(x - at) \qquad (9.7)$$

$$\text{Forward signal, } S_2(t) = \Delta v(x, t) - Z_0 \Delta i(x, t) = -2f_2(x + at) \qquad (9.8)$$

The signals $S_1(t)$ and $S_2(t)$ are used as the backward and forward relaying signals, respectively, in the travelling wave algorithm. The concept of travelling wave phenomena is utilised for developing the following three types of travelling waves relays:

(i) Amplitude comparison relay
(ii) Directional comparison relay
(iii) Phase comparison relay, which is also known as the *correlation technique relay.*

These relays are discussed in detail in the following sections.

9.4
AMPLITUDE COMPARISON TRAVELLING WAVE RELAY

The voltage and current at any point in the power network after the fault are regarded as the sum of the pre-fault and fault-generated components. The fault-generated components may be represented by travelling waves during the first few milliseconds after the occurrence of a fault. Internal faults are detected after the difference between the travelling waves at the local and remote ends is computed.

Three relaying signals are used for the detection of the three-phase faults. The relaying signals are zero if there is no internal fault and it has a value if there is an internal fault. The main demerit of this scheme is that it requires the exchange of quantitative information between the ends of the protected line.

Forward fault

The propagation and reflection of wave fronts in a forward fault is shown in Figure 9.3. The reflection of the wave fronts takes place at the points of impedance discontinuities like the fault point F, and the busbars A and B. For a derivation of the mathematical equations, the voltage and current signals measured at busbar R are considered to be the incremental components already described in equations (9.3) and (9.4).

The first backward travelling wave f_1 reaches R at τ_1 seconds after the fault occurrence, and after that, it reaches busbar A at time τ. This wave f_1 is reflected from busbar A and the travelling wave travels towards the fault location passing R at time τ_2. After reaching the forward travelling wave, f_2 at fault location F at time 2τ, a part of it is reflected from F while the other part passes towards busbar B. Then, f_{1r}, which is the reflected part of the travelling wave f_2 reaches busbar R at time τ_3 and is reflected from busbar A at time 3τ.

Let the pre-fault steady state voltage be $V_p \sin(\omega t + \phi)$. Now, a fault can be described as the introduction of a voltage source equal in magnitude and opposite in sign at the fault point, at the time of the fault.

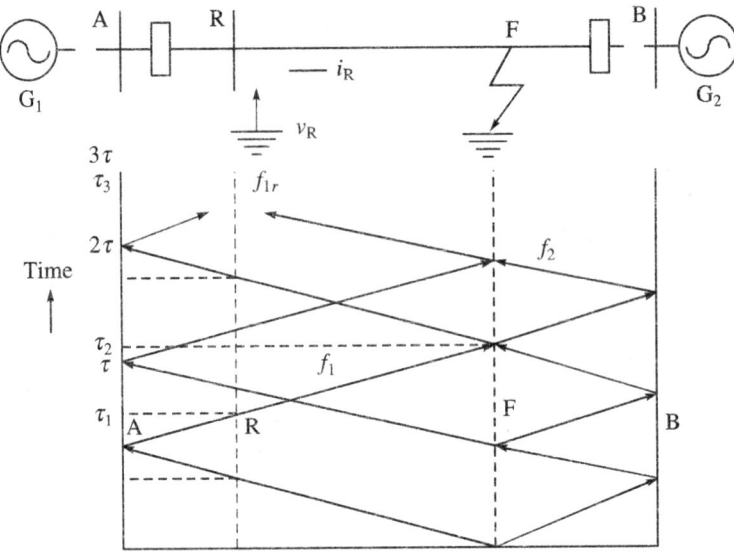

FIGURE 9.3   Wave fronts propagation and reflection in a forward fault.

The voltage deviation can thus be represented by the following expression:

$$\Delta v_F(t) = -V_p \sin(\omega t + \phi)$$

If the voltage and current changes produced by the fault point are $\Delta v_F(t)$ and $\Delta i_{RF}(t)$, respectively, then combining the voltage and current deviations seen at the relay point R, results in the following equations:

$$\Delta v_F(t) + Z_0 \Delta i_{RF}(t) = \Delta v_R(t + \tau_1) + Z_0 \Delta i_R(t + \tau_1) \tag{9.9}$$

$$\Delta v_F(t) - Z_0 \Delta i_{RF}(t) = \Delta v_R(t + \tau_1) + Z_0 \Delta i_R(t + \tau_1) \tag{9.10}$$

Thus, by combining equations (9.8) and (9.10), which is possible because at time τ_1, the forward travelling wave f_2 is equal to zero at the relaying point, we obtain the following equation:

$$\Delta v_R(t + \tau_1) - Z_0 \Delta i_R(t + \tau_1) = 0 \tag{9.11}$$

and

$$\Delta i_{RF}(t) = [\Delta v_F(t)]/Z_0 = [-V_p \sin(\omega t + \phi)]/Z_0$$

By putting the above values in equation (9.9), we obtain the following equation:

$$\Delta v_R(t + \tau_1) + Z_0 \Delta i_R(t + \tau_1) = -2V_p \sin(\omega t + \phi)$$

By introducing a time shift, we obtain the following:

$$f_1(t + \tau_1) = -2V_p \sin(\omega t + \phi)$$

$$f_1(t) = -2V_p \sin(\omega t - \omega t_1 + \phi) \tag{9.12}$$

For the time duration $\tau_1 < t < (2\tau + \tau_1)$, the backward travelling wave f_1 is initially independent of the terminal conditions. If, for the loss less line, τ_1 is small as compared with the period of 50 Hz or 60 Hz signal, f_1 may be approximated to have a constant value as given in equation (9.13) below:

$$f_1(t) \approx -2V_p \sin(\phi) \tag{9.13}$$

In case of an incoming wave on a loss less line with the surge impedance Z_0, its shape is determined by equation (9.13) when it is reflected from a source consisting of a pure inductance, say L_s, resulting in the following equation:

$$f_2(t) = V_p \sin(\phi)\,[1 - 2e^{-[(t-t_2)/L_s]Z_0}] \tag{9.14}$$

Equation (9.13) is for the time duration $\tau_2 < t < (2\tau + \tau_2)$. The low frequency signal is reflected with an exponential decay while the high frequency steep signal is reflected immediately.

Reverse fault

In a reverse fault, the forward travelling wave, say f_2, first occurs in the relays with respect to the relaying point R. The travelling waves are propagated past R along the line and reflected at the busbar B located at the remote end. The first forward wave front f_2 starts from the fault point F and reaches R in time τ_1. The same wave fronts reach busbar B at time τ. The wave front f_2 is reflected from busbar B, and then travels towards the relay location R at time τ_2.

The wave propagation for a reverse fault is shown in Figure 9.4. Let $V_p \sin(\omega t + \phi)$ be the pre-fault steady state voltage. The voltage deviation due to the fault Δv_F will then be as per the following equation:

$$\Delta v_F(t) = -V_p \sin(\omega t + \phi)$$

If we take the wave propagation times as shown in Figure 9.4 and the fault occurrence time as t, and then combine the voltage and current deviations seen at the relay point R, we obtain the following equations:

$$\Delta v_F(t) + Z_0\Delta i_{RF}(t) = \Delta v_R(t + \tau_1) + Z_0\Delta i_R(t + \tau_1) \tag{9.15}$$

$$\Delta v_F(t) - Z_0\Delta i_{RF}(t) = \Delta v_R(t + \tau_1) - Z_0\Delta i_R(t + \tau_1) \tag{9.16}$$

where τ_1 is the travel time for the waves from F to R and Δi_{RF} is the deviation in current with the current measurement taken as positive in the direction RB.

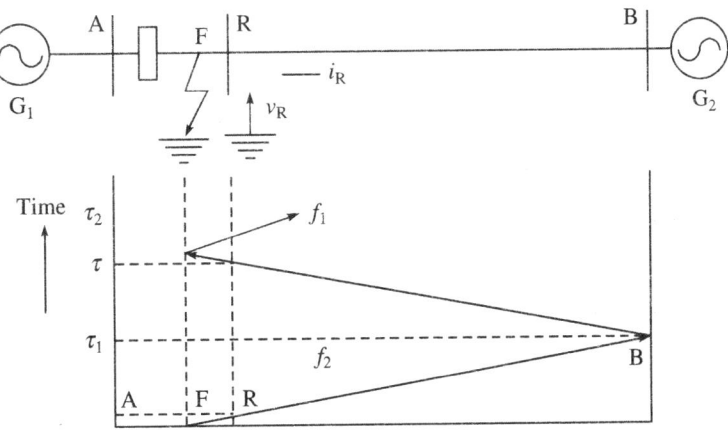

FIGURE 9.4 Wave propagation for a reverse fault.

In the beginning, however, the backward travelling wave f_1 is equal to zero at the relaying point.

Thus, from equations (9.7) and (9.15), we obtain the following equation:

$$\Delta v_R(t + \tau_1) + Z_0 \Delta i_R(t + \tau_1) = 0 \qquad (9.17)$$

By combining the equation $\Delta v_F(t) = -V_p \sin(\omega t + \phi)$ with equation (9.17) above, we get:

$$\Delta i_{RF}(t) = -\{\Delta v_F(t)/Z_0\} = V_p \sin(\omega t + \phi)/Z_0 \qquad (9.18)$$

Now by solving equations (9.16) and (9.18), we get the following:

$$\Delta v_R(t + \tau_1) - Z_0 \Delta i_R(t + \tau_1) = -2V_p \sin(\omega t + \phi) \qquad (9.19)$$

And by introducing the time shift in equation (9.8), we get:

$$f_2(t) = -2V_p \sin(\omega t - \omega \tau_1 + \phi) \qquad (9.20)$$

The time period for the above equation will be $\tau_1 < t < (2\tau + \tau_1)$.

Initially, the forward travelling wave f_2 is independent of the terminal conditions. For a very small value of τ_1 as compared to the time period of 50 Hz or 60 Hz, the value of f_2 in equation (9.20) can be approximated for the time period of $\tau_1 < t < (2\tau + \tau_1)$ to obtain the following result:

$$f_2(t) \approx -2V_p \sin(\phi)$$

9.5
PHASE COMPARISON TRAVELLING WAVE RELAY

Phase comparison travelling wave relays perform the correlation of a forward and backward travelling wave at the relay location. Correlation is a measure of a linear relationship between two variables and can be applied to both random variables and deterministic functions. The correlation technique is applicable to both the analog and discrete signals, and is widely used for detecting noise signals.

The sampled values of signals are used in the digital domain of the correlation. The cross-correlation function is used for comparing the relation between two time-shifted signals. The cross-correlation function of two signals $x(k)$ and $y(k)$ is defined by the following equation:

$$R_{xy}(t) = \left(\frac{1}{N}\right) \sum_{K=1}^{N} x(k\Delta t) \cdot y(k\Delta t + \tau) \qquad (9.21)$$

The cross-covariance function of the mean removed signals $x(k)$ and $y(k)$ is as follows:

$$\phi_{xy}(t) = \left(\frac{1}{N}\right) \sum_{K=1}^{N} [x(k\Delta t) - \overline{x}] \cdot [y(k\Delta t + \tau) - \overline{y}(\tau)] \qquad (9.22)$$

where

$$\bar{x} = \left(\frac{1}{N}\right) \sum_{n=1}^{N} x(n\Delta t)$$

$$\bar{y}(\tau) = \left(\frac{1}{N}\right) \sum_{n=1}^{N} y(n\Delta t + \tau)$$

Thus the cross co-variance function $\phi_{S_2 S_1}$ can be effectively used to find a match between the backward and forward relaying signals S_1 and S_2.

Superposition of the travelling wave signals takes place at the different mean values of the relaying signals S_1 and S_2. This helps remove the mean values in order to avoid complications in the cross-correlation output between the two signals. The discrete cross co-variance function between the stored signal S_2 and the reflected signal S_1 is given by the following equation:

$$\phi_{S_2 S_1}(t) = \left(\frac{1}{N}\right) \sum_{n=1}^{N} -[\bar{S}_2(k\Delta t) - \bar{S}_2] \cdot [S_1(k\Delta t + \tau) - \bar{S}_1(\tau)] \qquad (9.23)$$

where

$$\bar{S}_2 = \left(\frac{1}{N}\right) \sum_{K=1}^{N} S_2(n\Delta t)$$

and

$$\bar{S}_1(\tau) = \left(\frac{1}{N}\right) \sum_{K=1}^{N} S_1(n\Delta t + \tau)$$

Here, $\bar{S}_1$ and $\bar{S}_2$ are the mean values of the signals S_1 and S_2 in their respective sample groups. τ is the time shift corresponding to the peak value of $\phi_{S_2 S_1}$ and is generally used to estimate the distance to the fault.

9.6
DIRECTIONAL COMPARISON TRAVELLING WAVE RELAY

In this scheme, as in the extension of amplitude comparison relays, the relaying decision is based on two signals, i.e. the forward and backward travelling waves. The two signals may be taken as denoted in the following equation:

$$D_1 = |a(t)| - |b(t)|$$

$$D_2 = \left|\frac{da(t)}{dt}\right| - \left|\frac{db(t)}{dt}\right|$$

The signal D_2 helps address the possibility of a fault occurring near voltage zero. In the case of the forward fault, D_1 and D_2 can never be simultaneously negative. Similarly in the case of the reverse fault, both D_1 and D_2 can never be simultaneously positive. Qualitative information about the direction of the fault is exchanged through an ultra high-frequency carrier communication channel.

The directional comparison relay, when combined with the distance relay, is used for providing remote and back-up protection. The former integrates multiple functions to facilitate better performance through mutual support between the two relays.

9.7
FAULT LOCATION

In a power system under steady state operating conditions, the incremental components of the voltage and current signals ideally remain zero during an estimation of the travelling wave functions S_1 and S_2. In the event of occurrence of a fault, there is a significant deviation the magnitude of the voltage and current signals relative to their steady state values. The variation in the sequence in which the signals S_1 and S_2 become non-zero depends on the fault direction and the relay location. The *fault direction* is estimated by the sequence in which S_1 and S_2 exceed the pre-specified threshold value. For example, if S_2 exceeds a certain threshold value before S_1 the fault is said to be occurring in the backward direction, otherwise it would be occurring in the forward direction.

The fault inception angle ϕ plays very important role in this process because the backward and forward travelling wave functions, S_1 and S_2 are dependent on the angle ϕ. In the very rare situation wherein a fault occurs near a voltage zero, the fault inception angle ϕ would also be close to zero. In this situation, the signals S_1 and S_2 have small magnitudes and it is difficult to detect a fault by using the signals S_1 and S_2 alone. In this case, the use of a method independent of the angle ϕ is recommended as that can improve the identification of the fault conditions and reliability. The discriminant function independent of ϕ is thus used in order to improve the reliability of fault detection. Using the travelling wave functions and their differentiation generally facilitates an estimation of the discriminant functions. The backward and forward discriminant functions D_B and D_F, respectively are denoted by the following equations:

$$D_B = (\Delta v_R + Z_0 \Delta i_R)^2 + (1/\omega^2) [d(\Delta v_R + Z_0 \Delta i_R)/dt]^2$$

and

$$D_F = (\Delta v_R - Z_0 \Delta i_R)^2 + (1/\omega^2) [d(\Delta v_R - Z_0 \Delta i_R)/dt]^2$$

The discriminant functions D_B and D_F are equivalent to zero for a healthy line and show very high values in the inception of a fault. *If D_F exceeds a certain threshold value before D_B, the fault is said to be a backward fault, otherwise it would be a forward fault. Fortunately, these functions are of a highly decisive nature.*

For determining the *fault location*, it is imperative to ascertain whether the disturbance is occurring within the protected zone or outside it. Since it is simple to implement a correlation algorithm, a cross-correlation function is widely used to estimate the distance to the fault. The fault-generated wave fronts are propagated along both directions on the transmission line. The first wave front f_1 reaches the relay location R at time τ_1 [refer to Figure 9.3]. A forward fault will be detected in case the signal S_2 is low at this time τ_1. When the signal S_2 rises above a certain pre-specified value at time τ_2, the forward wave front will be detected at the relay location R. After detection of the forward fault, the relay starts storing the forward travelling wave signal S_2. This wave front signal S_2 can be stored in a window of N samples. The storage time is fixed in such a way that the stored signal contains both the pre-fault and post-fault information.

The reflection of the wave front f_2 takes place at the fault point and a portion of the wave front returns to the relay location at time τ_3. The identification of the reflected wave front is carried out by matching the backward relaying signal S_1 with the stored signal S_2. This happens through the cross-correlation of the stored and returning signals. The instant at which the cross-correlation of the two signals attains a peak value is the time τ_3 at which the reflected wave S_1 matches the stored signal S_2.

The distance to the fault will be as denoted in the following equation:

$$x_f = \frac{a(t_3 - t_2)}{2}$$

Here, a is the velocity of the propagation of the waves on the line. If, the distance x_f is less than the length of the line, the fault falls within the protected zone of the relay.

9.8
TRAVELLING WAVES IN A THREE-PHASE SYSTEM

Due to mutual coupling between two conductors of the three-phase transmission lines, even if a fault occurs in any one conductor, transient currents are induced in the other healthy conductors too. The induced waves on the healthy phase conductors travel between the two ends of the line and interact with the waves on the faulted phase line. Thus all the three phases get distorted. The set of equations describing the wave propagation on each phase therefore become interdependent because of the mutual coupling effect, and the theory of natural modes has to be applied to arrive at a solution. In this theory, a three-phase coupled line is decomposed into three independent modes of propagation or three single-phase lines. Now the travelling wave theory applicable to a single-phase circuit is applied to the independent modes separately to analyse the faults of a three-phase system. The manner in which a three-phase system can be transferred into three independent single-phase systems or modes is described below.

A line consisting of n conductors and ground has n modes of propagation. Each mode has a particular voltage-current relationship, velocity and attenuation constant at any given frequency. The modal transformation technique can be used to transfer the dependent n line system into n independent modes. The following equations show the relationship between the phase and modal quantities:

$$[v(t)] = [S]\,[v^{(m)}(t)] \tag{9.24}$$

$$[i(t)] = [Q]\,[i^{(m)}(t)] \tag{9.25}$$

Here, the inverse relationship can be described by the following equations:

$$[v^{(m)}(t)] = [S]^{-1}[v(t)] \tag{9.26}$$

$$[i^{(m)}(t)] = [Q]^{-1}[i(t)] \tag{9.27}$$

where $[v^{(m)}(t)]$ and $[i^{(m)}(t)]$ are the modal voltage and current matrices, respectively. $[S]$ and $[Q]$ are the voltage and current modal transformation matrices, respectively.

The modal transformation matrices $[S]$ and $[Q]$ are chosen in such a way as to facilitate independent modes of propagation. Karrenbauer, Clarke and Wedepohl transformations are the three constant transformation matrices used for perfectly transposed lines. These three transformations are detailed below:

Clarke transformation

$$Q = S = \begin{bmatrix} 1 & 1 & 0 \\ 1 & -1/2 & \sqrt{3/2} \\ 1 & 1/2 & -\sqrt{3/2} \end{bmatrix}$$

$$Q^{-1} = S^{-1} = (1/3)\begin{bmatrix} 1 & 1 & 1 \\ 2 & -1 & -1 \\ 0 & 1/\sqrt{3} & -1/\sqrt{3} \end{bmatrix}$$

Wedepohl transformation

$$Q = S = \begin{bmatrix} 1 & 1 & 1 \\ 1 & 0 & -2 \\ 1 & -1 & 1 \end{bmatrix}$$

$$Q^{-1} = S^{-1} = (1/3)\begin{bmatrix} 1 & 1 & 1 \\ 3/2 & 0 & -3/2 \\ 1/2 & -1 & 1/2 \end{bmatrix}$$

Karrenbauer transformation

$$Q = S = \begin{bmatrix} 1 & 1 & 1 \\ 1 & -2 & 1 \\ 1 & 1 & -2 \end{bmatrix}$$

$$Q^{-1} = S^{-1} = (1/3)\begin{bmatrix} 1 & 1 & 1 \\ 1 & -1 & 0 \\ 1 & 0 & -1 \end{bmatrix}$$

However, the performance of the correlation function method depends upon the fault resistance, the system configuration and the mode type. The evaluated modal quantities are of two modes-(Mode-1)- earth mode and (Modes-2 and 3)- aerial modes. These are described below.

[Mode-1] Earth Mode

The earth mode comprises the zero sequence components of the phase voltages and currents. It is frequency-dependent and its velocity and attenuation are affected by the resistivity of the earth. At low frequencies the velocity may be approximated to be about 75% of the speed of the light.

[Modes-2 and 3] Aerial Modes

The current in the three phases tends to be cancelled in this mode. Thus the effect of earth resistance in this mode is very small. Therefore, the aerial mode is told to be independent of the frequency and the velocities of the waves tend to approach the speed of the light.

The ground (earth) mode has lower velocity and greater attenuation and distortion than the aerial modes. The velocity and attenuation of the aerial modes for a fully transposed single circuit three-phase system are identical.

EXERCISES

1. Describe how superposition in a faulted network is helpful in a travelling wave protection scheme.

2. Explain how travelling wave propagation takes place in a power system network.

3. What are the different types of travelling wave-relaying schemes? Describe each of them.

4. How are forward fault and reverse fault identified in an amplitude comparison travelling wave relay?

5. What is correlation technique? How is this technique useful in the phase comparison travelling wave relaying scheme?

6. Write a short note on the directional comparison travelling wave relay.

7. Explain the method of fault location by using travelling waves.

8. Describe fault location determination by using travelling waves.

9. Describe how mutual coupling in the three-phase system affects the performance of travelling waves relays.

Chapter 10

Pilot Relay Protection

10.1

INTRODUCTION

'Pilot' means the existence of a communication channel of wire, carrier current or microwave between the two ends of a transmission system over which the exchange of information takes place. The pilot relaying scheme is the best choice for all types of short-circuits occurring at any fault location on the transmission and distribution line. This scheme permits high-speed automatic reclosing. Pilot relay protection schemes are of the following three types:

   (i) Wire pilot relaying
   (ii) Carrier current pilot relaying
  (iii) Microwave pilot relaying

Further, carrier current pilot relaying and microwave pilot relaying can be further divided into the following schemes:

   (i) Phase comparison scheme
   (ii) Directional comparison scheme
  (iii) A combination of the above two schemes
  (iv) Remote tripping schemes

10.2

REQUIREMENTS OF PILOT RELAYING

The fundamental principles involved and the basic equipments required for both carrier current pilot and microwave pilot relaying are more or less similar. In both cases, two principles, phase comparison and directional comparison, are used. It can thus be said that pilot relaying is an adaptation of differential relaying. The following equipment and relaying principles are needed for facilitating pilot relaying:

(i) Pilot wires, carrier wave generators, microwave generator

(ii) Phase comparison

(iii) Combined phase and direction comparison

(iv) Current transformers

(v) Blocking terminals

(vi) Attenuation

(vii) Quick and remote tripping

(viii) High-speed reclosing of circuit breakers

(ix) Automatic supervision of the carrier current channel

(x) Sensitivity and speed

(xi) Detection of sleet accumulation

(xii) Transient blocking

In a pilot wire channel, physical wires are run from one end to the other end of the line. In a carrier current pilot channel, low-voltage high-frequency (20 kilo-cycles to 700 kilo-cycles) currents are transmitted along a power line conductor. The ground wire generally acts as a return conductor.

Microwave pilot is an ultra-high frequency radio system. The use of a microwave pilot can help render many services which may not be technically feasible in the carrier current pilot.

10.3
WIRE PILOT RELAYING

Wire pilot for protection is in the form of either a buried cable or a pair of overhead wires other than power line conductors. This unit protection scheme is very similar to the percentage differential relaying scheme used for apparatus protection. When a carrier current pilot is found to be uneconomical, wire pilot relaying scheme is preferred for low voltage circuits as well as high voltage transmission line protection.

In cases wherein the attenuation caused in the power cable circuit by distributed capacitance and series resistance is found to be too high for carrier current, a wire pilot relaying scheme is used for the protection of certain power cables circuits. A wire pilot relaying scheme is generally the most economical form of high-speed relaying for short AC transmission line protection.

The technical limitations of using the pilot wire scheme include line resistance and shunt capacitance. Compensating reactors are generally required in order to compensate for the high shunt capacitance of long transmission lines. For this reason, wire pilot relaying is considered to be less reliable as compared to carrier current pilot relaying. Great care needs to be taken in the selection and subsequent protection of the pilot wire because its circuit is prone to troubles. It is important to maintain the requisite sensitivity and speed of protection in the case of wire pilot relaying equipments. In order to avoid an excessive burden on CTs, it is advisable not to adjust the equipment to obtain greater sensitivity.

Sometimes, the breakers of the schemes trip undesirably on a normal load current. In this situation, pilot wires become open-circuited. Such undesirable trippings of the breakers can be avoided by adjusting the pick-up value to make it at least 25 per cent higher than the maximum load current.

Amplitude comparison is preferred in most of the pilot wire protection schemes. Amplitude comparison is easier to apply in multi-terminal lines. Following are the two practical schemes used in wire pilot relaying:

(i) Circulating current wire pilot scheme, and
(ii) Opposed voltage wire pilot scheme.

The *circulating current wire pilot scheme* is similar to the percentage differential relaying scheme used for apparatus protection. A schematic arrangement of this scheme is shown in Figure 10.1.

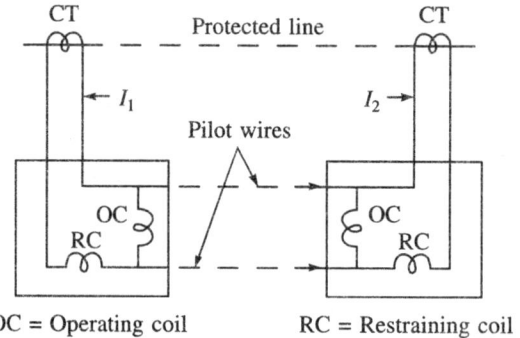

OC = Operating coil   RC = Restraining coil

FIGURE 10.1 Circulating current wire pilot scheme.

In the circulating current wire pilot scheme, two identical relays equipped with an operating coil and a restraining coil, are placed at both ends of the line. Under normal operating conditions, when there is no fault current and in case of an external fault, the output currents I_1 and I_2 of both the identical CTs connected at the two ends of the line are equal. These equal currents circulate through the restraining coil and no current flows through the operating coil. If there is a fault in the line, two different valued currents flow through the CTs and the difference of the current flows through the operating coil. This difference of the current generates a torque in the operating coil. If the latter exceeds the torque generated by the restraining coil, the relay passes on a trip command to the circuit breaker.

This circulating current scheme is generally suitable for pilot loop resistance of up to 1 mega ohm and an inter core capacitance of up to 2.5 µF. The practical scheme used in this type of relay is shown in Figure 10.2.

For economical reasons, this scheme is used in the case of line protection over a distance of 15–30 km as it becomes financially uneconomical when used for line protection over a distance of 30 km. This scheme is very popular for protecting AC lines for the following reasons:

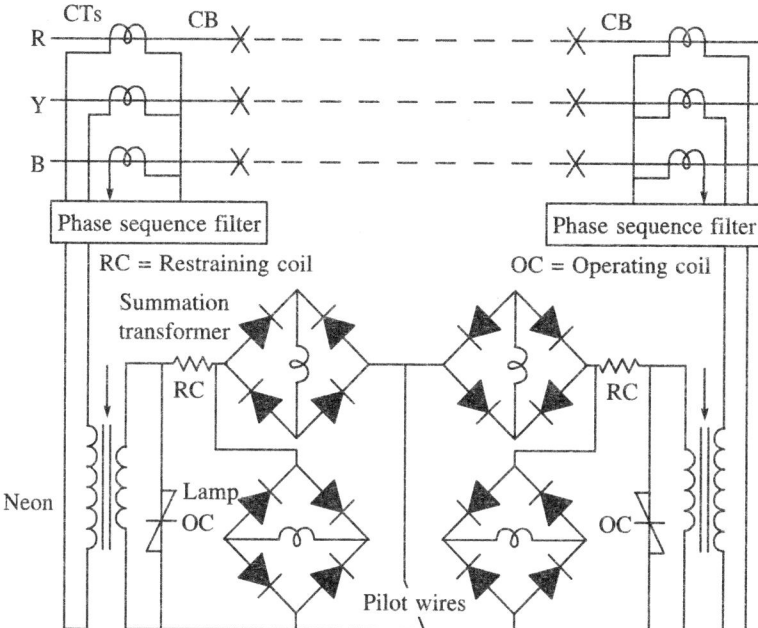

FIGURE 10.2   Practical scheme of a circulating current wire pilot relay.

(i) An AC transformer can reduce the magnitude of the fault current flowing through a pilot wire to a very low value.

(ii) The phase sequence filter connected at both ends of the line can reduce the three-phase fault current or ground current to single-phase current. This can also be achieved by mixing transformers. Thus two pilot wires are needed only for three-phase faults.

The restraining coil marked RC and the operating coil marked OC are energised by full wave rectifiers. A DC directional relay is used with the rectified AC quantities in order to obtain high sensitivity. Summation transformers limit the magnitude of the voltage impressed on the pilot circuit. Neon lamps are used to limit the peak of surge voltage caused by switching actions.

An *opposed voltage wire pilot scheme* does not have the provision for fault current circulation through pilot wires. In this type of scheme, the connection is reversed because of which there is no flow of normal current through the pilot wire. Since the operating coil is connected in series with the pilot wire, no current flows in the operating coil in normal conditions in the case of external faults. In case of internal faults, on the other hand, the polarity of the remote end CT gets reversed and the fault current finds a path to flow through pilot wires while the difference current flows through the operating coils of the relays. Then the relay passes a trip command to the circuit breaker. An opposed voltage scheme is shown in Figure 10.3.

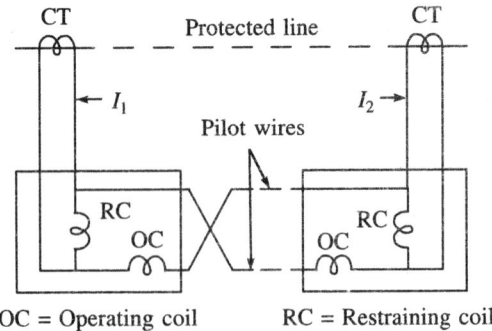

OC = Operating coil   RC = Restraining coil

FIGURE 10.3 Opposed voltage wire pilot scheme.

It should be noted that short circuit and open circuit have just the opposite effects on opposed voltage wire pilot and circulating current wire pilot schemes.

A short circuit in the wire pilot blocks tripping in the circulating current wire pilot scheme but causes tripping in the opposed voltage wire pilot scheme. And, an open circuit in the wire pilot causes tripping in the circulating current wire pilot scheme while it blocks tripping in the opposed voltage wire pilot scheme. This scheme is suitable for pilot loops up to 400 ohm. Figure 10.4 shows the practical connection diagram of the opposed voltage wire pilot scheme.

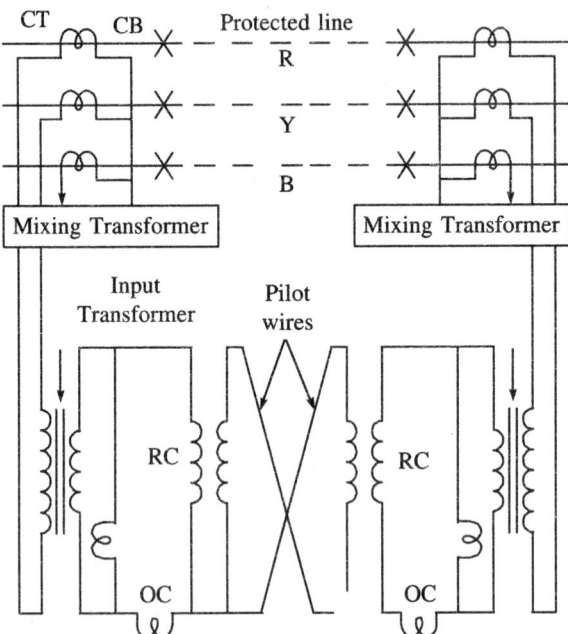

FIGURE 10.4 Opposed voltage wire pilot relaying.

The relay at each end is an AC directional type relay. The purpose of using mixing transformers is to provide single-phase quantity for all types of faults. The main limitations of an AC wire pilot relaying are as follows:

(i) If a fault occurs very near one end allowing only a very small current to flow at the other end, it is quite possible that the circuit breaker at the low current side may not trip.

(ii) The charging current of pilot wires makes the equipment less sensitive.

Multi-terminal lines protection

If more than one electric power generating station is connected in the multi-terminal lines, the protection schemes become less sensitive because of the complexity of the networking. Even though it is impossible to achieve perfect phase fault protection, a multi-terminal line with wire pilot relaying may sometimes be well-protected against ground faults.

Figure 10.5 shows the simplest multi-terminal lines system, which is a system with two terminals. In this system, the terminal, which has no source of generation, is treated/known as the "blocking" terminal.

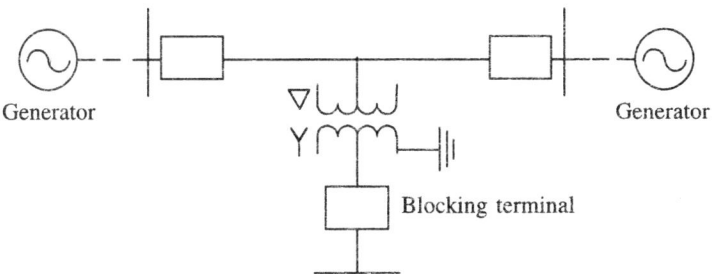

FIGURE 10.5   Illustration of a blocking terminal.

Figure 10.6 depicts the connection of an insulating transformer over the current relay across pilot wires in the blocking terminal technique. In order to protect the pilot wires, from either an open circuit or a short-circuit, instantaneous overcurrent relays are connected on the high voltage side. These relays are energised by the secondary of the CTs. The equipments have an adjustment facility to compensate for the nominal CT ratio. The CT ratio may be different for different terminals. But transient differences in different CTs with different ratios may cause undesired operations, especially in the presence of external fault currents. This is the reason why the same CT ratios are generally preferred at all the terminals.

In this case, if the transmission line terminates to a set of parallel-operated power transformers without high voltage breakers, the CT on the high-voltage side of the bushing energises the relaying equipments. The low-voltage sides of power transformers are generally connected in star to facilitate the neutral point. Therefore, if the low-voltage side of the transformer is used for a CT connection, then it should be arranged in a manner so as to compensate for the phase shift caused by the Δ-Y connection of the transformer and also to remove the zero-phase sequence components.

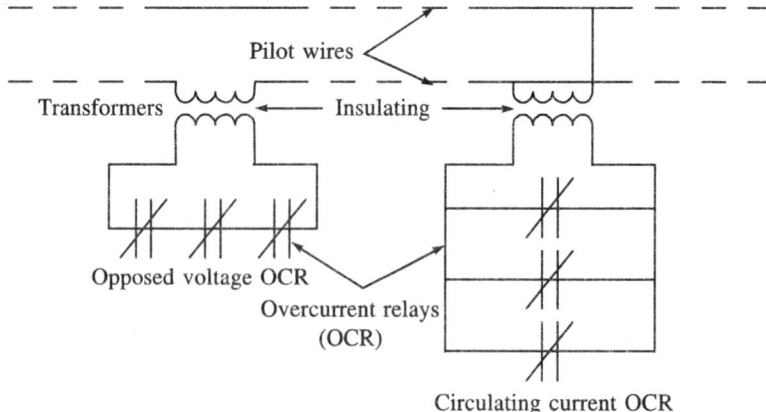

FIGURE 10.6   Blocking terminal technique.

Additional relaying equipments are also required to take care of undesirable tripping on a magnetising current inrush either at the time of energisation of the transformer or when disturbances occur in the system.

The tripping operating time of the main terminals is coordinated with that of the blocking terminal relays. The total current of all the blocking terminals on the line must be less than the current required to operate the wire pilot relays at one source terminal of the line with the breaker at the other source terminal open.

Wire pilot relaying does not have the provision of back-up protection. The existing overcurrent or distance relay equipments may be used as back-up protection if wire pilot relaying is applied as the main protection scheme.

If the wire pilot relays are arranged to receive only the CT neutral current, this type of multi-terminal line may be treated as a two-terminal line. Thus the impedance of the transformer is so high that the distance relays can usually be adjusted to protect nearly 90 per cent of the line without reaching through any of the transformers.

10.4
CARRIER CURRENT PILOT RELAY

When a voltage of positive polarity is employed on the control circuit of the transmitter, it generates a high frequency (50–500 kilo-cycles) output voltage. This output voltage is then employed between one phase conductor of the transmission line and the earth.

Carrier current pilot relaying is the most widely used kind of relaying for high voltage lines. This is used in lines transmitting voltage of 33 kV and above. This scheme is preferred over the wire pilot relaying scheme because of its following merits:

(i)   Carrier current pilot relaying consists entirely of terminal equipments.

(ii)   It is completely under the control of the user.

(iii)   Carrier current pilot relaying lends itself more conveniently to joint usage by other services such as emergency telephony and remote trip.

(iv) For lines protected with the distance scheme of protection, the main problem pertains to the non-simultaneous opening of the circuit-breakers at both ends of the lines. This leads to instability in the entire power system network. The carrier current pilot relaying scheme ensures the simultaneous and fast opening of circuit-breakers at both the ends.

(v) Above all, carrier current pilot relaying is cheaper, more reliable and widely applicable than other forms of relaying.

At each end of the line, there is a transmitter-receiver unit, coupling capacitor and a line trap. Each carrier current receiver receives the carrier current from the transmitter at either end of the line. The carrier signal is generated by the transmitter, consisting of a master oscillator and a power amplifier unit with an output of 15–20 W at a frequency of 50–500 Hz. The receiver converts the received carrier current into DC voltage. This voltage can be used in a relay or any other circuit to perform the desired function. In the absence of carrier current, this DC voltage will be zero.

The schematic illustration of a carrier current pilot channel is shown in Figure 10.7. Line traps are parallel resonant circuits and have negligible impedance to the power frequency (50 Hz or 60 Hz) currents. But they have very high impedance to the carrier frequency currents. This is explained below.

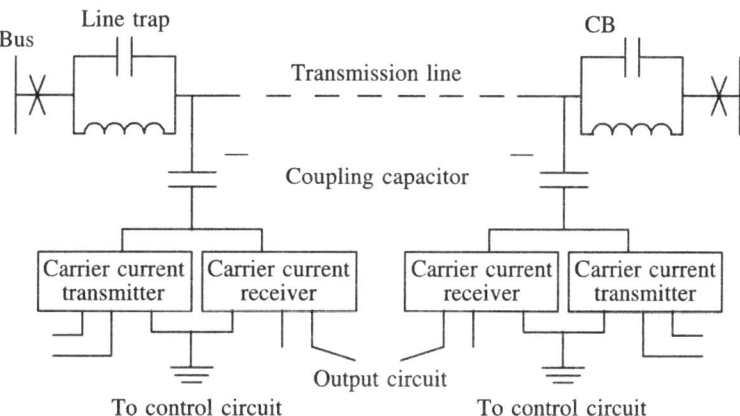

FIGURE 10.7   Carrier current pilot scheme.

Let X_C be the capacitive reactance.

$X_C = 1/(2\pi f C)$. For higher values of the signal frequency f, the capacitive reactance will be less, which means that the capacitance offers very low impedance to the carrier signal and for the power frequency (which at 50 Hz, is much less in comparison to the carrier frequency), the capacitance offers very high impedance. The capacitance thus allows the transmission of carrier signals but blocks power signals.

Similarly; let $X_L = 2\pi f L$. For higher values of the signal frequency f, the inductive reactance will be very high, which means that the inductive coil offers very high impedance to the carrier signal and for the power frequency, (which at 50 Hz is much less in comparison to the carrier frequency), the inductive coil offers very little impedance. The inductance thus allows the transmission of power signals but blocks carrier signals. The line trap thus consists of a parallel combination of inductance and capacitance, and it is tuned to the carrier frequency and connected in series at each end of the transmission line. Installation of the line traps helps confine the carrier to its own zone, i.e. only the line, and also prevents disturbances such as faults outside the protected zone from affecting the carrier. The value of inductance is in the order of 100 mH and while that of the capacitance is 0.001 μF.

In addition to relaying, the carrier signals are also used for communication. The lower value of the frequency, say 50 Hz, is chosen because below 50 Hz, the cost of the terminal equipment is high while above 500 kHz, the signal attenuation is too high.

In addition to the usual carrier equipments, the carrier current pilot relay system is also equipped with a fault detector and a directional relay. When carrier current pilot relaying was first introduced, the reliability of vacuum tubes was not up to the mark. Thus there was a perceived need for automatic equipments to supervise the pilot channel. But with the development of supervisory control and automation, the carrier current channel has become a very reliable element of protective equipments.

Each proposed application should be studied to ensure that the attenuation or the losses in the carrier current channel remain within the permissible limits. In their literature for users, manufacturers specify these limits and describe the calculations of attenuation in each element of the channel. It is not possible to implement a carrier current scheme in underground cables because of the high capacitance, which attenuates the carrier signal to a very low value. It is for this reason that the carrier current scheme is mostly preferred in overhead line protection. *Carrier currents are also used for detecting sleet accumulation, and accumulation of the dirt or salt on the line insulators and power conductors.* The method of detecting sleet accumulation is based on the fact that the attenuation of a transmission line increases as sleet accumulates on the line. Figure 10.8 describes the effect of attenuation on the magnitude of the output from the carrier current receiver.

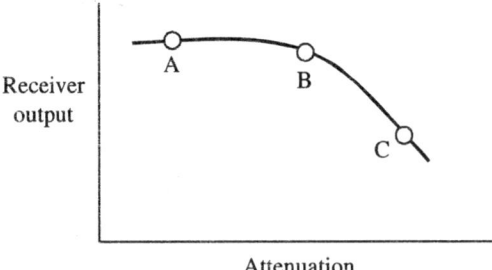

FIGURE 10.8   Attenuation effect on the receiver output signal strength.

Point A in Figure 10.8 represents the normal operation. In order to ensure safety from adverse atmospheric conditions such as sleet, the attenuation should not greatly exceed that represented by point B. For detecting the sleet accumulation, the operator at one end of the line generates the carrier current to be transmitted. Attenuation can be introduced into the transmitter or receiver circuit by pressing the button so as to advance the normal operating position from point A to point B. In this situation, the receiver output decreases rapidly for any other attenuation caused by the sleet. This information regarding the degree of attenuation must be coordinated with visual observation and experience before any useful meaning can be imparted to the receiver output readings. The same method of sleet detection also helps detect the accumulation of dirt or salt on the line insulators and power conductors.

The following three operating techniques are used for carrier current protection:

(i) Phase comparison technique
(ii) Directional comparison technique
(iii) Combined phase and directional comparison carrier current relaying

10.4.1   Phase Comparison Carrier Current Relaying

In this technique, there is a comparison of the phase relationship between the current entering one terminal and that leaving the other terminal of the transmission line. The current magnitudes are not compared. In this scheme, the transmission line current transformers feed a summation network that transforms the CT output currents into a single-phase sinusoidal output voltage. This sinusoidal output voltage is applied to a carrier current transmitter and to a comparer. The output of the carrier current receiver is also applied to the comparer. The comparer controls the operation of an auxiliary relay for tripping the transmission line circuit breaker. These components provide the means for transmitting and receiving carrier current signals for comparing at each end the relative phase relations of the transmission line currents at both ends of the line. A schematic representation of this scheme is given in Figure 10.9.

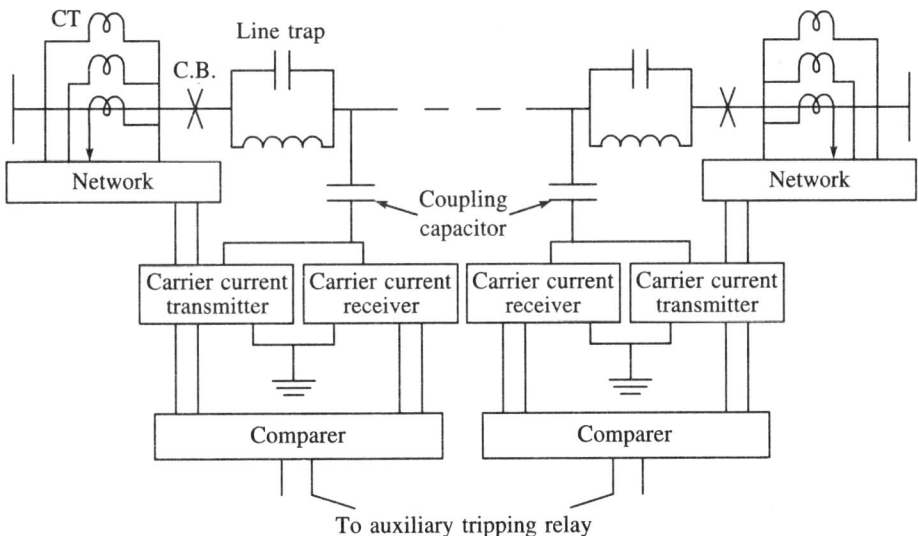

FIGURE 10.9 Phase comparison carrier current relaying schematic diagram.

The comparer may be called the heart of the phase comparison system. The schematic representation of the comparer is shown in Figure 10.10, which shows a vacuum tube equipment. The conduction of the tube takes place, when the voltage of positive polarity is impressed on the "operating grid" of the comparer by the local network. At the same time, the voltage of negative polarity should not concurrently impress on the "restraining grid" by the local carrier current receiver. With the conduction of the tube, an auxiliary tripping relay picks up and trips the local breaker. The voltage from the carrier current receiver impressed on the restraining grid causes the tube non-conducting wherever the carrier current is being received, irrespective of whether the operating grid is energised or not.

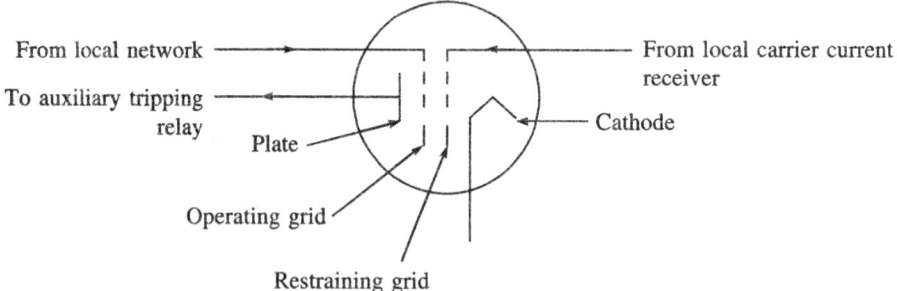

FIGURE 10.10   Schematic diagram of the comparer.

Figure 10.11 helps us examine the relations between the network output voltages at both ends of the line. The figure also shows the carrier current signals that are transmitted during external and internal fault conditions. If a fault occurs, a carrier current is produced at a very high frequency and is transmitted to the other end through the injection of this carrier into the power line through the coupling capacitor; normally the ground forms the return path. Due to the presence of the line trap, the carrier signals remain confined to their own zone, which means that they stay within the protected line. For an external fault at point 2, the network output voltages at stations A and B become 180° out of phase. This phase difference is because the CTs at the two stations are connected in reverse. Since an AC voltage is used to control the transmitter, it only transmits the carrier during the period when the voltage wave is positive. Thus the blocking carrier blocks the relay operation only in the presence of a continuous carrier. This is the reason why in case of an external fault outside the protected zone at point 2, the relay does not cause the circuit-breaker to trip.

However, when an internal fault takes place at, say at point 1, the voltages at both the stations A and B are in phase. Since the two voltages are in phase, there is carrier transmission from both the ends during the same period. This implies that there is a carrier in the positive half-cycle while there is no carrier during the negative half-cycle. In the absence of a carrier signal, there will be a relay operation that will cause tripping. It is thus observed that the phase comparison pilot is a blocking pilot because the carrier signal is not required for tripping. The failure of the pilot will not prevent tripping.

The phase comparison relaying is inherently immune to the effects of switching surges and loss of synchronism between generation sources beyond the boundary of the protected zone. Similarly, the induction current flow in the line caused by mutual induction from the

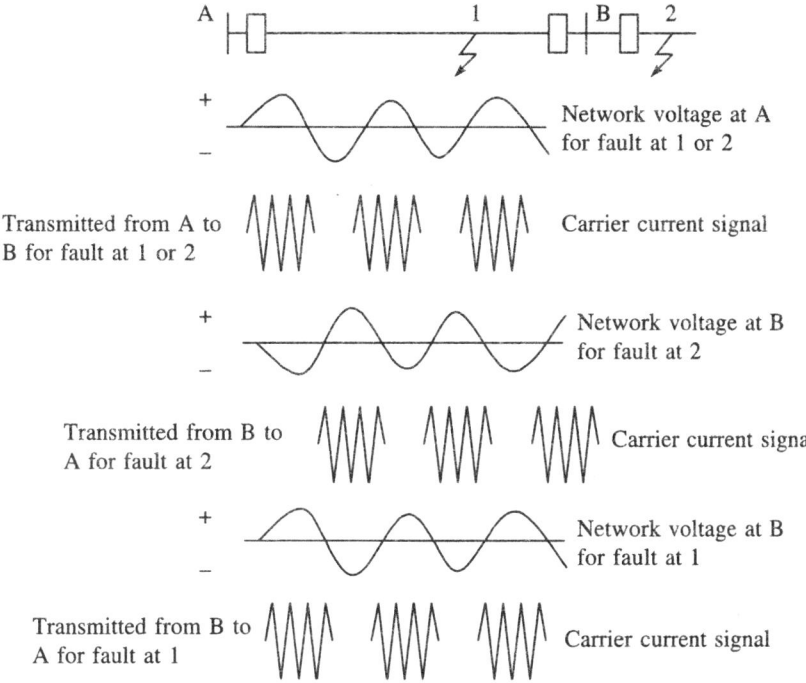

FIGURE 10.11   Relations between network output voltages and carrier-current signals.

nearby circuits does not affect the operation of the phase comparison equipments. The protected length of the transmission line in this scheme is limited by the phase shifts caused by the following factors:

(i) The time taken by the carrier signal to travel from one end to the other end of the section of the protected line, which is also known as the "propagation time"—this is generally 0.06° per km.

(ii) The band pass filter time response—this is about 5°.

(iii) The phase shifting caused by the transmission line capacitance (which is up to 10°).

Multi-terminal lines protection

The sensitivity of the protective scheme depends upon the number of terminals linked with the source of generation. The increase in the number of terminals with source generation decreases the sensitivity of the protective scheme. For example, in Figure 10.12, three terminal lines with three generators are shown. It should be noted that the phase-comparison relaying scheme is a non-directional scheme. In this scheme, any one terminal will operate to trip, whenever its current crosses the preset value unless it receives a blocking carrier current signal from any other signal. The worst case would be when equal currents enter buses 1 and 2. Thus, for equal magnitudes of fault currents fed into terminals 1 and 2, for an external fault beyond terminal 3, the pick-up current of the tripping fault detectors at terminal 3 would have to be more than twice the carrier current starting, or blocking, the fault detectors at terminals 1 and 2. If the currents

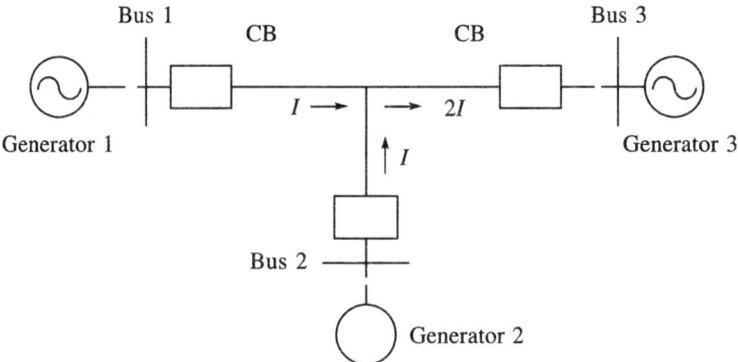

FIGURE 10.12 Reduction of sensitivity of phase comparison relaying on multi-terminal lines.

from the terminals 1, 2 and 3 are not equal, then the tripping fault detector at terminal 3 can be lowered otherwise the tripping fault detector at terminal 3 will have to be adjusted to make it three times the maximum load.

10.4.2   Directional Comparison Carrier Current Relaying

The most widely applicable type of relay is the directional comparison relay. This relay is, however, not suitable, if there is sufficient mutual induction with another line. This is also not applicable when directional ground relays are used instead of ground distance relays. In order to overcome these problems, the 'combined phase and directional comparison' technique, which is described in the later part of this chapter, is preferred.

Modern directional comparison relaying schemes operate in coordination with distance relays because distance relays make the scheme eligible for providing back-up protection. In this scheme, the pilot carrier signal informs the equipments at one end of the line as to how a directional unit at the other end responds to a short-circuit. A pilot signal transmission takes place from the terminal wherein the short-circuit current flows out of the line, i.e. in the non-tripping direction. The pilot signal in this scheme is a blocking pilot because the reception of a pilot signal is not required to permit tripping. In this scheme, the pilot signal, once started, remains steady, and is not converted into frequent half cycles as in phase comparison relaying technique.

This scheme may be applied to any multi-terminal line. The equipment type and relay adjustment for tripping and blocking need to be chosen with extreme care. The following are examples of situations when simultaneous high-speed tripping may not be obtained at all terminals:

(i) A situation may develop wherein there are phase faults or ground faults in which any internal fault current flowing from the line, say at A, in Figure 10.13 is higher than the blocking relay pick-up at that point. If there is no provision to increase the blocking relay pick-up, this situation can be avoided if the tripping at A takes place only after the back-up relays at B trip their breakers. After that, high-speed tripping will occur at the other two terminals.

(ii) Another similar situation can arise when the high apparent impedance to the fault causing the current is too low to prevent or delay the tripping. Referring to Figure 10.13, in case the fault is closer to the junction, the current at A would either flow in the blocking direction but would be too low to operate a blocking relay, or it would be zero, or would flow in the tripping direction but would be too low to operate a tripping relay. But with this tripping, the other ends will not be blocked.

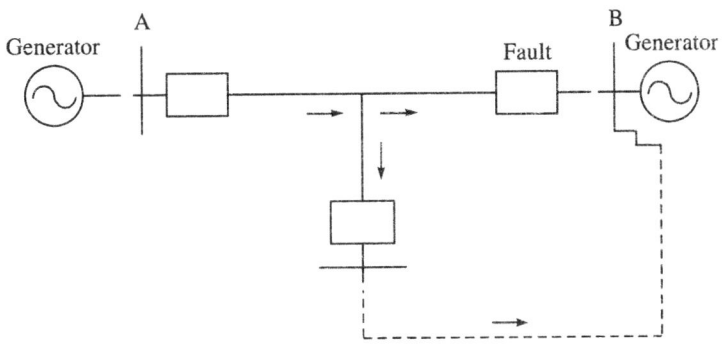

FIGURE 10.13   Directional comparison relay blocks tripping for an internal fault.

The chances of malfunctioning of directional comparison relaying using high-speed ground relays energised from zero phase sequence quantities are more than in the case of phase-comparison relaying. In order to minimise the tendency of disoperation, certain features are generally added in the directional comparison-relaying scheme. These features include:

(i) Limited sensitivity
(ii) Slight time delay in auxiliary relays
(iii) Transient blocking

The problem of malfunctioning on transients can be eliminated by using ground distance relays which, respond to positive phase sequence impedance.

10.4.3   Combined Phase and Directional Comparison Carrier Current Relaying

Enough induction with a nearby power circuit may cause the undesirable operation of directional comparison carrier current relaying through the use of directional ground relays. The maloperation of the directional ground relay is because of its adversely affected polarisation. The entire equipment used in the combined phase and directional comparison carrier current relaying technique is only slightly more expensive than that used in directional comparison relaying because some of the equipments such as transmitters and receivers are common to both. Fortunately, the phase-comparison ground fault equipment is relatively less affected by most of the transient conditions that affect directional ground relays.

The basic principle involved in the undesired operation of directional ground relays due to mutual induction is illustrated in Figure 10.14. The fault current, I_F, which flows in the nearby

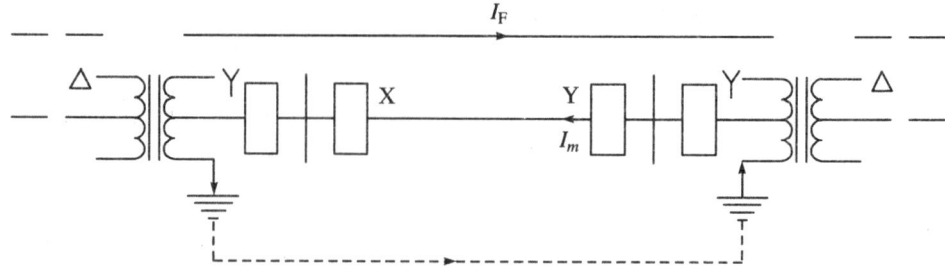

FIGURE 10.14 Undesired directional ground relay operation resulting from mutual induction.

line induces a current, say I_m, by mutual induction in the protected line. This induced current circulates through the grounded neutral power transformer banks at the ends of the line and the earth. Directional ground relays at both ends of the line find this situation conducive for operation. The polarising current flows from the ground into the neutral at location Y. Both the currents I_F and I_m at the end of location X get reversed with respect to the direction at Y causes tripping of the relay. Voltage polarised relays also have the same operating tendencies. It can therefore be said that the direction of the current flow in the line is dependent on the phase of the polarising quantity. The solution to this problem lies in using combined phase-comparison and directional comparison carrier current relaying techniques.

10.5
CARRIER-AIDED DISTANCE PROTECTION

Unlike the phase-comparison scheme, the distance protection scheme is eligible for providing back-up protection. But distance protection does not provide high-speed protection for the entire length of the line. The circuit-breakers thus cannot trip simultaneously at both ends for the end-zone faults, which is a highly desirable situation.

(i) There are a few schemes which provide both instantaneous tripping for the entire length of the line as well as back-up protection. Power line carriers, radio link or microwave may obtain the channel for signalling. These schemes are known as carrier-aided distance protection schemes. There are mainly two such schemes, which are: Carrier blocking scheme, which blocks the tripping, and

(ii) Carrier transfer or carrier inter-tripping scheme, which is used to trip the breaker.

The carrier transfer tripping schemes can further be divided into the following sub-schemes:

(a) Direct underreach transfer tripping scheme

(b) Permissive underreach transfer tripping scheme

(c) Tripping by carrier acceleration

(d) Permissive overreach transfer tripping scheme

10.5.1   Carrier Blocking Scheme

In the case of an external fault, the carrier blocking scheme blocks the operation of the relay by using the carrier signal. This scheme is suitable for the protection of multi-ended lines. A block diagram of this scheme is shown in Figure 10.15.

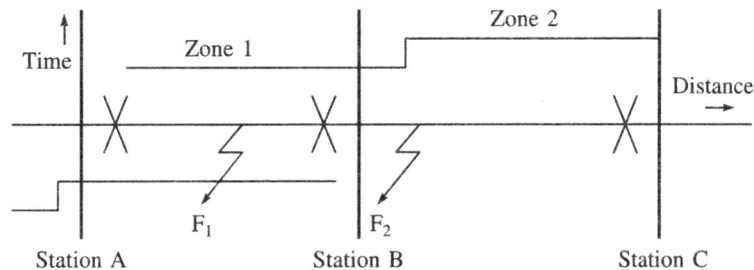

Stepped time-distance carrier blocking characteristics of the relay

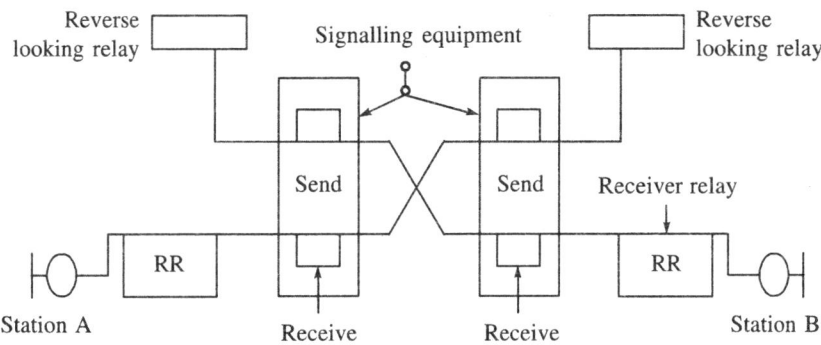

FIGURE 10.15   Block diagram of carrier blocking protection scheme.

Generally, in this scheme, zone 1 extends up to a range of 120 per cent beyond the principal section. The scheme thus covers the maximum underreach conditions such as 100 per cent of the protected line. A reverse looking directional relay installed with distance relays on both ends of the line detects all types of phase and ground faults. If the fault is detected in the nearby line, it is blocked by the carrier signal. The reverse looking directional relay generates a signal, which is then transmitted to the other end in order to prevent tripping.

The distance relays need to be operated at stations A and station B if a fault occurs, say at F_1, anywhere in the protected zone. But if the fault is external to zone 1 and occurs in the forward direction of the reverse looking relays, the relays will not operate. The relay contacts on both ends will remain closed thereby keeping the fault isolated from both the stations.

When a fault occurs at F_2 in zone 2, it will be seen by the forward-looking zone 1 at A and by the reverse looking zone 2 at B. It will be an external fault for zone 1. Since the fault is located near station B, it has to be cleared by the unit of zone 2. If the fault is not instantaneously cleared by the relays installed at station B for zone 2, the relay at A will trip after the time lapse in zone 2. This works as a back-up protection.

10.5.2   Carrier Transfer or Carrier Inter-tripping Scheme

The inter-tripping scheme, which is also known as *transferred tripping* undertakes controlled tripping of the circuit-breaker. This scheme completely isolates the circuit simultaneously from both the ends. The two main modes of operation of the inter-tripping scheme are as follows:

  (i) Underreach transfer tripping
 (ii) Overreach transfer tripping

Some extra precautions need to be taken to enhance the protection level under these schemes. Firstly, two-phase coupling needs to be employed and secondly, in the multi-purpose equipment, all the other facilities have to be cut off during a signal transmission at the time of occurrence of a fault.

Direct underreach transfer tripping scheme

The scheme in which the directly incoming carrier signal trips the circuit-breaker is known as a 'direct underreach transfer tripping scheme'. For effecting direct tripping, the receiver relay is bypassed in order to prevent the possibility of accidental operation of the carrier signalling caused by noise frequencies that are generated by switching surges, or similar occurrences. It is advisable to use certain methods to overcome the problems of accidental operation in the direct underreach transfer tripping scheme. These methods, which are expensive to use but which enhance the reliability of the scheme, are as follows:

  (i) Two separate channels are used and tripping commands are simultaneously transmitted through both these channels. The contacts of the receiving relays are connected in series. For tripping the circuit-breaker, both the channels must transmit the trip command simultaneously. In this, noise-signals do not cause malfunctioning of the circuit-breaker.
 (ii) The use of coded signals is another solution for preventing accidental operation. In order to prevent the transmission time from being too long, the code words used must be as small as possible.
(iii) In case of single-phase auto-reclosure, which is needed for phase selection at the receiving end, one channel must be employed for each phase.

The direct underreach transfer tripping scheme is illustrated in Figure 10.16.

Considering the protection for line AB, if the fault occurs at F_1 within the first zone of both the relays, the latter will instantaneously pick up and clear the fault. If the fault occurs at the end zone, say at F_2, the station B distance relay will trip its circuit-breaker instantaneously and immediately send an inter-tripping carrier signal to station A for tripping the circuit-breaker after sometime delay. The underreach phenomenon can therefore be described in terms of the failure of the relay to operate even when the fault point, even though it may be at the far end of the protected line, is within the reach of the relay.

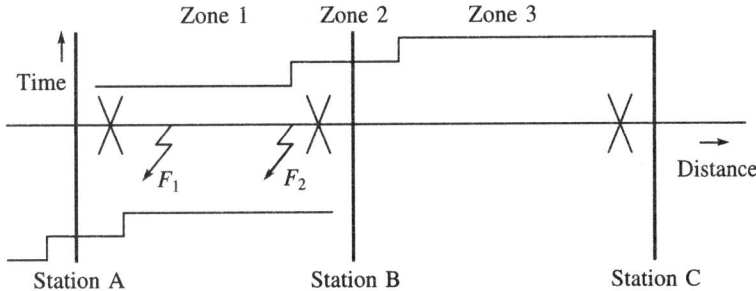

Stepped time-distance carrier inter-tripping characteristics of the relay

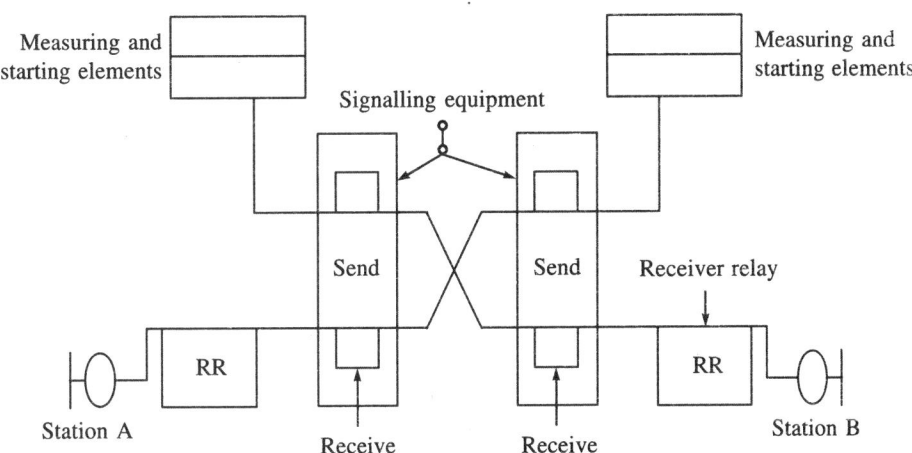

FIGURE 10.16   Block diagram of carrier inter-tripping relaying scheme.

Permissive underreach transfer tripping scheme

The scheme wherein the incoming signal is utilised to trip the circuit-breaker at the receiving end after the local distance relay is operated, is known as the *permissive underreach transfer tripping scheme*. In this scheme, the execution of the transmitted command takes place at the receiving end wherein the starting elements of the local distance relay have already picked up and confirmed the existence of the fault. This scheme overcomes the possibility of undesired tripping due to the prevalence of noise in the line. This is achieved by interlocking the received signal with the starting elements of the distance relay as shown in Figure 10.16. Hence, this scheme fulfils the requirement of rapid auto-reclosure.

Tripping by carrier acceleration

The scheme in which the incoming signal extends up to the measuring range of the first zone of the distance relay at the receiving end, is known as the *carrier acceleration scheme*. The basic difference between the underreach transfer and carrier acceleration schemes is that unlike the former, the latter accepts a certain extra time lag. In general thus, in the accelerated carrier scheme, the signal received from the opposite end extends up to the reach of zone 1 from about

80 per cent to 120 per cent of the length of the protected section. When an end zone fault occurs, the relay trips at that end and sends the carrier signal to the remote end. On receipt of the carrier signal from the other end, the range change relay extends the reach of the relay (mho unit) from zone 1 to zone 2 immediately, thereby accelerating the clearance of the fault at the remote end.

Permissive overreach transfer tripping scheme

The 'permissive overreach transfer tripping scheme' comes into effect when the relay operates to tackle the faults occurring beyond its zone of protection. In this scheme, the receive relay contact is supervised by a directional relay which is why this scheme is also known as a *directional comparison scheme*. Unlike the blocking carrier scheme, this scheme does not require any reverse looking directional relay.

The stepped time-distance permissive overreach inter-tripping characteristics of the relay and block diagram of the carrier inter-tripping relaying scheme are given in Figure 10.17. In this scheme, the zone-2 unit sends the carrier signal to the remote end of the protected line. The zone 2 relay monitors the RR contact of the receiver relay. The first zone of both the distance relays covers the fault at F_1, anywhere in the protected line while the tripping command of both the distance relays sends a carrier signal to the opposite end to close the normally open 'NO' contact of the receiver relay.

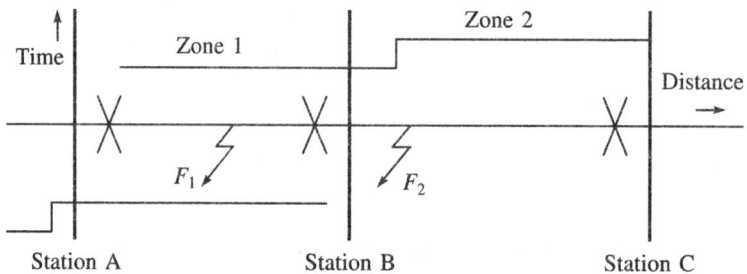

Stepped time-distance carrier inter-tripping characteristics of the relay

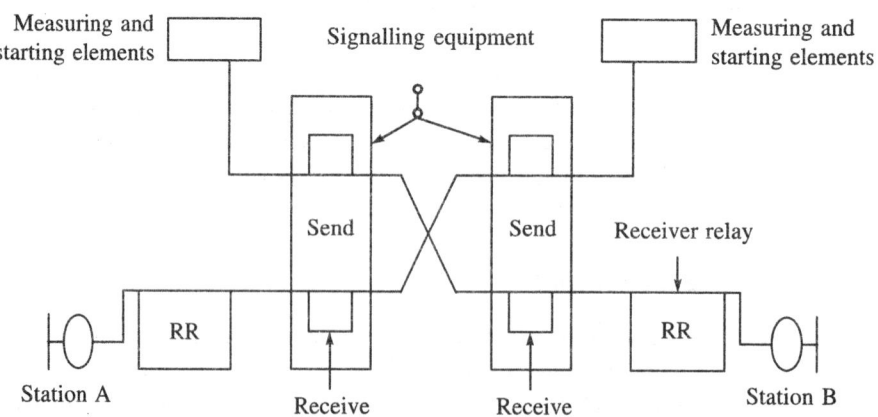

FIGURE 10.17   Block diagram of overreach inter-tripping relaying scheme.

The fault at F_2 will be seen by the distance relay at station A in zone 1 and carrier signal will be sent to close the receiver relay contact at station B. But if the fault in the reverse zone at station B does not pick up, no signal will be transmitted from station B to station A. The advantage of this is that no impulse signal will be transmitted and passed over to the circuit-breaker, even after the local circuit-breaker has operated.

This method is, however, not usually preferred for two reasons. Firstly, it takes much longer to isolate the faulty section in comparison to the corresponding time taken by the underreach schemes. Secondly, it is difficult to correctly determine the direction of the fault in this method, especially in the presence of a capacitor in the nearby stations.

10.6
COMPARISON OF THE TRANSFER TRIP AND BLOCKING SCHEMES

1. The lesser the attenuation, the more reliable will be the transmission. The attenuation due to a fault in the 'transfer tripping scheme' is variable and unpredictable. This attenuation is absent in the 'blocking scheme'.
2. In case of the transmission line with auto-reclosing, the blocking scheme is a better option than the 'transfer trip scheme' because in the latter, the relay does not operate instantaneously for tackling the end zone faults. One of the major advantages of the transfer trip scheme is the speed at which operates.
3. The 'carrier acceleration scheme' is more economical to use than the other schemes.
4. The cost of coupling equipments in the transfer scheme is more than that in case of the blocking scheme, because the latter utilises a healthy line for sending the signal and a single-phase coupling is enough for its operation.
5. The distance relays behave differently in the case of carrier channel failure. In the underreach transfer trip, the end zone fault will be cleared after the time delay corresponding to the nearest zone tripping. In the over reach transfer trip scheme, if the distance relay fails to isolate the fault in the principal section from both ends, say in zone 1, then all the faults will be cleared after a delay according to the zone 2 setting. Blocking schemes operate effectively and correctly for dealing with faults in the principal section but not for faults occurring in the adjoining line sections.

EXERCISES

1. What do you understand by 'pilot'? What is a pilot relaying scheme?
2. Discuss the various pilot relaying schemes with the help of neat diagrams.
3. What are the various requirements of a pilot relaying scheme. Describe each of these requirements in brief.
4. Discuss with the help of neat diagrams the circulating current wire pilot scheme and the opposed voltage type wire pilot type schemes.

5. What is a carrier current pilot scheme? With the help of a neat and complete diagram showing its components, explain the working principle of the carrier current pilot scheme.

6. Explain the carrier tripping and carrier blocking schemes.

7. Discuss with the help of a neat diagram the theory of the carrier current scheme based upon phase angle comparison.

8. Highlight the merits and demerits of the carrier blocking schemes as compared to other types of carrier-aided distance protection schemes.

9. Explain when 'combined phase and directional comparison carrier current relaying' is required.

10. What do you mean by 'permissive underreach' and 'permissive overreach transfer tripping schemes' of protection?

11. What do you understand by unit protection? In what ways does carrier-aided distance protection show better performance than carrier current protection?

Chapter 11

Apparatus Protection

11.1

GENERATOR PROTECTION

11.1.1 Faults in Generators

The protective schemes for generating units need to be designed carefully. Generators and their associated parts are subjected to various pressures and hazards.

At present 500 MW capacity generators are in use and loss of any unit of a generating station may cause overloading of the other units which subsequently makes the system unstable and reduces the reliability index. High capacity modern generators are interconnected with the transformers. Opening of the circuit-breaker to isolate the faulty generator is not enough like other apparatus because to protect the generator from damage is not guaranteed. Until field excitation is suppressed, the generator will continue to supply power to the stator winding fault. So, while deciding the type of protective system for generators, it is important to consider the origin and effect of the faults. The following provisions are desired:

- To open the field
- To stop the fuel supply to the prime mover
- To apply braking, if requirement arises.

The major faults and possible abnormal conditions occurring in the synchronous generators are listed in Table 11.1. In the second half of the table, protective schemes that are generally employed to protect the different sections of a generator are also listed. The possible hazards in operation of synchronous generators are generally categorized into two major sections such as internal faults and abnormal system operating conditions.

TABLE 11.1   Faults and Protective Schemes of Generators

Faults and Abnormal Conditions	Protective Schemes
Stator winding insulation failure	Differential protection
Rotor winding insulation failure	Inter-turn fault protection
Field winding failure	Stator earth fault protection
Overloading	Rotor earth fault protection
Unbalanced loading	Overcurrent earth fault protection
Overvoltage	Negative phase sequence protection
Prime mover failure	Field failure protection
Overspeeding	Overload protection
Loss of synchronism	Overvoltage protection
Overheating	Reverse power protection
Underfrequency	Back-up impedance protection
	Underfrequency protection
	Pole slipping protection

The generator faults can be summarized broadly as follows.

Stator faults include the following:

(i)   Phase-to-earth fault

(ii)   Phase-to-phase fault

(iii)   Inter-turn fault

Inter-turn fault and phase-to-phase fault are developed rarely but once these are developed cause more damage. Majority of the faults developed in the stator of a generator are earth faults. Following are the effects of earth fault in the stator:

(i)   Arcing to core develops welds in laminations. Welding the laminations together causes eddy current hot spots on subsequent use. Repairing in this case is costly and time consuming affair.

(ii)   Excessive heating in the conductor damages them along with insulation. There are also fire risks.

Rotor faults in the rotor circuit may be either earth faults or inter-turns develop due to severe mechanical or thermal stresses acting on the winding insulation. Generators are operated with its field winding isolated from earth and therefore a single fault between field winding and rotor body due to insulation breakdown gets avoided. Observing the incipient fault in the rotor should stop the generator otherwise it may short circuit some part of the field winding. Short circuit in some part of the field winding causes asymmetry of the air gap fluxes which further causes severe vibration of the rotor.

Abnormal running conditions include the following:

(i)   Loss of excitation

(ii)   Unbalanced loading

(iii)   Overloading

(iv)   Failure of prime mover

(v)   Overspeeding and

(vi)   Overvoltage

Loss of excitation may occur due to the exciter failure or improper operation of the field-breaker. Failure of field excitation causes slight speeding up of the generator and generator starts acting as an induction generator which derives power for excitation from the system and supplies power at a leading power factor. Further fall in voltage may result in loss of synchronism and system instability. Induction current in rotor and damper windings may also cause overheating.

Unbalancing occurs due to:

(i) Single phasing and associated faults
(ii) Unbalanced loading
(iii) Open circuits due to broken lines
(iv) Failure of circuit-breaker pole

Unbalancing gives rise to negative sequence currents producing armature reaction field which rotates in a direction opposite to that of the rotor and hence produces flux sweeping through the rotor with twice the rotational speed. This flux induces current twice the machine frequency in the body of the rotor in the field and damper windings. An appreciable unbalancing causes overheating.

Overloading of the generator overheats the stator which damages the insulation and creates complications.

Prime mover failure transfers generator to motoring mode. If it is not a single-generator system in the motoring mode, the generator takes power from the system. How much it adversely affects the system depends on the type of drives used for the generator.

Loss of load suddenly causes the machine to overspeed. This is prominent in the hydro-generators because of the mechanical and hydraulic inertia.

Overvoltages occur due to defective voltage regulators or due to overspeeding.

11.1.2   Classes of Generator Protection

Protection schemes of synchronous generators are divided in the three classes: class A, class B and class C, according to the severity of the faults and abnormal conditions. Table 11.2 shows the classes of the protection schemes and the actions which are to be initiated for operation, when corresponding sections show faults.

TABLE 11.2   Classes of the Protective Schemes and Actions to be Initiated

Class A Protections	Class B Protections	Class C Protections
'Appearance of Class A Trip' annunciation	'Appearance of Class B Trip' annunciation	Generator-breaker tripping only
Auxiliary unit of the transformer tripping	Turbine tripping	
Generator-breaker tripping	Time delayed of Class A relays	
Generator field-breaker tripping	Boiler tripping	
Prime over tripping		
Boiler tripping		
Closing of tie-breakers		

11.1.3   Conventional Protection of Generators

Stator protection

The earth fault current through stator is usually limited by inserting resistance in the neutral of the generator. Resistance earthing can limit the earth fault current between 5A and 10A and it is not practicable if the stator winding is directly connected to the delta winding of the main transformer. A current transformer is mounted in the generator neutral and connected to either an inverse time relay or an instantaneous attracted relay, if the stator neutral is earthed through a resistor. It depends whether the generator is directly connected to the station busbar or with a delta/star transformer. The inverse time relay requires grading with earth fault relays in the system. If the generator is connected to the station busbar through delta/star transformer and because the earth fault loop is restricted to the stator and transformer primary winding no discrimination with other earth fault relays is necessary. Resistance earthing does not provide 100 per cent protection of stator winding. The percentage of winding protection depends on the value of the neutral earthing resistor and the relay setting. High-speed relays and breakers are required to be installed if the resistance earthing is provided. To avoid any other winding fault due to the resonance generated overvoltage, the distributed capacitor in between stator and ground is installed. An earth fault relay in a resistance earthed generator is shown in Figure 11.1.

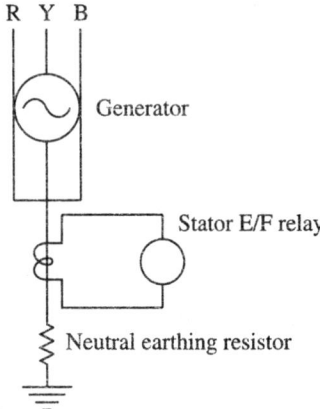

FIGURE 11.1   Earth fault relay in a resistance earthed generator.

Stator protection from phase-to-phase and phase-to-ground fault protection

Longitudinal differential relaying scheme is said to be the best scheme for stator protection from phase-to-phase and phase-to-ground fault protection. An instantaneous attracted armature-type relay with setting 10 per cent to 40 per cent is used. This scheme is unaffected by ac transients and has high-speed feature. In the case, if the characteristics of the current transformers on the line side do not match with those mounted at the neutral end of the winding, a relatively large spill current flows through the relay's operating coil. A percentage biased differential relaying scheme is used in the place of a longitudinal differential relaying scheme which is shown in Figure 11.2.

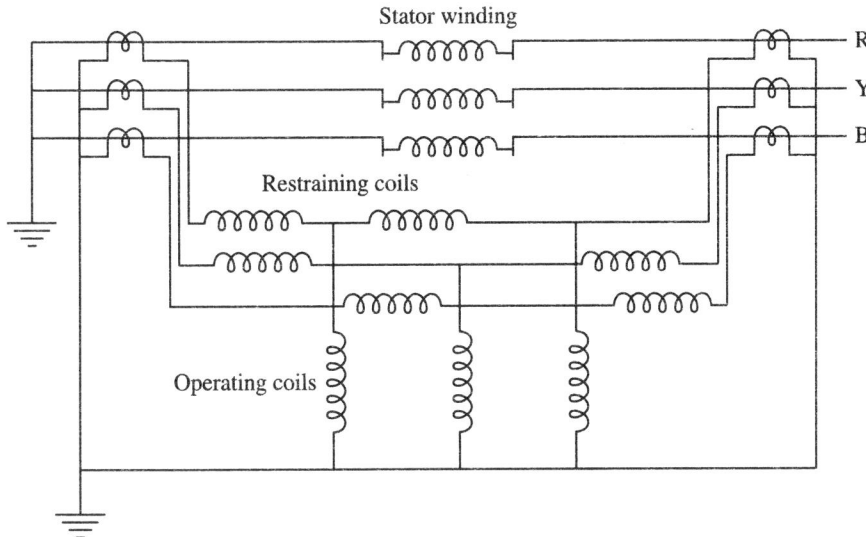

FIGURE 11.2   A percentage biased differential relaying scheme.

Stator inter-turn fault protection

Balance between the currents in the neutral and the high voltage CTs do not get disturbed, if inter-turn fault on the same phase of the stator winding occurs. This is the reason why inter-turn fault can not be detected by longitudinal differential protection. A biased transverse differential protection as shown in Figure 11.3 is used for protecting the inter-turn faults of the generator.

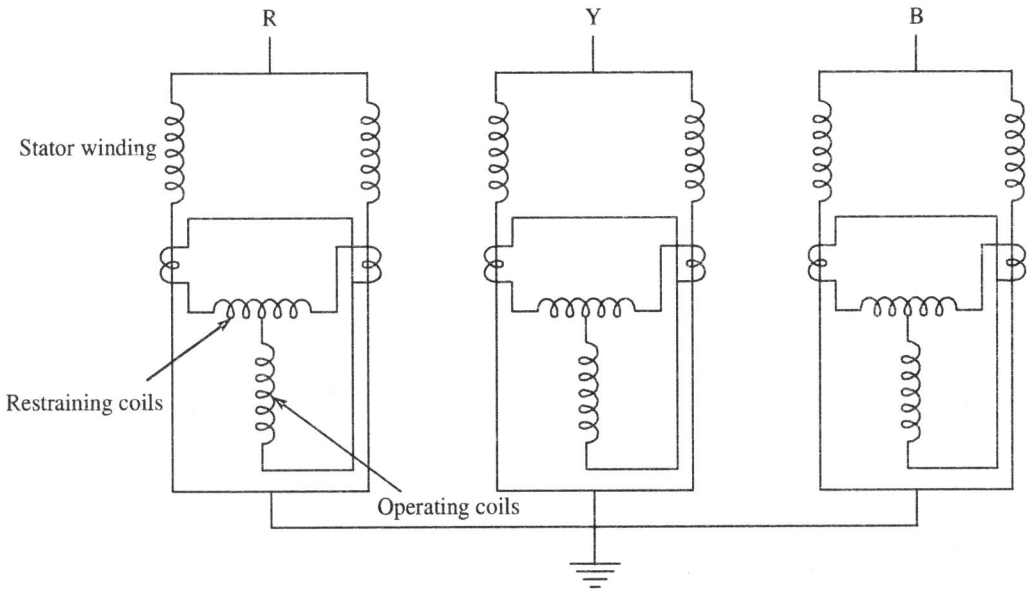

FIGURE 11.3   A biased transverse differential protection scheme.

Rotor protection

Rotor winding can be damaged by open circuits or earth faults. In the earth fault protection scheme of the rotor, the field is biased by a dc voltage which causes current to flow through the relay R for an earth fault anywhere on the field system. Method of detection of the rotor open circuit is the same as that for detecting loss of excitation described in the next section. A method of rotor earth fault detection is shown in Figure 11.4.

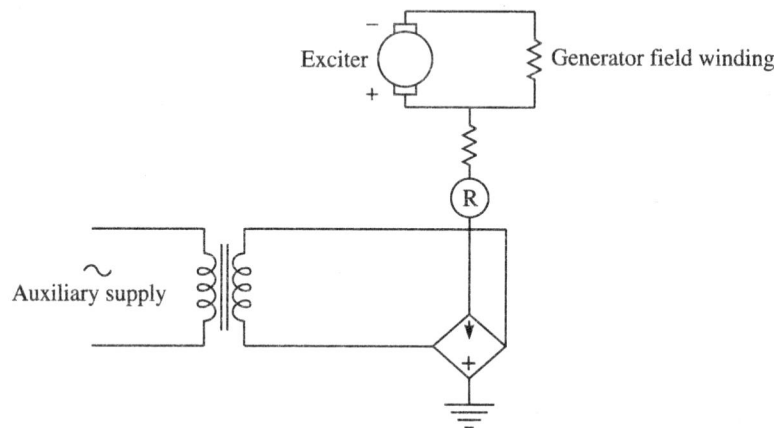

FIGURE 11.4   Rotor earth fault detection method.

Loss of excitation (field failure) protection

Loss of excitation mainly causes two adverse effects and both cause overheating of the rotor:
   (i) Machine starts drawing large magnitude magnetizing current from the system
   (ii) Slip frequency emf gets induced in the rotor circuit

Indication of loss of synchronism is an importation of excessive value of VAR, which, in turn, indicates field failure. A fixed time delay in between one and five seconds in the tripping sequence of the relay is usually incorporated. This causes a momentary reversal of VAR component.

An offset impedance or mho measuring relay at the generator terminals is applied for protection of the field failure. Operating characteristics of this scheme are required to be arranged as shown in Figure 11.5. With these characteristics in the event of complete loss of excitation or extremely low excitation, generator equivalent impedance falls within the tripping zone.

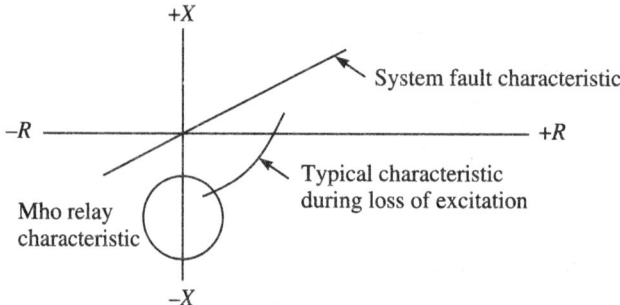

FIGURE 11.5   Loss of excitation characteristics.

Unbalanced loading protection

During unbalance, the negative phase sequence current causes heating of the stator. The negative sequence heating follows a normal linear resistance law and hence is proportional to the square of the current.

The heating quotient depends on the cooling system employed. An inverse definite minimum time delay relay connected to a network which segregates the negative sequence current from the positive and zero sequence current is shown in Figure 11.6. This relay has a long operating time and has setting range which allows its characteristic to be accurately matched to those of the machine.

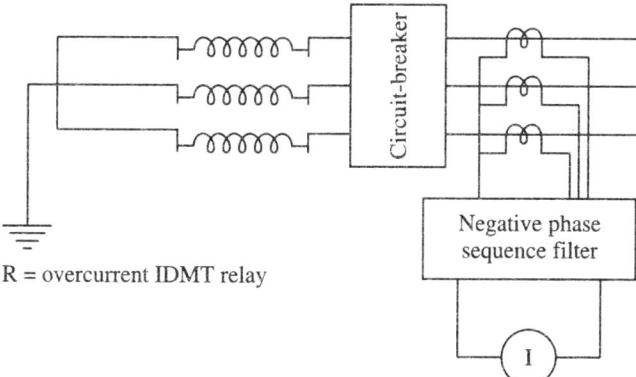

FIGURE 11.6 Protection against unbalanced load.

Overload protection

Stator winding gets overheated, if the generator is overloaded. Temperature detector may be of the types: thermistor, resistance temperature detector or thermocouple, which are used to sense the temperature of the winding. Temperature detector coils are embedded at various points in the stator winding, which provides indication to the temperature detector. The temperature detector coils form one arm of the Wheatstone bridge of the circuit.

Small generators of rating below 30 MW are not required for the provision of embedded temperature detectors. Thermal relays are provided for such small generators. Thermal relays have a bimetallic strip heated by secondary current from the stator. The housing of the bimetallic strip is designed to have heating and cooling characteristics similar to that of the machine. In the event of failure of the cooling system, thermal relay does not provide protection against overheating.

Prime mover protection

Relay working on load balance principle with directional characteristic is used for protection of prime mover. Actually, if the prime mover fails, the generator goes in motoring mode, means it draws electrical power from the system and starts driving the prime mover. This situation imposes a balanced load on the system, which is detected by the relay with directional characteristic as shown in Figure 11.7. This relay can detect the power balance condition over the full power factor range. Auxiliary time-lag is provided to prevent operation by synchronizing surges and power oscillations following system disturbances.

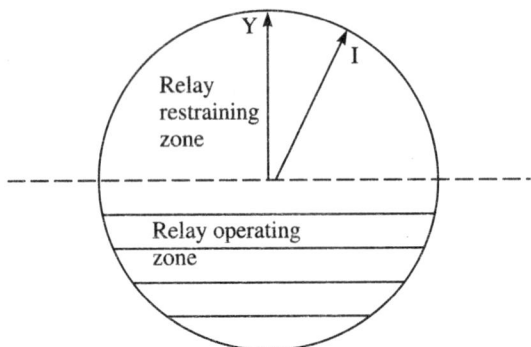

FIGURE 11.7   Reverse power relay operating characteristic.

Back-up reverse power protection is also required in unattended stations. A sensitive directional relay with low setting of about 2 per cent of rated power in conjunction with timing relay serves the purpose.

Overspeed protection

Unlike steam turbines, hydraulic turbines speed up rapidly, if the load is suddenly lost because steam sets are easily controlled by fast-acting governor. For hydro-generators a reverse or under-power interlock relay is used. A reverse or under-power interlock relay prevents the main generator circuit-breaker being tripped under non-emergency conditions. A mechanical overspeed device which is in the form of centrifugally-operated rings mounted on the rotor shaft is also used. This mechanical overspeed device mounted on the rotor flies out and closes the stop valves, if the speed of the generator increases by more than 10 per cent.

Overvoltage protection

An overvoltage relay is provided to protect the generator, if the voltage exceeds the pre-specified value of the generator. Inverse definite minimum time relay is preferred. Relay with IDMT characteristics opens the main circuit-breaker and the field switch, if the overvoltage persists.

11.1.4   Digital Protection of Generators

Synchronous generators are among the vital equipment of the power systems. Availability of the electric power totally depends on the health of the synchronous generators. Generating units are very expensive so their protection is also very important. Selection of the protective schemes for generating units are again done very carefully because malfunctioning of any sections of the generating unit may cause overloading of the rest of the system leading to oscillations and instability in the power system. So, to avoid all these, high-speed protective relays and associated switchgear having the simplicity, flexibility, sensitivity and high-grade reliability become necessary for the synchronous generator protection. Electromechanical, static and electronic relays are not found capable to provide the required high speed, simplicity, flexibility, sensitivity and high-grade reliability. These features are available in the digital protection schemes i.e. microprocessor-based relays, mini computer-based relays or PC-based relays. Digital relaying schemes are *the best* because the threshold characteristics of these relays can be easily changed by modifying the hardware and minor change in the software.

Microprocessor-based unit protection scheme

The basic percentage differential protection scheme as shown in Figure 11.8 is used in this unit protection scheme.

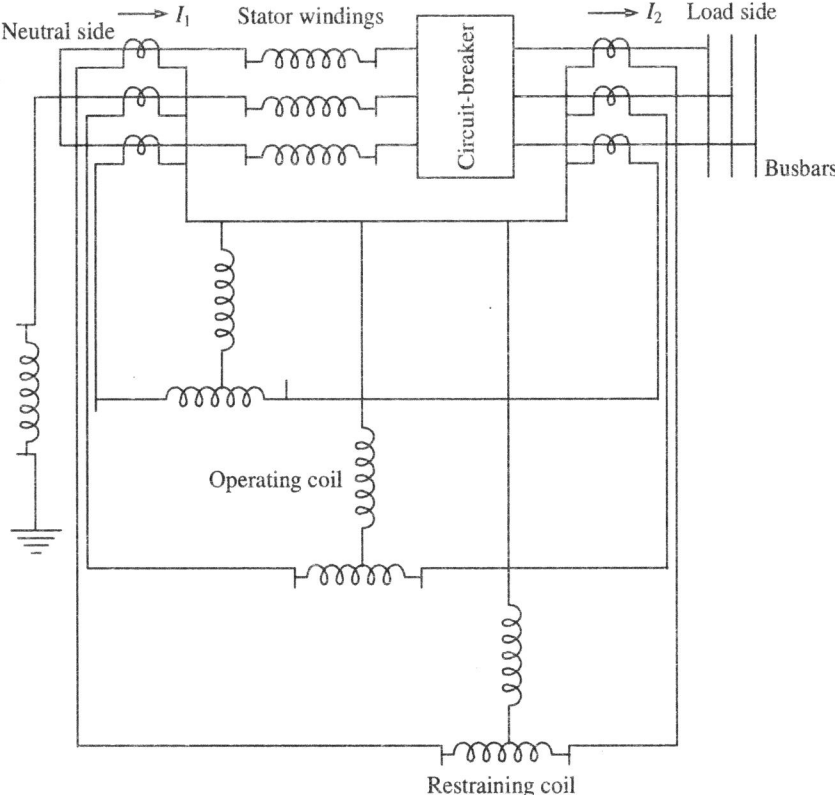

FIGURE 11.8 Percentage differential relay.

In this scheme, I_1 is the current entering in the stator winding and I_2 is the current leaving. These currents are stepped down by the exactly similar CTs. In the percentage differential relaying schemes magnitude of the difference of the currents I_1 and I_2, i.e. $(I_1 - I_2)$ is the operating signal and $(I_1 + I_2)/2$ is the restraining signal.

The condition for tripping is:

$$|(I_1 - I_2)| > S \left| \frac{I_1 + I_2}{2} \right|$$

Here, S is known as bias factor of the percentage biased differential relay. The value of S lies between 0.05 and 0.15.

During external fault or normal load conditions, the phasor difference of the currents flowing in both the windings is negligibly small, and this is the operating signal. But during internal fault in the windings, there is considerable change in the magnitude and phase difference between the currents at two ends of the windings. So, during internal fault condition, the operating signal $(I_1 - I_2)$ is sufficiently high to exceed S (bias factor) times the mean value of the restraining signal, i.e. the current $[(I_1 + I_2)/2]$. Then, the circuit-breaker will get the tripping command.

The block diagram of this relaying scheme is given in Figure 11.9. The output of current transformers, I_1 and I_2 are fed into the current-to-voltage converter for converting the current signals into voltage signals. Current-to-voltage conversion is done by the transactor. In the next step these signals are fed into the multiplexer for selection of the signals one by one. After selecting the signals, sampling is done. Sampled signals are then fed into the analog-to-digital converter for conversion of the analog fault signal into digital signals. After this, the digital converter filters the dc components and harmonics of the signals and send them to the memories of the microprocessor. This way only rms value of the operating signal $(I_1 - I_2)$ and restraining signal $[(I_1 + I_2)/2]$ reach to the microprocessor. Sometimes, unit protection scheme for the protection of synchronous generators is not preferred because of the non-accessibility of the ends of the neutral windings.

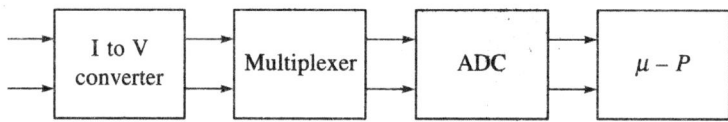

FIGURE 11.9   Block diagram of microprocessor-based unit protection scheme.

Second harmonic current induced in the rotor field circuit-based digital protection

Any abnormality in the synchronous generator is indicated by internal unbalancing. The stator field generated due to the positive sequence stator current rotates at the same speed and in the same direction as that of the rotor. Due to this, no induced emf generation takes place in the rotor circuit because the stator field remains decoupled with respect to the rotor.

Negative sequence currents in the stator winding cause stator field, which rotates at the synchronous speed in the direction opposite to the rotation of the rotor. This induces an emf which is two times the synchronous frequency in the rotor circuit. This is the reason why the negative sequence current in the stator winding produces second harmonics in the rotor field circuit.

The second harmonic component present in the field current is utilized to detect the presence of fault or abnormality in the stator winding. Descrimination between external fault and internal fault is made by direction of the flow of positive sequence current at the generator terminals. Detection of the magnitude and phase difference between sequence currents decides the fault types.

Sampling of the field current monitors the loss of excitation. With the application of suitable filter, the second harmonic component of the field current is obtained. Generally, the second harmonic component presence exceeding the 0.2 pu is taken as an indication of abnormality in the stator circuit or fault external to the stator circuit. A reverse power relay is also required to show the direction of the current such as the power flow direction at the stator terminals. Steps of the scheme are shown in the form of a block diagram in Figure 11.10.

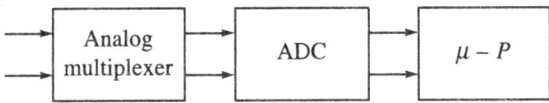

FIGURE 11.10   Steps of the scheme second harmonic current induced in the rotor field circuit-based digital protection.

The scheme can be realized by using either an 8085 or 8086 microprocessor. In order to make the relay reasonably fast, eight samples per cycle is preferred. The programmable timer can help achieve simultaneous sampling of (+ve) and (–ve) sequence currents in the armature and field circuit. The output of the timer is then fed into the interrupt, say RST 7.5. In the next step, the sequence current, be it armature or field current, is selected through an analog multiplexer. The trip signal may be the output of the microprocessor of suitable port.

The following logical possibilities help in detecting the types of faults:

All the sequence currents are equal in magnitude	Line-to-ground fault
+ve and –ve sequence currents are equal in magnitude and the zero sequence current is absent.	Line-to-line fault
Only +ve sequence current is present, –ve and zero sequence currents are absent.	Triple line fault or triple line-to-ground fault

Sub-synchronous component injection in rotor circuit scheme

This scheme is used to isolate the synchronous generator from the rest of the system immediately on the occurrence of a fault. The maximum single line-to-ground fault current is very small because of the high ground resistance. Consequently, the ground fault current may also be too small for the relay to pick up. However, a single line-to-ground fault is very common in the synchronous generators, which essentially need to be isolated immediately to safeguard the lives of workers. If a ground fault occurs, its detection in the full length of the stator winding depends on the short-circuiting of the generator capacitance. A change in the capacitance reactance can be detected at the sub-harmonic frequency.

The injection of the voltage of suitable frequency (say 12.5 Hz) between the neutral and ground takes place in this scheme, and this voltage facilitates the monitoring of the resultant current by the scheme. This injection of the neutral voltage sub-harmonic frequency detects ground faults. This scheme works on the principle that when the ground fault occurs, the total (12.5 Hz) current increases because of the shorting of the generator capacitance, thereby causing the relay to operate. The 12.5 Hz frequency current path is shown in Figure 11.11.

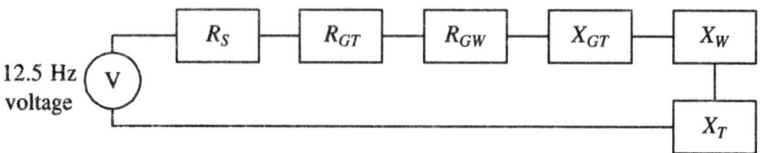

FIGURE 11.11   Low frequency (12.5 Hz) current path.

Here:

V = 12.5 Hz generated voltage

R_S = secondary resistance referred to primary

R_{GT} = grounding transformer resistance

R_{GW} = injection transformer resistance

X_T = capacitive reactance (3-phase) to ground of protected zone

X_{GT} = leakage reactance of generator grounding transformer

X_W = leakage reactance of injection transformer

From the above circuit as shown in Figure 11.11;
Minimum theoretical setting current;

$$I_{min} = \frac{V}{R + j(X + X_T)}$$

Maximum theoretical setting current

$$I_{max} = \frac{V}{R + jX}$$

The following precautionary measures must be taken:

(i) Injection of the voltage in unexcited generator while on turning gear should turn off.

(ii) A voltage supervision relay is used to deactivate the injection scheme when generated voltage is below 0.4 pu.

The schematic diagram of the scheme is shown in Figure 11.12.

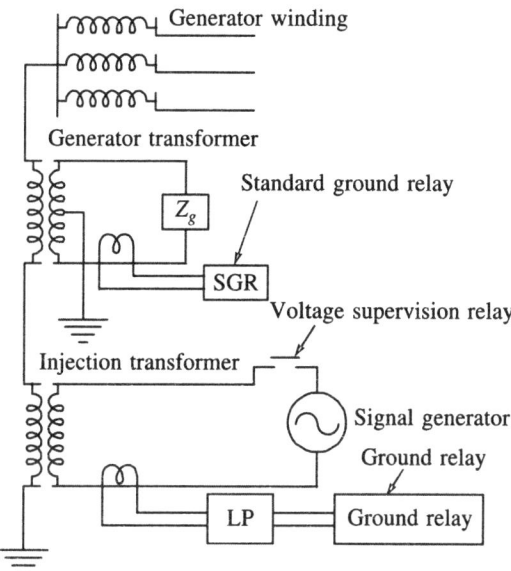

FIGURE 11.12 Schematic diagram of stator ground fault protection.

11.2
ELECTRIC MOTOR PROTECTION

There are several types of electric motors which are used for various purposes. The motor protection philosophy remains same because the fundamental problems affecting the choice of protection are independent of the type of motor and the type of load to which it is connected.

11.2.1 Types of Motor Faults

Types of motor faults are similar to those of generators. Motors need to be protected against the following faults:

    (i) Stator faults
    (ii) Rotor faults
    (iii) Overload
    (iv) Unbalanced voltage supply
    (v) Undervoltage
    (vi) Open phase or reverse phase starting
    (vii) Loss of synchronism in synchronous motors

11.2.2 Stator Protection

Like stator short circuits in the generator, it may also happen to the stator of the motors. Short circuit faults may be of the type either phase-to-ground or phase-to-phase. Protection from these faults is provided with the help of overcurrent tripping devices giving an inverse time current characteristic with the provision of instantaneous tripping at high current. Larger motors of rating 50 hp or above need to be protected by the instantaneous overcurrent relays. The setting of the instantaneous overcurrent relays is so chosen that it is well above the maximum starting current.

Earth fault protection for a motor operating on an earthed neutral system is provided by means of a simple instantaneous relay having a setting of approximately 30 per cent of the motor full load current in the residual circuits of three current transformers. Operation of the relay due to current transformer saturation during initial high starting current should be avoided by increasing the voltage setting of the relay by inserting a stabilising resistance in series with it.

Differential protection scheme is useful and provided on very large capacity motors applied at important applications, if the system is unearthed. Induction motor protection scheme having instantaneous overcurrent relays and inverse time-lag overcurrent relays is shown in Figure 11.13.

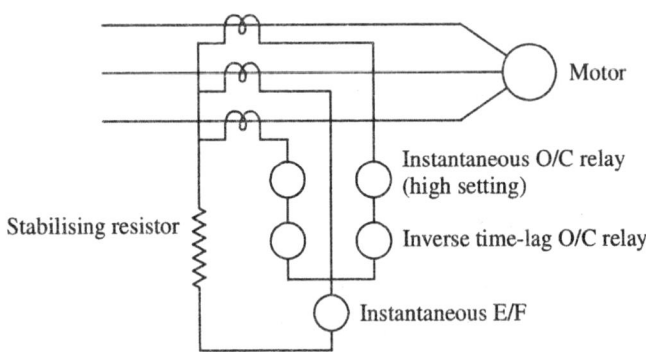

FIGURE 11.13   Induction motor protection scheme.

11.2.3   Rotor Protection

Negative sequence current flow induces high frequency currents in the rotor and is produced from unbalance either in the supply voltage or in the loading pattern. Frequency of these currents in the rotor is (2-S) times the normal frequency of supply. Here S is slip. Heating effect on the rotor winding of the negative sequence component is proportional to $(2-S)fR_{ac}$, while rotor heating due to positive sequence component of the stator current is proportional to R_{dc}. Here, R_{ac} and R_{dc} are the resistance values of the rotor winding for ac and dc supplies respectively. Obviously, the heating by negative phase sequence current is more than that of the positive phase sequence current. So, while deciding the protection scheme for motor, it is important to take into account what load the motor can stand for a given degree of unbalanced voltage.

11.2.4   Overload Protection

The overload protection is required to be designed in such a manner so that it matches as closely as possible with the heating curve of majority of the motors. The protection characteristics for overloading should lie just below the heating curve of the motor protected. The characteristics should be of adjustable nature so that different designs of motors with different duties can incorporate it in their respective protection scheme. During winding temperature is still high and just after tripping, the motor should not be restarted. So, it is said that an ideal protection should therefore not only match the heating characteristics of the motor but also its cooling characteristics. Relays employed for overload protection should have a time constant slightly lower than the thermal time constant of the motor. Closer the characteristic of overload relay matches the starting current curve the better is the motor protected against damage due to flow of longer time starting current. Induction type over current relays with setting for overload 120 per cent of full load current are employed for protecting the motors against overloading.

11.2.5   Unbalance and Single-phasing Protection

For star-connected motors, overload and single-phasing protection is provided by fitting two overload (overcurrent relays) elements as shown in Figure 11.13. The characteristics of the overload elements are chosen such as the motor is permitted to run with supply only on two phases till such time as there is risk of thermal damage is reached. Motor running with more than 70 per cent of full load can be considered to be protected by such arrangement for delta-connected motors.

A better scheme for detection of single-phasing provides a phase balance relay or bimetal relay. Bimetal relay with single-phasing trip feature is shown in Figure 11.14(i, ii and iii) in the different operating modes. Figure 11.14(i) shows the assembly of the bimetallic relay when motor is in cool state. Figure 11.14(ii) shows the position of both slides S_1 and S_2 of the bimetallic relay when symmetrical loading is applied. On symmetrical loading, all the three bimetallic strips a, b and c bend equally and both the sliders S_1 and S_2 move in the same direction by distance d thereby initiating tripping.

On single-phasing, only the two bimetal strips say a and c bend whereas the third strip b remains cold. The slider S_1 is accordingly moved by the heated bimetal strips a and c whereas slider S_2 is kept in position by the cold strip b. Thus, this differential motion of the two sliders causes the tripping lever T to move the distance d required for initiating tripping, which is shown in Figure 11.14(iii).

Large capacity motors are provided with thermal protection with thermistors. Occurrence of excessive heating due to overloading or single-phasing, the thermistors embedded in the stator causes tripping as a result of change in the resistance.

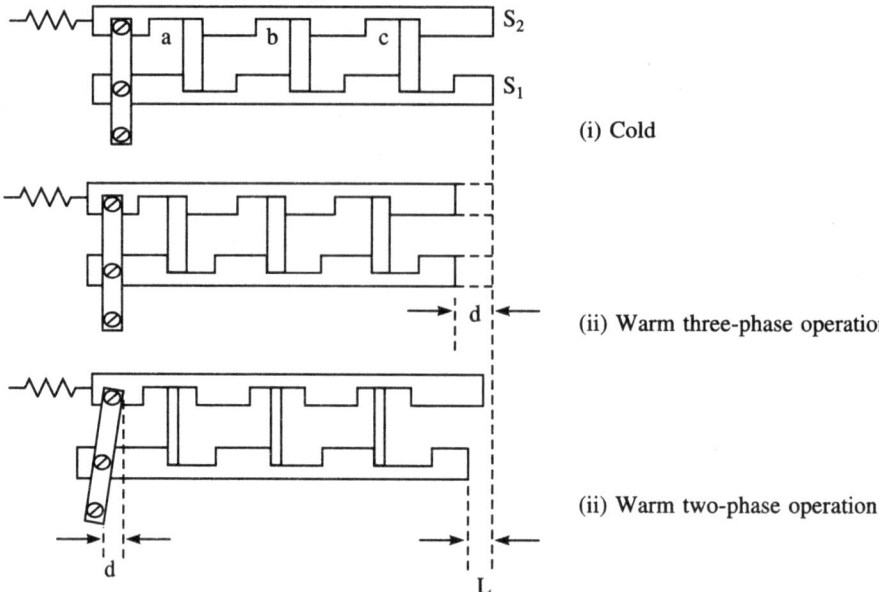

FIGURE 11.14   Bimetal relay with single-phasing trip feature.

11.2.6   Undervoltage Protection

While delivering the same power, operation of motor on undervoltage generally causes overcurrent and thus motors are protected by overload devices or temperature-sensitive devices. A separate single element undervoltage relay energized with phase-to-earth or phase-to-phase voltage can be provided to protect against a three-phase drop in voltage or in attempt to start the motor with low voltages on all the phases. A time delay is required to be incorporated in order to prevent tripping due to transient voltage drop.

11.2.7   Reverse Phase Protection

Improper phase sequence causes change in the direction of rotation of the motor. Movement of the motor in the opposite direction may cause severe damages and hence in the some motor applications reverse phase protection becomes an essential requirement. Generally, an induction disc-type polyphase voltage relay is used to protect motors from starting with one phase open or with reversed phase sequence. Torque produced by the relay is proportional to the sine product of the two line-to-line voltages. The relay will not close its contacts and hence the motor will not start unless all the three phases are present and in the correct sequence.

11.2.8   Loss of Synchronism Protection

A synchronous motor may get pulled out of the step due to the following reasons:
   (i) Severe overloading
   (ii) Reduction in supply voltage

A relay may detect the above conditions, if it responds to the change in power factor when there is pole slipping. In this scheme, comparison of voltage between two phases with the current in the third phase is required to be done. An attracted armature relay which gets energized from a full wave rectifier bridge, is connected differentially and remains in operation so long as motor remains in synchronism. A non-linear resistor is used to protect the rectifiers and extends the operating range of the relay. An out-of-step protection relay is shown in Figure 11.15.

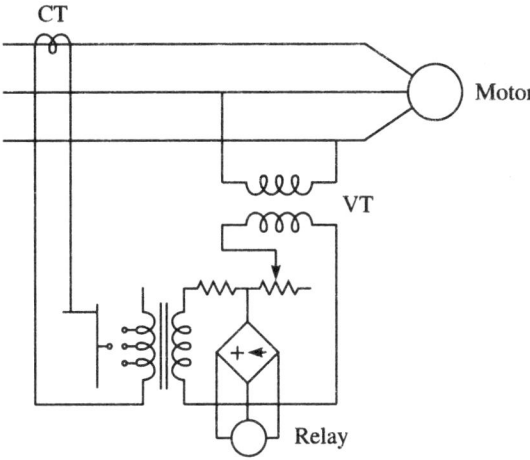

FIGURE 11.15   An out-of-step protection relay.

11.3
TRANSFORMER PROTECTION

11.3.1   Introduction

The traditional percentage differential relays are used for all kinds of internal transformer faults. But this scheme sometimes fails in case of magnetizing and overvoltage inrush current generated by interturn faults, switching, etc. Use of current transformers with different turns ratios, different ratings of the either sides of the main transformers and also tap changer in the main transformer are the associated phenomenon that add up the problems of malfunctioning of the transformers. Digital protection of the transformer is the ultimate solution to these problems.

Whether it is internal or external fault both cause the malfunctioning of the transformers. Protective schemes for the transmission line take care of the external faults while various schemes are employed for protecting the internal faults. These are gas actuated relays (Buchholz relays), overcurrent relays, restricted earth fault relays, percentage differential relays, etc. Selection of the protection scheme depends upon the types of faults as well as type, rating, size and importance of the transformer.

11.3.2 Faults in Transformers

The various faults that occur in transformers in a transformer can be classified as follows:

(i) *External faults,* i.e. appearance of overcurrent for short duration, short circuit outside the transformer, overloading, etc.

(ii) *Internal faults,* i.e. incipient faults, terminal faults and winding faults, etc.

Incipient faults cause overheating and pose consequent hazard, if not attended in time. The various incipient faults are:

(i) Core faults due to the breakdown of insulation of the lamination, of bolts or of clamping rings, which may cause limited arcing inside the oil.

(ii) Earth fault very near the neutral point of the star-connected transformer.

(iii) Poor electrical connection of conductors.

(iv) Coolant failure, i.e. failure of the cooling arrangement.

(v) Improper oil flow, loss or blockage of oil causing development of local hot spots.

(vi) Regulator failure and improper load sharing between transformers operating in parallel, causing circulation current.

Terminal faults include the following:

(i) HV or LV terminals phase-to-phase fault.

(ii) HV or LV terminals phase-to-earth fault.

(iii) HV or LV terminals three-phase faults.

Winding faults include the following:

(i) HV or LV windings phase-to-phase fault.

(ii) HV or LV windings phase-to-earth fault.

(iii) Short circuit between turns of HV or LV windings.

(iv) Short circuit between turns of tertiary windings.

(v) Earth fault on tertiary windings.

Terminal faults as well as winding faults are very serious in nature and can damage the transformer.

11.3.3 Electromagnetic Protection of Transformers

Introduction

The percentage differential relaying scheme provides reliable and fast operating relaying. The digital-based unit protection scheme is reliable and very fast in operation, possesses several additional features like high flexibility and self-checking provision. Certain phenomena associated with the transformer protection affect the reliable operation of the protection scheme are discussed as follows.

Magnetizing inrush current

Magnetizing inrush current refers to the current which flows in the primary winding of the transformer while the secondary winding is open-circuited just after the switch is closed on the primary side. The magnitude of magnetizing inrush current depends upon several factors such as steel used for core construction, magnitude and polarity of residual magnetism, instant of switching and so on. The initial asymmetrical flux requirement forces core deep into saturation, which leads to an abnormally high magnetizing current. The most power transformers operate fairly close to saturation level. Magnetizing characteristic determines the exact shape of the magnetizing inrush current, which sometimes goes up to 10–15 times the rated value. In many cases, this current takes nearly 1–2 seconds to decay owing to a large magnetizing inductance. This effect is more severe in a three-phase transformer as the current flowing in a winding is dependent upon currents of the remaining windings as well. The magnetizing inrush current, which is rich in second harmonics may cause false tripping of the relay. It is observed that the second harmonic content is always more than 16 per cent and is found to be very useful in detecting the difference between fault current and magnetizing inrush current for relaying purposes.

Overvoltage inrush current

During abnormal system conditions, a short duration voltage increase may take place, which, in turn, may cause saturation of transformer resulting in high differential currents. Increased current rises the magnitude of internal fault current causing an erroneous operation to the protective relay. It is observed that a 20 to 50 per cent overvoltage in grain steel may cause an increase as high as 10 to 100 per cent in excitation currents. The overvoltage magnitude is determined by several factors such as:

(i) Generator excitation

(ii) Remaining line length connected to a generating station

(iii) System shunt reactance

(iv) Generating capacity

An increase in the magnetizing current due to an overvoltage causes an increase in the third and fifth harmonic components, which, in turn, leads to malfunctioning of the relays.

Differential protection of transformers

Transformers rated 5 MVA and above are protected by differential protection scheme against internal phase-to-phase and phase-to-earth faults. Working principle of differential relay is already described in earlier chapter. Unbalanced fault current is applied for indicating and tripping the fault current. It needs highest selectivity with the lowest tripping time.

An ordinary differential protection for a three-phase star–delta power transformer is shown in Figure 11.16. The load currents in the two windings of the star–delta transformers are not in the direct opposition, but are displaced by 30°.

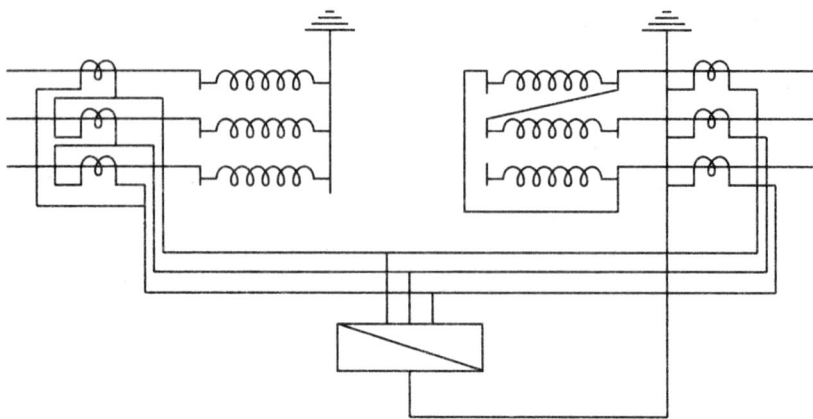

FIGURE 11.16   Differential protection for a three-phase star–delta power transformer.

The secondary of the current transformers are connected in delta on the star side and in star on the delta side of the transformer. Following problems are generally associated with the differential protection of the transformers:

(i) Magnetized inrush current

(ii) Unmatched characteristics of the current transformers

(iii) Turn ratio change due to tapping

Wherever through faults occur, unless saturation is avoided, the difference in current transformer characteristics due to different ratios being required in circuits of different voltage may cause appreciable differences in the respective secondary currents. This trouble gets aggravated in the case of transformers due to unequal ratio current transformers being employed on either side of the protected transformer.

For altering the turn's ratio effectively, tap changes are the common feature of a power transformer. Biased or percentage differential relays ensure stability with the amount of unbalance occurring at the extremities of the tap-change range.

Percentage biased differential protection of transformers

A restraining winding which gets energized by the passing current is provided with the differential protection scheme. Restraining winding helps in avoiding the undesirable operation on heavy external faults caused by turn ratio change due to tap changing or error in current transformer. In other words, the operating winding is biased and operated by some percentage of the passing current. Relay becomes more sensitive at low current without tripping for external fault.

If the restraining and operating magnets are identical with the turn ratio say T, static comparator operates, if the following criteria meet:

$$\left| \frac{I_1 - I_2}{T} \right| \geq \left| \frac{I_1 + I_2}{2} \right|$$

And a criterion for electromagnetic comparator is:

$$\left(\frac{I_1 - I_2}{T}\right)^2 \geq \left(\frac{I_1 + I_2}{2}\right)^2$$

Recommended value for T is 0.1 to 0.4. A biased differential protection scheme for the transformer is shown in Figure 11.17.

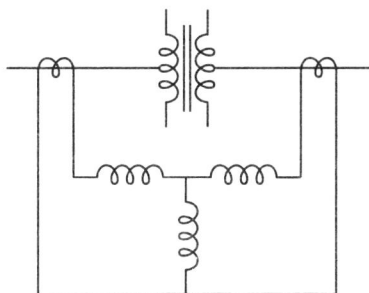

FIGURE 11.17   Biased differential protection scheme for transformer.

Methods for preventing operation on inrush current

During energization of the transformer, the transient inrush current may be ten times the full load current which decays slowly and its magnitude is the function of flux trapped in the core of the transformer and the voltage at the point of switching ON the supply. Immunity to operation by magnetizing surges is developed by many ways. One of them may be by setting the relays higher than the maximum inrush current. The other provision may be long time setting so the magnetizing current falls below the primary operating current before the relay operates.

Internal fault currents are generally sinusoidal but magnetizing inrush current contains appreciable harmonics. For high speed and low primary operating current, the harmonic content of the current is first filtered and then applied for restraint. Even harmonic cancellation and harmonic restraint are two methods of preventing the inrush currents in the transformer. Amplitudes of the harmonics in percentage of the wave shape of the magnetizing inrush current are shown in Table 11.3.

TABLE 11.3   Amplitudes of the Harmonics in a Typical Wave Shape of Magnetizing Inrush Current

Harmonic component	Fundamental	DC	2nd	3rd	4th	5th	6th	7th
Typical value (%)	100	55	63	26.8	5.1	4.1	3.5	2.5

Even harmonic cancellation: This method assumes the following:

  (i) Third harmonics and its multiples do not appear in the leads of the current transformers. These components circulate in the delta windings of the transformers and connected current transformers.

 (ii) Weak and small fifth and seventh harmonics can be ignored.

Even harmonics and dc components can get cancelled in the operating circuit of the rectifier bridge relay and in the restraint circuit.

Harmonic restraint relay: Harmonic restraint method is applied to make the differential relays immune to magnetizing inrush currents. A basic circuit of the harmonic restraint relay is shown in Figure 11.18.

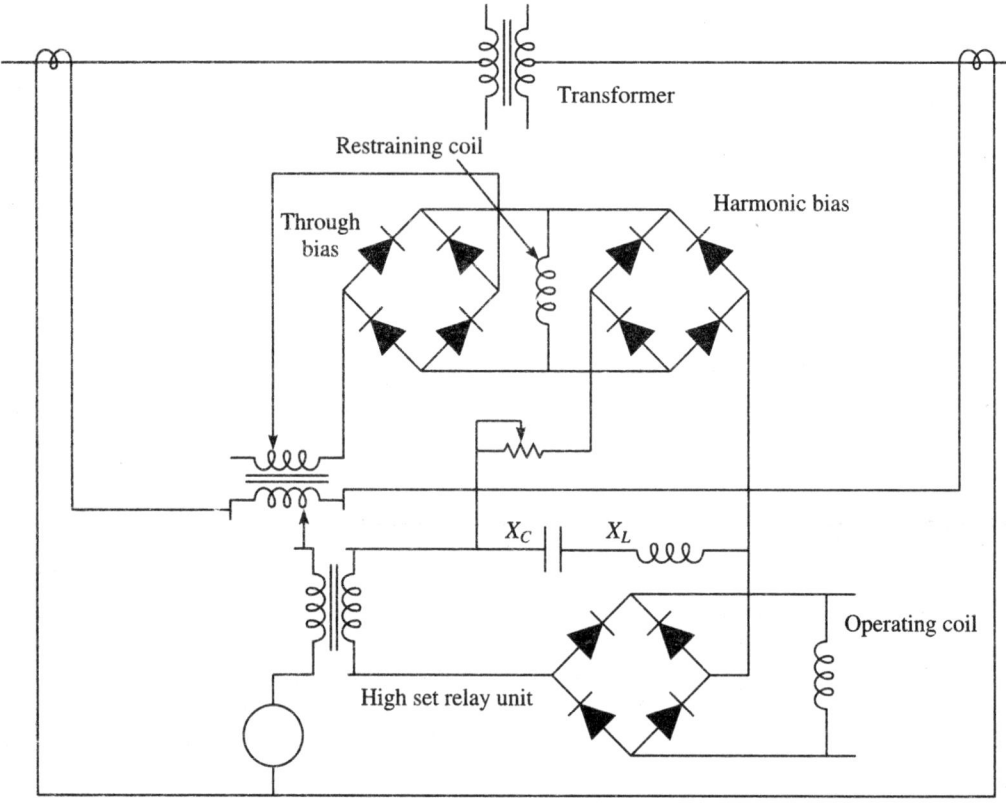

FIGURE 11.18 A basic circuit of the harmonic restraint relay.

The restraint coil is energized by the direct current proportional to the biased winding current and direct current due to harmonics. Tuned X_C and X_L permits only currents of fundamental frequency to enter the operating circuit and hence harmonic restraint gets obtained. The dc and second harmonics get diverted into the rectifier bridge feeding the restraining coil. Relay is adjusted in such a manner that it will not operate when the restraining harmonic current exceeds 15 per cent of the fundamental operating current.

Tank leakage protection

Earth fault protection for the transformer tank is provided by connecting a relay between the transformer tank and the earth. This is necessary, if the transformers are banked and are provided with single overall differential protection because it is difficult to detect the faulty transformer if they are banked together.

Restricted earth fault protection

Restricted earth fault protection is provided to detect earth fault within the protected zone of the transformer. A current transformer is fitted in each connection to the protected winding. Secondary windings of the current transformer are connected parallel to the relay. The circuit connections for restricted earth fault protection for star and delta windings are shown in Figure 11.19.

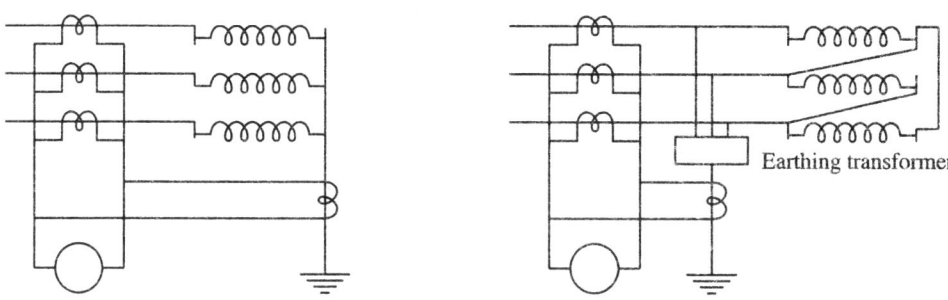

FIGURE 11.19   Restricted earth fault protection for star–delta windings.

For internal earth faults sum of the zero sequence currents in the line and neutral earth connection is equal to two times the total fault current. Zero sequence currents are either absent or sum to zero in the line and the neutral earth connection. Minimum primary current at which the earth fault current operates is the criterion for determination of the portion of the winding to be protected against each fault. If the neutral point of star winding of a transformer is earthed through a resistor, the rating of the neutral earthing resistor and relay setting decides the variations in the percentage of the winding to be protected.

Gas-actuated relays

Gas generation inside the transformer tank takes place when internal fault occurs. It is slow for the incipient fault but violent for severe faults. The heat produced by the high local current causes the transformer oil to decompose and produce the gas which is used to detect the winding faults. Impulses developed between the winding turns cause short circuit, which, in turn, heat up the interior of the tank. Following types of relays are available:

   (i) Buchholz relay actuated by the gas produced which is also known as the gas accumulator relay.

  (ii) Rate of rise of pressure relay which works on measuring the rate of gas formation.

 (iii) Pressure relays and pressure relief devices work on the total accumulated pressure.

 (iv) Gas analyzers, which act on the analysis of decomposition of the products.

A Buchholz relay is commonly used in all types of transformers. It is the simplest form of protection. It is provided with the conservator. This relay consists of a chamber connected in the upper side of the pipe run between the oil conservator and the transformer tank, and contains two cylindrical floats. One float is placed near top of the chamber and the other opposite the orifice of the pipe to the transformer. Floats are at upper position under normal conditions. But under occurrence of fault current, gas bubbles try to flow out of the transformer tank towards the conservator and in the path they are trapped by the Buchholz relay and cause upper float to fall. Under low-level fault a pair of contacts is controlled by the float and with its movement buzzer gets activated to produce audible signal as well as glowing a bulb for giving visible warning. A Buchholz relay is shown in Figure 11.20.

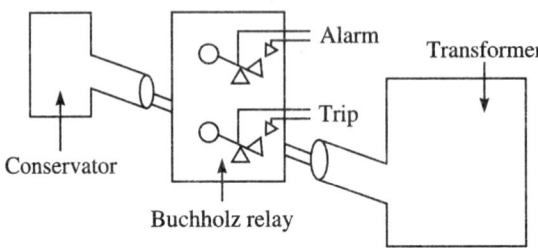

FIGURE 11.20   A Buchholz relay.

If the fault is severe, rush of the gas and oil come upper side of the pipe towards the conservator. Now the gas engages the lower float, which is pushed over instantaneously and engages its associated contacts, which, in turn, trip the circuit-breaker. Oil leakage causes the upper float to operate and if it persists, it also causes the lower float to operate.

11.3.4   Digital Protection of Transformers

Filtering schemes

Analog and digital filtering are basically two types of filtering schemes. Analog filters are not so reliable because of the hardware complexity. Analog filters are complicated circuitry because separate filters are to be provided for each input signal. Digital filters are preferable because of the following merits:

(i) High accuracy

(ii) High reliability

(iii) Non-critical component tolerance

(iv) Performance unaffected by component aging and greater flexibility

There are following two classes of digital filters namely:

(i) *Non-recursive filters*, whose output is a function of previous and present inputs, generate a finite impulse response and therefore, it is called *FIR filters [Finite impulse response filters]*.

(ii) *Recursive filters*, in which the output becomes a function of the past and present inputs and outputs, through the use of feedback. Due to the feedback, the recursive filters, have an infinite impulse response and are known as *IIR filters [Infinite impulse response filters]*.

In order to meet the prescribed requirements, transferring of the analog filters to the digital filters is required. This is achieved by:

(i) Impulse invariance

(ii) Direct transformation using S to Z- plane transforms

(iii) Conversion of differential equations to difference equations

(iv) Direct synthesis.

A *legendre filter* of the fourth order gives minimum settling (approx. 56.2 ms) time for a step input and tripping decision can be reliably taken only when the filter outputs have settled down. IIR filters do not perform well if they are desired to trip within a cycle after the fault occurrence. This is the reason why the use of FIR filters over IIR filters is a better option. FIR filters are digital equivalent of a tapped delay line analog filter or transversal filters. Walsh transform-based filters are simpler and better in usage, described below.

Walsh transform based filter

The Walsh function, denoted by Wal (k, t), forms a complete set of orthonormal functions and bears a number of resemblances to the foregoing sines and cosines, appearing as their square version. They take only the values ± 1 and change the sign only when t is a multiple of a power of $\frac{1}{2}$.

The synthesization of a continuous function with a periodic time period $T = 1$, is achieved by the Walsh series in the following way:

$$f(t) = \sum_{n=0}^{\infty} W_n \text{Wal}\,(n, t)$$

where

$$W_n = \int_0^n f(t)\,\text{Wal}\,(n, t)$$

For the function $f(t)$ specified by N samples such as $(X_1, X_2, ..., X_N)$ during a time period, the use of a Walsh transformation is not possible. However, the discrete Walsh transform and its inverse from the above two equations are as follows:

$$X_n = \frac{1}{N} \sum_{k=0}^{N-1} W_k \text{Wal}\left(n, \frac{k}{N}\right)$$

$$W_k = \sum_{k=0}^{N-1} X_n \text{Wal}\left(n, \frac{k}{N}\right)$$

For $k, n = 0, 1, 2, ..., (N - 1)$

The settling time of this filter is less than 20 ms.

Percentage digital differential relaying

Two output currents, I_1 and I_2 which are obtained from the current transformers, first need to be converted into voltages by being passing through transactors. After this, these two voltage signals, S_1 and S_2, as shown in Figure 11.21, are selected one by one with the help of a multiplexer and then sampled. The sampled frequency has to be two times the highest frequency so that it can be detected by the sampling process. This process, however, depends on the type of microprocessor, being used.

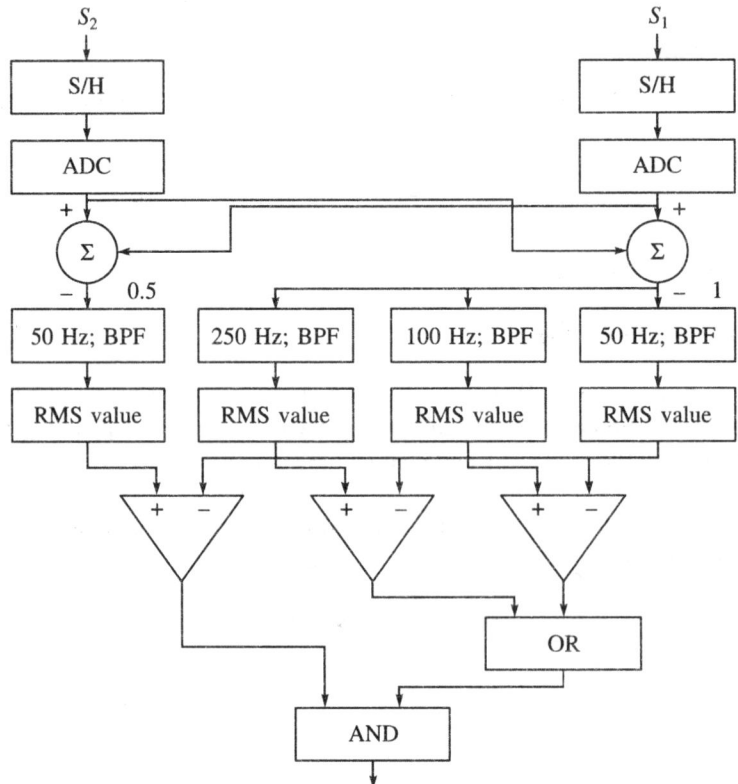

FIGURE 11.21   Percentage digital differential relaying.

The condition for tripping of the percentage differential relay is:

$$|I_1 - I_2| \geq S \left| \frac{I_1 + I_2}{2} \right|$$

Here, S is a bias factor whose value varies in the range of 0.15 to 0.6. I_1 and I_2 are the fundamental components of the pure sinusoidal waveforms. But in actual practice, after the fault, the current waveform consists of fundamental as well as several harmonic components.

The filtering of the harmonics is thus necessary otherwise these interfere with the fundamental component of the signal. The signals, which are achieved for tripping by filtering, are depicted in the equations given below.

For fundamental frequency:

$$|I_1 - I_2|_1 \geq S |I_1 - I_2|_1$$

For the second harmonic component, on the basis of earlier results wherein an overvoltage inrush condition exists:

$$|I_1 - I_2|_2 \geq 0.16\,|I_1 - I_2|_1$$

And for fifth harmonic component:

$$|I_1 - I_2|_5 \geq 0.08\,|I_1 - I_2|_1$$

Here, 1, 2, and 5 refer to the fundamental, first and second harmonic components respectively.

11.4
BUS-ZONE PROTECTION

11.4.1   Introduction

Bus-zone failure may take place due to the following reasons:
- (i)   Accidental fall of the foreign objects on the busbars
- (ii) Insulation failure
- (iii) Failure of circuit-breakers
- (iv) Errors in the operation and maintenance of the switchgear
- (v)   Flashover caused by overvoltages

The operating principle of the differential current protection scheme of the bus-zone is based on the Kirchhoff's law, i.e. the algerbraic sum of all the currents entering and leaving the busbar is zero. However, if there is internal fault, algebraic sum of the current meeting at bus-zone will not remain zero and relay operates to send the signal to the circuit-breaker connected for tripping and isolating the faulty section.

Faults on bus-zone rarely occur and the zone includes the bus, circuit-breakers, disconnecting switches, instrument transformers, bus-sectionalizing reactors, etc. It is essentially needed to provide protection to the bus-zone, otherwise once the fault occurs, the system will face the following adverse effects:

- (i) MVA loading increases due to addition of fault MVA which may damage the bus
- (ii) Loss of zone causes supply interruption

Routine testing to ensure the proper working of the bus-zone protection is also essential. Protection scheme must be fast, stable and most reliable.

11.4.2   Bus Back-up Protection

For the small switchgear installation, busbar protective zone may be covered within the reach of the distance protection scheme provided to the feeders. But for the large and strategically important installations, a separate bus-zone protection scheme is required. A back-up scheme is provided with appropriate time delay through a timer.

11.4.3   Differential Protection Scheme for Busbar

Circulating current entering the busbar is equal to the currents leaving the busbar under normal operating condition and for external fault. For any given conductor, if the sum of currents entering the busbar is not equal to the currents leaving the busbar, that may be due to L–G or L–L fault. This condition suits the differential relaying scheme.

As shown in Figure 11.22, relay detects the current in the protect zone and sends tripping signal to the associated circuit-breakers. To avoid the false operation due to the difference in the magnetic conditions of the iron-cored current transformers, current transformers without iron cores are used. Current transformer without iron core is also known as *linear coupler*.

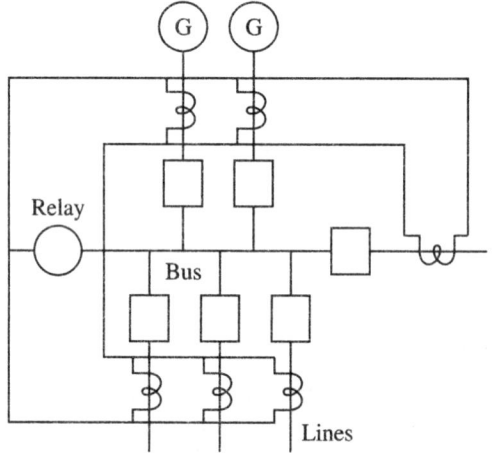

FIGURE 11.22   Differential protection of a bus section.

EXERCISES

1. What are the different types of faults, which occur in synchronous machines? Discuss briefly the protection schemes to protect the synchronous machines against these faults.

2. Describe the merits of microprocessor-based relaying scheme.

3. Explain the theory and working principle of microprocessor-based unit protection scheme of a synchronous generator.

4. Discuss, with the help of neat diagram, the microprocessor-based relaying scheme for the protection of a synchronous generator by monitoring its field current.

5. When an unbalanced current caused by a fault or external disturbances is produced in the stator, which harmonic component is induced in the field circuit? How do you distinguish between an internal fault and an external disturbance in this scheme?

6. Discuss briefly the different types of faults in a transformer. Name the protection schemes to protect the transformer against these faults.

7. Discuss, with the help of a neat diagram, the theory and principle of operation of a unit protection scheme (i.e. percentage differential relaying scheme), using the static components for the protection of a transformer.

8. What do you understand by inrush magnetizing current in transformer? Discuss the factors upon which the inrush magnetizing current depends.

9. Discuss the theory and principle of operation of a microprocessor-based percentage differential relay for the protection of a transformer.

10. Discuss briefly the reasons why a digital filtering scheme is preferred. What do you understand by recursive filter and non-recursive filter?

11. Describe a Buchholz relay and discuss its merits and demerits.

12. Draw neatly the differential protection scheme of an alternator. Discuss its limitations and suggest remedies to overcome them.

13. Distinguish between overload protection and short circuit protection of a motor.

14. Describe the different types of faults which occur in a three-phase induction motor.

15. Write a protection scheme for large induction motor.

16. Describe the method of protecting busbars by applying differential relaying scheme.

Chapter 12

Recent Developments in Digital Protection

12.1

FIBRE OPTIC-BASED RELAYING

12.1.1 Introduction

Direct fibre optic cable communication is a straightforward and simple medium media for facilitating relay-to-relay logic communication. It is preferred because of its immunity against electrical interference. It has a very small bit error rate below 10^{-9}. In this relay, direct digital relay-to-relay logic communication takes place via a direct fibre. This overcomes the problem of ground potential rise and interference problems encountered when a metallic cable is used. In fibre optic relays, the data transfer rate between relays is very fast and any delay in data transmission in the fibre optic transceivers and optical cable is typically only a fraction of microseconds. This delay in data transfer is negligible as compared with the corresponding delay in the case of other relays.

In direct fibre optic cable communication, a fibre optic transceiver is used at each relay terminal. This fibre optic transceiver converts the digital relay signal into an optical signal which is transmitted over the fibre optic cable. The use of single mode optical cable and transceivers facilitates transmission up to a distance of 90 km. Multi-mode fibre optic cable and transceiver technology supports optical signal transmission up to a distance of 3 to 5 km.

The development of new generation fibre optic voltage and current measuring systems is expected to facilitate the wider application of travelling wave protection using fibre optical-based communication, in the near future.

12.1.2 Fibre Optic-based Direct Relay-to-Relay Digital Logic Communication

Interfacing of the network communication multiplexer with the relay-to-relay logic communication is achieved through a channel card inserted in the multiplexer rack, as shown in Figure 12.1. The relay serial communication port is connected to the channel multiplexer interface card with a shielded fibre optic cable equipped with fibre optic transceivers.

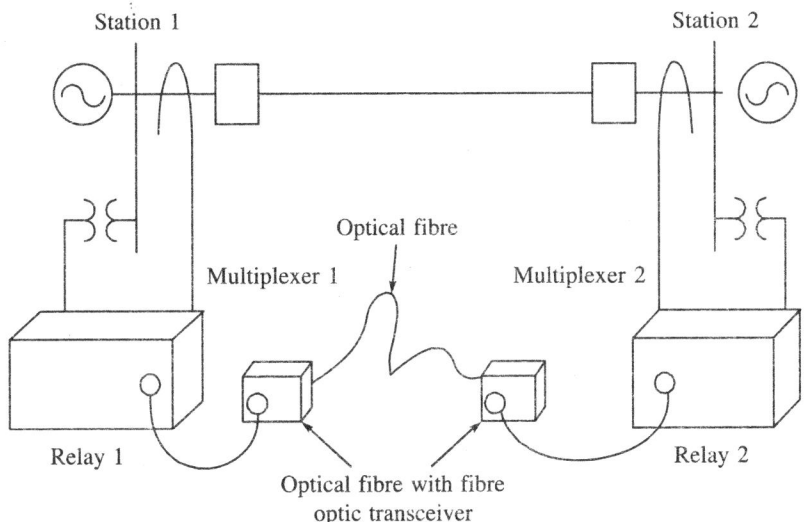

FIGURE 12.1   Fibre optic-based direct relay-to-relay digital logic communication.

Network topology has several communication nodes wherein channels are inserted or dropped. There is a facility for making loops in order to provide alternate paths, which can be used when any one of the network segment fails or is taken out of service for maintenance.

Fibre optic communication is recommended between the relay and the multiplexer to eliminate the effect of any electrical interference emanating from the sub-station environment.

The various tasks undertaken by the network multiplexer are detailed below:

(i) Error checking is performed by the multiplexer.

(ii) The network multiplexer handles the task of channel addressing and synchronising.

(iii) Re-synchronisation of signals takes place in the multiplexer whenever the channel breaks down due to failure or a data error. The multiplexer takes a few milliseconds to re-synchronise the channel.

(iv) The network multiplexer also performs the task of error detection. This may delay the data communication and affect the end-to-end relay logic response time.

Figure 12.2 shows the hardware connection of the fibre optic-based direct relay-to-relay digital logic communication scheme.

It is thus clear that optical fibres, which are fine strands of glass, behave as wave-guides for light. They are immune to electromagnetic interference. At the sending end, the transmitter converts the electrical pulses into light pulses. The light pulses are then transmitted through the optical fibres and subsequently decoded at the other remote terminals. At the receiving end, the light pulses are converted into electrical pulses. Digital techniques of pulse code modulation are used in the terminal equipments.

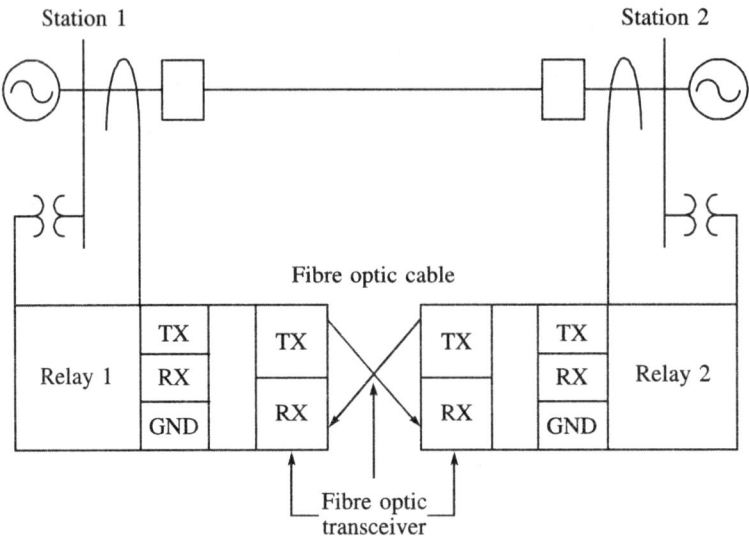

FIGURE 12.2 Hardware connection of the fibre optic-based direct relay-to-relay digital logic communication scheme.

Analog relaying quantities are sampled and transmitted as coded information to the remote terminals by the analog-to-digital converters. The decoded information is then compared with the locally derived signal at the remote signal.

12.2
MICROWAVE RELAYING

12.2.1 Introduction

Microwave relaying is applicable in power line protection. In this scheme, an antenna is erected and installed at the line of the sight. The gap between the two sights of antennas is generally kept at 90 km. This scheme is more or less similar to the career current pilot scheme except that in this case, signals are not transmitted through the same power line as in the case of the career current pilot scheme. In the microwave relaying scheme, a point-to-point digital radio provides stand-alone communication between two sites. The line of sight antenna helps generate and transmit in space a carrier of very high frequency in the range of 900 MHz to 6000 MHz, with relatively low power ratings.

In this scheme, a digital radio transceiver is needed to interface with the relay serial communication port at rates of up to about 9,600 baud. This microwave-relaying scheme does not require a coupling capacitor and line traps.

Relay-to-relay logic communication works well without in-built radio error detection because the latter adds only two or three milliseconds to the overall relay-to-relay data delay. The radios with in-built error detection may introduce data delays of 60 milliseconds or more. Since speed is vital in the most pilot communication schemes, it becomes imperative to check the radio specifications carefully for the radio system data delay characteristics.

12.2.2   The Microwave Pilot

The microwave pilot is an ultra-high frequency radio system operating in allotted bands of above 900 MHz. In this line, coupling and trapping are not required and are hence eliminated. The requirement for other equipments and their operations are the same in the microwave pilot as in the carrier current pilot. Line of sight antenna equipments are also required in this case.

Due to certain disadvantages, the microwave relaying scheme should not, however, be used for relaying alone if the carrier current pilot relaying scheme or wire pilot is available for usage. These disadvantages are listed below:

Disadvantages of microwave pilot

(i) The microwave pilot scheme is not entirely reliable because of the presence of complex circuitry, and the involvement of a large number of parameters as also a large number of services on the same microwave channels.

(ii) In cases when repeater stations are required, the complexity of the networks further affects their reliability adversely.

(iii) It is sometimes felt that the availability of microwave signals is more important for communication purposes than for the relaying scheme. And, in the presence of a fault, the unavailability of the microwave signals affects the system adversely.

The connection diagram of microwave direct relay-to-relay digital logic communication is shown in Figure 12.3.

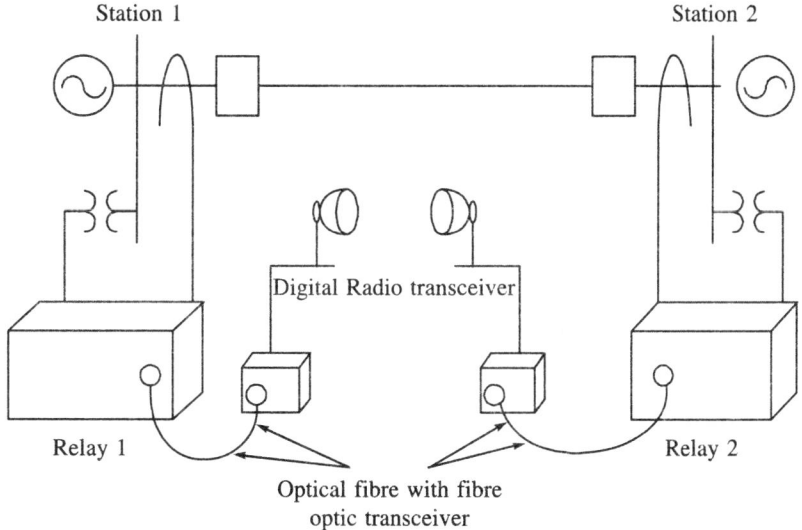

FIGURE 12.3   Microwave direct relay-to-relay digital logic-communication.

The microwave relaying scheme also has several advantages. These are listed below.

Advantages of a microwave pilot

(i) The microwave relaying network is dissociated from the power line, which negates the chance of occurrence of a fault in the line length.

(ii) The presence of a fault on the protected line does not interfere with the transmission of a remote tripping signal of the microwave relay.

(iii) The same relaying equipments that are used with a career current pilot can also be used with a microwave pilot.

12.2.3   Microwave Channel

A line of sight radio system, which operates on a frequency band, is known as a microwave channel. This system requires a straight line from one antenna to the other above the intervening objects. The space requirement for this is about 50 feet. The intervening objects limit the distance between antennas to about 30 to 90 km. The distance between antennas depends upon the topology of the land. It may be necessary to set up one or more repeater stations to establish a longer channel. Each repeater station, however, increases the number of base equipments and the antenna tower, which, in turn, leads to an increase in the cost of the microwave channel.

Operation from a power system AC source is not acceptable for protective relaying, as the latter cannot tolerate a moment's outage. It therefore becomes necessary to ensure stand-by power supply in the form of an AC generator or a battery bank. This requirement may at times pose problems at a repeater station where a suitable battery source may not be available.

In the microwave channel also, there is a provision to use stand-by equipment, which is automatically switched into service in the event of the failure of regular equipment.

Microwave frequency is generally modulated directly by the usual methods, such as by a so-called "tone". Such a tone is a single-frequency voltage in the audio range or above. Since the time constants of the filter circuits of the tones above the audio range are shorter, the tones above the audio range are preferred.

12.2.4   Remote Tripping

The main advantage of using microwave protection for relaying is that it precludes any interference of the fault with the remote tripping signal. In the case of microwave protection, the first relays to operate can cause the transmission of a tripping signal to another terminal and thereby eliminate part of the time delay in the sequential tripping of the other terminal.

If each terminal is arranged to transmit a trip signal to any other signal, simultaneous high-speed tripping occurs at all terminals. The relays set at remote places may trip after delay of about 2–3 cycles.

This type of relaying has the widest application in cases wherein remote tripping is required for multi-terminal applications.

12.3
UNDETECTABLE ELECTRIC POWER SYSTEM FAILURES

12.3.1   Introduction

The failures, which occur at random and do not lead to any change of the state of the power system network, are known as *hidden failures* in electric power systems. These failures are rare but, if the power system is stressed, their effects on the power network may be catastrophic because the undetectable hidden fault can occur in any equipment placed at any place in the entire network.

The causes of such catastrophic failures in power mechanisms have been a matter of concern for power engineers in recent years. In the present competitive deregulated electric power supply system, power utilities are being forced to improve their productivity and efficiency while offering a greater degree of reliability. This is only possible in a dedicated, sophisticated and efficient control and protection system.

While catastrophic failures in power grids are at times unavoidable, power engineers and researchers still need to find answers to the following questions:

(i) What causes catastrophic failures?

(ii) Are any defensive measures available to reduce the likelihood of such failures?

(iii) Is there any mechanism that can reduce and limit the extent to which these unpredictable faults can spread?

One of the defensive measures used against catastrophic failures may be to make the power system more robust by using modern computer-based protection schemes and adapt newer theoretical ideas in the field of protection. Communication links play a key role in such defensive schemes, along with improved monitoring through synchronised phasor measurements and adaptive relaying.

12.3.2   Hidden Failures in the Power System

Scheduled or unscheduled breakdowns or outages of main generators and other high-rated transmission apparatus put pressure on the power system network. Such pressure conditions increase the chances of occurrence of catastrophic faults. In recent times, there has been a significant increase in catastrophic failures in the grid systems. In the beginning, generating stations were not interconnected. With the advent of reserve capacities, isolated stations became capable of catering to the entire load of the area for which they were built. Today, however, the interconnection of generating stations in a grid has become a popular option for optimising the availability of electrical power.

The main advantage of a grid system is that it can supply the excess power kept in reserve to its neighbouring sub-stations during emergencies. Duties for load frequency regulation can also be divided equally among all the generators of the interconnection, which, in turn, reduces the burden on each region.

The hidden failures are failures which cannot be detected when they occur, because they do not cause any problems when the system is normal. The reasons for unavoidable hidden failures in power systems are as follows:

(i) The power network may get stressed due to unscheduled events like fires, floods, hurricanes, faults, etc. This causes generators, power lines and transformers to reach dangerously close to overshooting their safe operating limits.

(ii) System operators may commit human mistakes like failure to make real-time measurements, false assessment of the system conditions, keeping inefficient or inadequate safety margins, etc.

(iii) If the relaying scheme does not clear the fault well within time, the fault may start a chain reaction under this condition.

(iv) The initiation of incipient faults may lead to the development of instabilities, which, in turn, may further weaken the network.

(v) The separation of grids into sub-regions of load and generation imbalance may cause their eventual collapse in some regions.

(vi) System failure is also inevitable if the settings of the protection systems are inappropriate.

(vii) The failure of components due to wear and tear or damage caused by adverse environmental conditions may also lead to a failure in the network.

(viii) Incorrect and non-judicial human intervention could also be another reason for failure in the power system.

12.3.3   Countermeasures to Avoid Hidden Failures

Hidden failures, which are often unavoidable, play an important role in catastrophic failures. The only way of dealing with these failures is by taking remedial measures, and considering countermeasures that reduce the likelihood of hidden failures and prevent them from spreading in the network.

The following three approaches may be adopted as countermeasures to prevent hidden failures:

(i) Adaptive relaying

(ii) Hidden failure monitoring and control

(iii) Special protection system

Adaptive relaying implies making an assessment of the state of a power system and then automatically making adjustments in protection systems in order to ensure their correct settings for the prevailing conditions. Adaptive relaying concepts are generally implemented in digital relays. The relay settings in digital relaying schemes are generally very close to appropriate relay settings, which help prevent hidden failures caused by incorrect settings.

One example of an adaptive relaying scheme is that of adaptive out-of-step relaying. These relays are designed to detect conditions when a group of machines or one or more portions of the power system are about to go out of synchronism with the rest of the network. It is the duty of the relay to detect such a condition of non-synchronism and separate the affected machines from the network before the occurrence of a catastrophic failure in the entire network.

A *hidden failure monitoring and control* [HFMC] system is shown in Figure 12.4. In this principle, the first two relays are critical from the point of view of the catastrophic failures. Input signals supplied to the HFMC (Hidden Failure Monitoring and Control System) emanate from the critical relays. It should be noted here that the input signals are common for both the critical relays and HFMC, to facilitate duplication of the relay functions as back-up relay. The output of the HFMC system is used to supervise the output of the two critical relays.

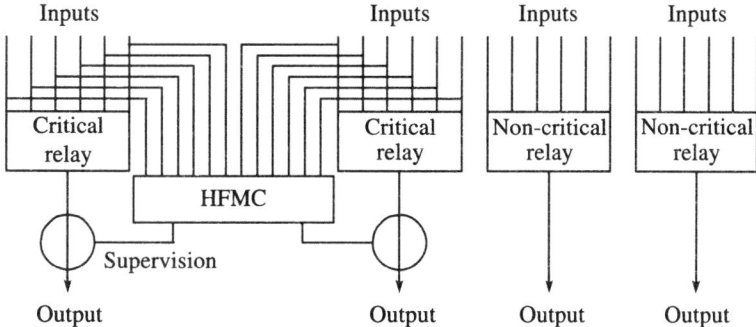

FIGURE 12.4   Basic principle used for the supervision of hidden failures in a protection system.

According to the same principle (see Figure 12.4), at any critical location, if the three protection systems are in use on a critical line at any critical location, the normal mode of operation is for any one of the relays to trip the breakers. This is an 'OR' function applied to the output of the three relays. In order to determine the emergency state of the power system, the logic of the protection system can be changed to 'AND' or a '2-out-of-3' logic. This change in logic will ensure that at least two relays confirm the occurrence of a fault. This increases the reliability of the system and greatly reduces the likelihood of a relay maloperation caused by hidden failures.

Special protection system (SPS) refers to the protection systems that trigger protection and control actions, which reduce the probability of a disturbance turning into a catastrophic event. Several tables stored in the SPS system meet the requirements of varying system conditions. In a communication network, data regarding the status of the key circuit breakers and switches, system loading, loading on key transmission facilities, etc., are brought to the SPS. A table look-up determines the appropriate response for the prevailing system conditions.

12.3.4   Communication Infrastructure Integration

Communication plays an important role in all defensive strategies. However, in order to prevent complete dependence on the communication system in a power network, it is essential to have in place an alternative protection system, which does not utilise the communication system.

In *primary protection*, the communication channels used are generally dedicated channels. Primary relaying is the fastest acting protection, usually operating within 10-30 milliseconds after the inception of a fault.

Back-up protection relaying schemes are slower than primary relaying systems. The most traditional back-up functions do not use communication lines. The overcurrent and distance relaying functions provide sufficient reach, precision and speed in the network. Some *adaptive protection* systems do not need inputs from remote portions of the transmission lines. But, inputs from remote sub-stations are needed in the most adaptive protection systems. Adaptive protection systems also require some data from the system control centre, or from the several remote sub-stations.

Special protection systems require alternative data paths in order to avoid dependence on a single failure mode. These systems also require data from remote sites. *Hidden failure monitoring systems*, *however*, do not require data from remote sub-stations.

The power system engineers have the option to use any of the following communication systems:

(i) Power line carrier
(ii) Pilot wires
(iii) Microwaves
(iv) Leased telephone circuits
(v) Fibre optic links

All the above communication systems are used in protection systems for preventing catastrophic failures. In all cases of failure, the security of the communication channel must be the top priority. Under any condition of stress in the network, failure in any one link should not lead to failure in any of the other links. It is also important to keep communication systems safe and secure from the intervention of hackers. Thus the adoption of computer-interfaced relays, synchronised phasor measurements, and adaptive relaying ideas can help in developing protection systems that are suited to handle stresses and reduce the likelihood of catastrophic failures.

12.4
MAINS PROTECTION

12.4.1 Introduction

Supply failures always disrupt industrial production and entail economic losses. The parallel operation of smaller and medium level electric power generation is gaining importance in the power sector as it helps prevent longer duration disruptions of electric power supply mains. Besides, in the industrial sector, there is need to install captive power plants as alternative energy sources to fill the gap between the demand and supply of energy from the power grid. Thus Combined Heat and Power Plants (CHP) are in use because they employ alternative energy sources.

In the event of a grid failure, the inter-tie circuit-breaker must be disconnected immediately, which means that the power generating system should be decoupled automatically from the faulty section of the grid. *Decoupling of the mains implies opening of the closed inter-tie circuit-breaker wherein the private power system and public grid are connected. In some cases, the generator circuit-breaker may serve the purpose of decoupling.* In the decoupling process, the main focus is on the measurement of the surge and evaluation of the rate of change of the frequency (df/dt).

The following parameters need to be studied before precautionary measures can be established to facilitate the early detection of incipient failure:

(i) Detection of the overstressed mains whose failure is possible.
(ii) Identification of the switching operations of the mains, which may have dangerous side-effects.
(iii) Need to protect the faulty sections.

(iv) Identification of individual failures in the system.
(v) The prevalent state-of-the-art protection schemes.
(vi) Prioritisation of either system protection or supply continuity.
(vii) Necessity for additional protection measures.

The inter-tie circuit-breaker is kept open for eliminating the possibility of any external fault. This measure also protects consumers from voltage breakdowns.

12.4.2   Loss of Mains Protection

Routine checking or detection of an earth fault in the grid necessitates the simultaneous decoupling of all the private power suppliers and captive power plants. Decoupling, in turn, needs to be undertaken with proper co-ordination and planning, otherwise manual re-connection of the faulty section can prove to be extremely dangerous. In such a situation, there is a possibility of loss of synchronisation at the opened isolating point and it is not possible to close the opened circuit-breaker without risk.

Another reason for the loss of mains protection may be the short-term interruptions of the mains with automatic reclosing, which sometimes activates the clear faulty phase following which reconnection can be done. However, the electric arc ignited during short-circuiting keeps on burning and weakens the insulation. Failure of the insulation, in turn, causes the loss of mains protection.

The after-effect of a blown fuse or a broken conductor may be the cause of a single-phase conductor failure. Even the out-of-phase voltage recovery endangers the main parallel running generator. Generally, overcurrent protection becomes effective and results in the disconnection of the faulty outgoing section just after the fault has occurred. The disconnection of the faulty section helps the voltage on the line to start recovering. The mains voltage and the generator voltage are now positioned directly opposite each other. The transient processes come into effect if the two voltages are not synchronous at the instant of disconnection. In such a situation, the mains and the generator are extremely stressed.

12.4.3   After-effects of Mains Disconnection

The mains can be decoupled without altering the system voltages, which means that the mains disconnection takes place without a preceding change in voltage. This is possible because of the provision of a variable load at the generator end as shown in Figure 12.5. The circuit-breaker of the interconnection line isolates the captive power plants from the grid. In this case, it is possible for the residual load to remain at the generator side while the generator continues to supply power on this load.

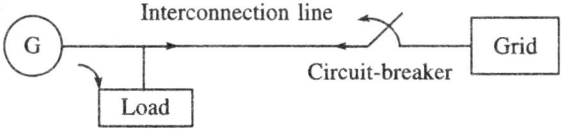

FIGURE 12.5   Isolation after switching off manually.

If the active power load increases during the isolation process, there is a decrease in the generator speed. And if there is a reduction in the active power requirement of the isolation system, the speed of the generator increases, which, in turn, leads to an increase in frequency.

Now, if during the formation of the isolation, there is an increase in the reactive power load requirement, the terminal voltage of the generator decreases. If, on the other hand, there is a reduction in the requirement of reactive power supply, the terminal voltage increases.

In case of occurrence of a short-circuit with or without automatic reclosing (AR), there will be a change in the voltage. The formation of an isolated system through a short-circuit and automatic reclosing is shown in Figure 12.6. Here, the transient for the generator and the driving generator set is the result of a short-circuit, followed by its quick clearance.

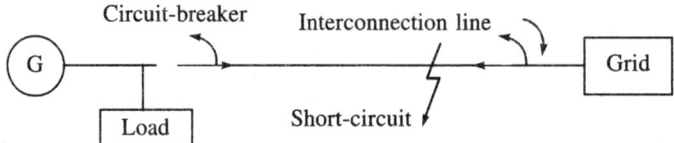

FIGURE 12.6   Isolation after short-circuit and automatic reclosing.

By opening the circuit-breaker, the failures in the mains enter the second phase wherein the faulty section of the line is switched off from the generator and forms the remaining load, after which the voltage recovery takes place.

12.4.4 Decoupling of the Mains

Decoupling of the mains may depend on several factors including the variation in voltage, voltage unbalancing, reverse power flow, variation in frequency, the rate of change of frequency, change in load, etc. These factors are discussed in detail below.

Voltage monitoring

The most suitable criterion for decoupling is measurement of the voltage variation because the voltage is supposed to remain identical at all measuring points. By using the voltage measurement criteria, it is also possible to monitor a complicated and branched grid right into every feeder.

If there is any sign of overvoltage or undervoltage at the decoupling point, it is necessary to go in for decoupling in order to safeguard against any risk to either the system or the consumers connected. An example of the need for decoupling may be that of motors on load, which need higher current. If there is a reduction in the mains voltage for any reason, thermal overloading will take place.

Voltage unbalancing may also be a criterion for decoupling. It can be monitored by evaluating the following symmetrical components:

(i) Positive phase sequence voltage (V_1)

(ii) Negative phase sequence voltage (V_2)

(iii) Zero phase sequence voltage (V_0)

Deviation from the original voltage phasor takes place when a single-phase or double-phase fault occurs. It is reflected by an increase in the amplitude of the negative phase sequence system and a drop in the voltage of the positive phase sequence system. The zero phase sequence system provides information regarding earth faults in isolated and compensated systems.

Reverse power monitoring

Another method of protection is the monitoring of the reverse power flow. In this method, the direction of energy flow at the mains coupling point is monitored. The only precondition in this method is that energy import should be permitted during the normal operation. Thus if the own generator supplies power to external consumers during a fault, the coupling switch has to be operated.

Frequency monitoring

If the load on the generator is increased, both its speed and the frequency of the system decrease. It is therefore said that the failure of the mains supply can be recognised by frequency supervision. Additional frequency supervision elements are required for realising the loadshedding scheme. In some situations, the frequency changes to such a degree that it crosses the absolute limit. And if there is no frequency, what can one measure? In such a situation, the detection of the rate of change of frequency (df/dt) comes into use. The df/dt relay analyses the tendency of the change in frequency.

Voltage vector surge monitoring

The amplitude of the vector surge is defined by the active load change at the generator during a failure in the mains. The vector surge relay is suitable for handling strong active load changes as it ensures rapid and reliable decoupling. This is delineated in the following example.

Let a motor load be connected with the own generator. Another load Z_L is connected to the consumers. The consumers' load Z_L imports power from the own generator as well as from the mains grid. This process is shown in Figure 12.7.

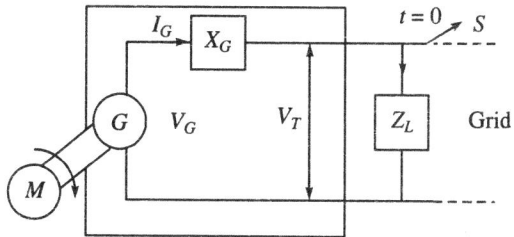

FIGURE 12.7   Simulation circuit diagram.

Here:

V_T = generator terminal voltage

V_G = generator internal voltage

X_G = generator reactance

Z_L = load of the connected consumers

θ = rotor displacement angle, the angle between V_T and V_G

Thus, prior to the decoupling, Z_L is partly fed by the mains and partly by the own generator. In other words, both the voltage sources, i.e. the own captive generator or the standby generator and the grid mains, are available to the Z_L load. In case of a mains failure or auto-reclosing, the generator suddenly feeds a very high consumer load. In other words, if the supply voltage gets interrupted, say at time $(t > 0)$, the power supplied by the mains will also suddenly be supplied by the generator, thereby constituting an additional load on the generator. Due to the inertia, however, the speed of the rotor will not change abruptly. The rotor displacement angle θ will be changed due to a change in the current from I_G to I_L. Simultaneously, there is also a change in the phase angle of the terminal voltage.

Figure 12.8(a) shows the phasor diagram before the decoupling at time $(t < 0)$, while Figure 12.8(b) shows the phasor diagram after the decoupling at time $(t > 0)$. *The rotor displacement angle between the rotor and the stator depends on the mechanical moving torque of the generator shaft. The mechanical shaft power is balanced with the electrically fed mains power, which therefore keeps the synchronous speed constant.* The change of angle takes place very quickly. Another effect of this load change is a drop in speed, which attains a measurable value only after some time due to the effect of inertia. The minus sign (–) before I_m in the phasor diagram denotes the opposite direction of the current I_G.

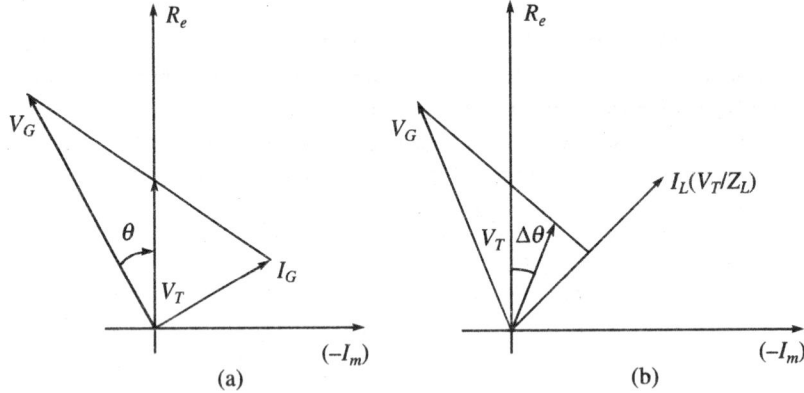

FIGURE 12.8   (a) Phasor diagram before decoupling, (b) Phasor diagram after decoupling.

The oscillogram of the generator terminal voltage at the instant when the mains are switched off as shown in Figure 12.9. In the oscillogram, it can be seen that the generator terminal voltage V_T jumps to a different instantaneous value. When the voltage jumps to another value, the change in the phase value is known as *voltage* or *vector surge*.

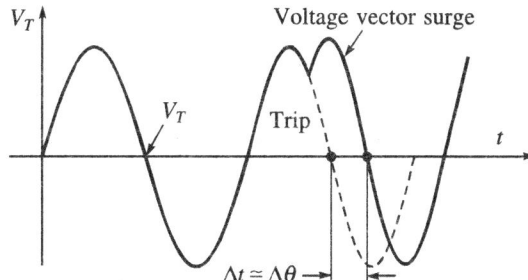

FIGURE 12.9   Oscillogram of the terminal voltage at the instant of decoupling showing a voltage vector shift.

Change in load (ΔP) monitoring

Monitoring the change in load is an essential part of the power system. A sudden electrical deloading of the generator causes overspeeding of the turbine. The mains protection has a direct effect on the regulation of the generator and the relay reduces the steam power in case of protection. In this method, the load is measured by measuring three different voltages and three different currents for each phase.

Combined criteria

It is possible to apply a combination of the different criteria discussed earlier for mains protection and decoupling. For example, a short-circuit or excessive output load is an indication of an overcurrent and undervoltage condition. In this case, an overcurrent relay having inverse current-time characteristics is used wherein the characteristics are decided by the undervoltage relay. If the mains voltage drops, say below 60 per cent of the phase to neutral at a coincidental current which is greater than the pick-up value, a close short-circuit can be assumed and the relay starts operating. Selective shutdown realisation takes place through a relay located closer to the actual fault. Since this method is very slow, it is not advisable for quick mains decoupling at automatic reclosing.

Another example of combined criteria for mains decoupling is the detection of underfrequency and undervoltage conditions. The underfrequency element of the relay trips if due to the disconnection of the mains, the remaining generators get overloaded to such a degree that their speed drops considerably. This method is used as a reliable indication for decoupling during short-circuits.

12.4.5   Mains Decoupling Relays

The basic functions of the relay which disconnects the captive power plant supply from the grid in the event of grid failures are to:

    (i)   Provide over- and undervoltage protection

    (ii)   Provide over- and underfrequency protection

(iii) Supervise phase sequences

(iv) Monitor the phase-to-neutral voltages

(v) Decouple the generator in the case of mains failure

Digital mains decoupling relays are superior to conventional relays because of their following features:

(i) Compact size

(ii) Ability to measure RMS value

(iii) Extremely short response time

(iv) Ability to indicate faults via LEDs

(v) Use of digital data processing for measuring, which gives high accuracy

(vi) Scope for fine graded wide setting ranges

The relay is supplied directly from the measuring quantities itself by connecting it in the line as shown in Figure 12.10. The analog input signals of the voltages are connected to the protection device via terminals R, Y, B and N. Generally, the supply range of the relay is kept at +25 per cent of the phase-to-phase voltage. The highest value of the voltage is evaluated for over voltage protection while the lowest value of the voltage is evaluated for under voltage protection. The line and phase voltages are compared with their pre-specified values respectively. Tripping is indicated by flashing the LEDs meant for each condition, be it overvoltage or undervoltage.

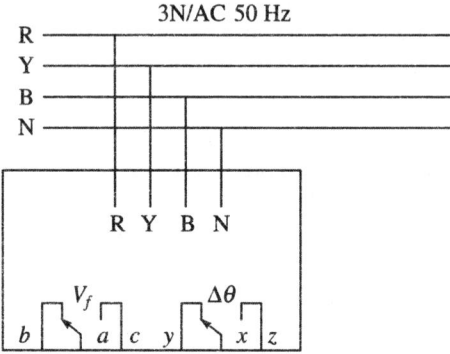

FIGURE 12.10   Connection of the relay with four-wire system.

Change of frequency relay

In digital relays, the evaluation of the cycle time is done for *frequency monitoring* and this measurement is independent of the harmonic effects. Independence in terms of the harmonic effects should thus be a strict criterion. In order to prevent tripping during normal operation due to voltage transients and phase transients, a fixed measuring repetition is therefore used. In the three-phase system, each of the phases is individually monitored. If at least one phase exceeds the set reference value, tripping takes place. Tripping is indicated by the flashing of the LEDs under each condition, be it overvoltage or undervoltage. Flashing of the LEDs also indicates the conditions of overfrequency and underfrequency.

A change of the frequency relay works on the basis of the following two separate measuring methods:

(i) Mean value detection method

(ii) Frequency gradient (df/dt) detection method

Mean value detection method: Suppose:

f_{th} = threshold value of the frequency

f_t = tripping value of the frequency also known as the second threshold value

If there is a change in frequency:

$$\Delta f = f_{th} - f_t$$

In the mean value detection method, the relay is activated only when the frequency reaches above or below the adjustable threshold value f_{th}. When this situation arises, a time counter (setting the value dt in cycles) starts. As the measured system frequency also reaches above or below the second threshold value that is the tripping value f_t, within time Δt, the relay trips immediately.

Tripping value depends on the setting df, f_{th}, and dt.

$$df = \frac{f_{th} - f_t}{\Delta t}$$

So

$$f_t = f_{th} - df \cdot \Delta t$$

The tripping value f_t must reach within time Δt, otherwise tripping will not take place.

Frequency gradient (df/dt) detection method: In this method, each new value of the cycle duration measured is detected via df/dt, the frequency gradient of the mains voltage. In order to evaluate the rate of change of frequency, the relays consider more than two values of the frequencies. Multiplex evaluation provides more accurate tripping. However, it is possible to parameterise the multiplex evaluation duration. Accuracy and speed can be adjusted optimally to the prevailing mains conditions. There should be provision for a high degree of tripping reliability and consequently, high stability during the switching actions. Short-term oscillations in the mains frequency created by the switching actions should not result in trippings.

The frequency gradient (df/dt) detection method is highly resistant to switching transients and associated short-term frequency oscillations.

Vector surge relay

The auto-reclosing of the mains is very dangerous for alternators. The returning mains voltage can hit the alternator in the asynchronous mode. In the case of mains failure, the *vector surge monitoring* provided by the relay protects alternators in a parallel operation. Short-circuits in the mains can also create conditions leading to the tripping of the relay because a relay can detect a vector surge higher than the pre-set threshold value. The value of the vector surge depends on the distance of the short-circuit from the generator.

The relay measures the cycle duration and at each zero crossing of the voltage, a new measure is started. The relay reproduces an internal frequency replica of the mains voltage to predict the time of the next voltage zero crossing.

In the presence of the vector surge (see Figure 12.9), the reference replica and the mains frequency are not identical and the next zero passing occurs either earlier or later. The angular shift of the mains voltage can be fixed on the basis of the difference between the reference time and the measured time. Generally, relays detect vector surges after a few milliseconds, say about 10 ms. Two or three successive angular measures give a tripping impulse. This relay therefore ensures high tripping reliability is possible and it is also fast enough to be used for quick mains decoupling.

The relay tripping adjustment is preferably done for either the one-phase or the three-phase vector surge. The vector surges are supervised by the vector surge functions of the relay in all three line-to-line voltages and phase conductor voltages simultaneously. In the event of a one-phase failure, i.e. a phase conductor-to-ground fault, the single-phase variant adjustment is chosen to be activated in order to trip the circuit. The three-phase variant adjustment is chosen if the three-phase failure implies that the complete loss of mains in all the three phases needs to be detected. False tripping due to angular difference must be prevented by disconnecting the respective mains sections instantly.

In a single operation, any load change generates a voltage surge. The vector surge operation is thus advisable for the parallel operation of mains and generators only. This means that the vector surge function must only be activated after the generator has been synchronised. Generally an active residual load of at least 15–20 per cent of the rated load is necessary for identifying the vector surge at the mains parallel operating generators. A failure with insignificant load changes will cause neither a vector surge nor a change in the load and frequency and this failure will not be recognised.

One example of the relay operation for a long transmission line closed loop network such as a meshed overhead line is shown in Figure 12.11.

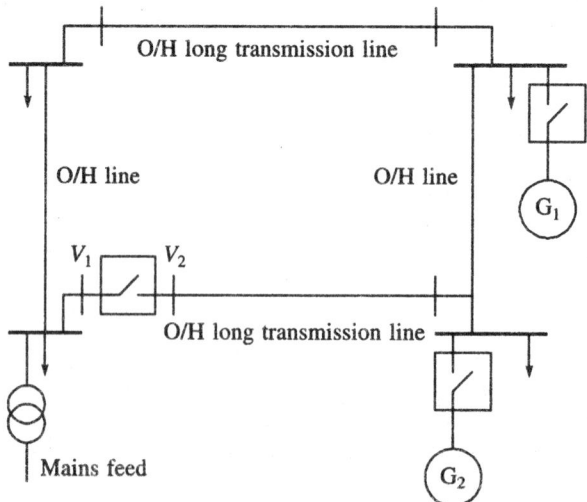

FIGURE 12.11 Interconnected long transmission network.

In this type of network, the consumers are spread over the entire ring, which may encompass a distance of 1000 km or even more. If the relay operates, i.e. the decoupling point opens, V_1 and V_2 turn by some angular degrees due to line impedance. Here, the vector surge relay connected to V_2 may trip due to these switching actions. The rate of change of the frequency relay is found to be more suitable for the loss of mains protection.

EXERCISES

1. Discuss the applicability of the optical fibre link for power line protection.
2. What are the merits of the optical fibre channel which enable it to be used as a communication link?
3. Compare the optical fibre pilot-relaying scheme with the general wire pilot relaying scheme.
4. Explain the working of the optical fibre pilot-relaying scheme with the help of a hardware connection diagram.
5. Describe the requirements and roles of various terminal equipments required in a fibre optic-based direct relay-to-relay digital logic communication scheme.
6. How does the use of microwaves help in power system protection?
7. Describe microwave direct relay-to-relay digital logic communication. How does this help in power system protection?
8. Explain the advantages and disadvantages of the microwave-relaying scheme.
9. Write notes on:
   (a) Microwave channel
   (b) Remote tripping
10. What are catastrophic failures?
11. Describe the various hidden failures in power systems.
12. Give reasons for unavoidable hidden failures.
13. Explain some important countermeasures used to prevent hidden failures.
14. Discuss how integrating the communication infrastructures helps in preventing hidden failures.
15. What is mains decoupling? Describe it.
16. Discuss the after-effects of mains decoupling.
17. Describe the electrical parameters that need to be carefully monitored for mains decoupling.
18. What do you understand by voltage vector surge? Explain with the help of a simulation circuit.
19. Write a note on the following relays that are generally used for mains protection:
   (a) Change of frequency relay
   (b) Vector surge relay

Chapter 13

Network Relays

13.1
INTRODUCTION

Since the very inception of power system development, the automatic detection and removal of faults have been the preconditions for operating a power system smoothly and reliably. The early methods of operating a power system were based on electromechanical devices and relatively complex schemes. The next step in the development of power systems pertained to solid-state relays, which were implemented for detecting faults through electronic circuits. Currently, fault detection is performed in numerical relays, which, in many cases, consist of a digital signal processor (DSP) with additional measuring circuits and output circuits. The future holds promise of further improvement in terms of the speed and consequent processing capacity of computers used in power systems.

The ability of advanced microprocessors and microcomputers to communicate with each other and the advent of broadband communication links have facilitated the sharing of information that can be used for purposes of supervision, control and protection of power systems. These advances in the manufacture of microprocessors and computers have laid the ground for the development of a wide range of network relays like adaptive protection relays, protection based on artificial intelligence, protection based on wavelet transform methods, wide area protection and hybrid protection. These relaying schemes have become popular as ultra-high-speed protection methods, which need to be incorporated in the present complicated grid system of the interconnected power system network.

Some of these recent advances and developments are briefly described below.

13.2
ADAPTIVE PROTECTION

13.2.1   Introduction

Adaptive protection is a protection philosophy which permits and seeks to make adjustments to protection functions in order to make them more attuned to prevailing power systems conditions. This implies updating of the relay settings, which represent a change in the power system topology. This change in power system topology is mainly due to the following two reasons:

(i) Deliberate system switching, for example, isolation of a part of the equipment for maintenance.
(ii) Unplanned system switching, for example, removal of a fault by a relay from a healthy power system.

In these cases, the power system topology gets changed but the system protection settings remain the same as they were before the modifications, which is why the latter may not be able to adequately protect the current system. This is the main basis for using the adaptive protection scheme. Once the system topology is changed, the settings of the system devices should be updated as soon as possible to protect the system against new faults. This can be achieved by applying the adaptive protection scheme.

Adaptive relaying can be utilised in the following areas:

(i) Automatic circuit-breaker reclosing control
(ii) Power transformer protection
(iii) Multi-terminal transmission line protection
(iv) Relay settings

Specifically, adaptive protection is applied to dynamically determine the pick-up current and time multiplier in overcurrent relays. These parameters can be determined through optimisation techniques. Continuous real-time simulations of the power system also need to be done to reconfigure the protection system. Thus, as and when there is a change in the power system, continuous real-time simulations help protection devices to get updated automatically. Adaptive protection systems are expected to perform the following functions:

(i) Relays should only operate for faults in their respective zone of protection, as defined by the direction that the relay supervises.
(ii) If both the relays sense a fault in their respective zones of protection, the relay with the shorter time delay or the relay subject to the larger fault current should trip first.
(iii) As the power system topology changes, the relay will automatically switch to the appropriate relay group for proper power system protection.

It is obvious that adaptive protection is an old concept, which is substantiated by the fact that directional relaying schemes adapt the direction of the fault current while time-delay overcurrent relays adapt their operating time to fault current magnitude. But these old concepts represent the permanent characteristics of a relay or relay system. These characteristics are part of original designs of the installations used to perform a given function.

The concept of this relaying is based on the fact that many relay settings are dependent upon the assumed conditions in the power system. Presently, it is not possible to realise these assumed conditions in complex interconnected power system networks for reasons of accuracy. The relay setting has to adapt itself to the real-time system as and when the system conditions change. The advantages and applications of adaptive protection schemes are shown in the Table 13.1.

TABLE 13.1   Applications and Advantages of Adaptive Protection Schemes

Applications	Advantages
Adaptive system impedance model	Improved reliability
Adaptive sequential instantaneous tripping	Faster back-up protection
Adaptive multi-terminal relay coverage	Improved zone settings
Adaptive zone-1 ground distance	Greater sensitivity to high resistance ground faults
Adaptive response to defective relaying equipments	Minimisation of the need for second pilot scheme
Adaptive reclosing	Faster restoration following incorrect trips
Variable breaker-failure timing	Improvement in back-up timing margins and elimination of unnecessary tripping of back-up breaker
Adaptive last-resort islanding	Facilitation of load restoration
Adaptive internal logic monitoring	Improved relaying reliability
Relay setting co-ordination checks	Optimisation of the co-ordination

13.2.2   Fault Analysis and Protection Determination

Faults can be analysed by applying them to various locations in a modelled power system and then recording the resulting fault currents and voltages. The results of this fault analysis help decide the determination of the level of protection. Generally, for simulation purposes, fault locations are kept at each bus and the mid-point of each line. The fault currents and voltages for each type of fault and location represent the fault analysis results. In order to determine the adaptive protection settings, fault analysis results are found for all the possible system configurations.

In order to realise the system protection, the relay settings are selected on the basis of the power system configuration and fault conditions, once the fault analysis has been completed. A table of different power system topologies consisting of associated relay settings is organised in such a manner as to ensure that the sets of relay settings for a particular relay location are similar. Each group of similar relay settings are analysed to arrive at a set of common settings which are useful for that group and to manage all situations that need to be covered.

The adaptive relaying scheme operates very rapidly and the relays can communicate with one another to indicate failures, suppress operation and suitably isolate a fault.

These schemes also facilitate communication between devices other than relays. Local or remote computers can interface with digital relays and allow humans to easily interact with the relay. Any digital aspect of the relay including current settings, operation logs and event-reports can be downloaded by computers, which can also carry out the requisite changes in the current settings. The details can be saved in a separate file for future reference and use. Relay-to-relay and relay-to-computer communication paths for a digital relay are shown in Figure 13.1.

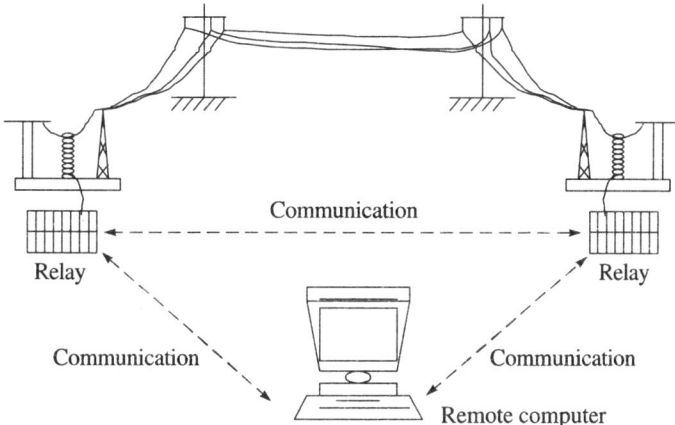

FIGURE 13.1   Relay share logic status in pilot logic communication.

13.2.3   Adaptive Techniques

The numeric techniques used for protective relays are based on protection algorithms, which make them adaptive to the power system. This protection algorithm increases the security and dependability of the protection. The adaptive techniques described here are the Adaptive Reactance Supervision technique and the Adaptive Polarising Memory technique.

Adaptive reactance supervision

The supervision of reactance characteristics is used to prevent operation under load conditions. One boundary of the quadrilateral characteristics of the reactance characteristic limits the resistive reach while using the resistive blinders. The setting of the supervising functions is done on the basis of the minimum load impedance which the relays are supposed to see in the service. The use of an mho distance function with an adaptive reach is shown in Figure 13.2. The application of digital relays facilitates modification in the supervising functions on the basis of the existing load flow on the power line. The maximum fault resistance for any given load is monitored by the adaptive function. The digital relays continually monitor the load impedance and adjust the reach setting of the supervising mho distance function to provide the maximum resistive coverage while maintaining a safe margin from the load impedance.

In Figure 13.2, the reach is limited by the minimum load impedance, which is plotted at point X. The resistive coverage in this case is limited by the expected maximum load flow.

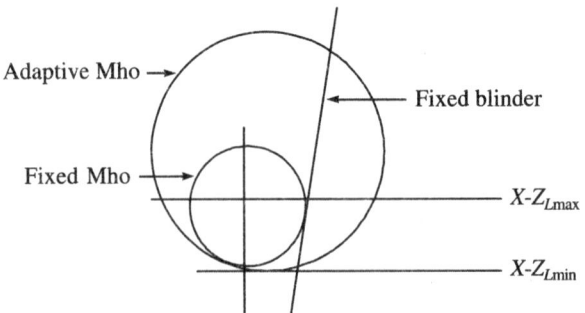

FIGURE 13.2   Adaptive reach using an mho distance function.

Adaptive polarising memory technique

The polarising voltage memory performs the following functions:

(i) Allowing the distance relays to operate for zero voltage three-phase faults at the front of the relay.
(ii) Preventing the relay from operating for zero voltage three-phase faults behind the relay.
(iii) Imparting a variable characteristic to the relay.

Unlike electromagnetic relays, in digital relays, memory time can be made adaptive. For example, if the positive sequence voltage is less than 6 per cent of the rated voltage during a fault, the relay continually uses the pre-fault memory voltage to polarise the distance functions. The pre-fault memory voltage is used until the positive sequence voltage increases to above 6 per cent. This change does not affect the performance of faults other than the three-phase faults, if the relay uses positive sequence voltage polarising for all distance units.

13.3
PROTECTION BASED ON ARTIFICIAL INTELLIGENCE

In a situation wherein the deterministic modelling for the non-linear system behaviour representation is difficult to achieve, an artificial neural network (ANN), commonly known as artificial intelligence (AI), provides a suitable solution. In order to determine the direction of the fault, direction discrimination can be achieved through a multi-layered feed forward neural network. The following four steps need to be taken in this scheme:

(i) Preparation of suitable training data
(ii) Selection of a suitable ANN structure
(iii) Training of the ANN
(iv) Evaluation of the trained network by using test patterns

Training data can be obtained by simulating the system on any software like EMTDC. The selection process of a suitable ANN structure necessitates the careful and appropriate choice of the number of layers, the transfer function, the number of neurons, and the number of inputs and outputs for an accurate identification of all fault conditions.

ANNs are being used in various industries and have proved to be vital tools in applications related to power systems. The main advantage of ANNs is that their usage does not necessitate a complete understanding of the system behaviour and they can therefore be used in extremely complex situations. However, the accuracy of the network depends on the size and accuracy of the test set. The training and testing of ANNs may take a long time.

The magnitudes of the three-phase voltage and current phasors can be utilised in an ANN-based protection relay to achieve fast and precise operation under different fault conditions and changes in the network. The back propagation algorithm may be used as a neural network works by adjusting the weights connected in successive layers of multi-layer perceptions. The ANN-based approach improves the security of the distance relays by extending the zone-1 reach.

13.4
PROTECTION BASED ON WAVELET TRANSFORM METHODS

Wavelet Transform is an operation that transforms a function by integrating it with the modified versions of some kernel function. The kernel function is called the mother wavelet, while the modifications are translations and compressions of the mother wavelet. This is a very powerful technique for the multi-scale representation and analysis of signals. Interestingly, Wavelet Transform (WT) is useful for the analysis of non-stationary signals. WT uses long windows at low frequencies and short windows at high frequencies. The property of the WT of localising the information in the time frequency plane enables it to trade one type of resolution for another, which, in turn, makes it suitable for analysing non-stationary signals. The mother wavelet must be admissible and a function $g \; \varepsilon \; L^2(R)$ is admissible if

$$c_g \equiv \int\limits_{-\infty}^{+\infty} \frac{|G(\omega)|^2}{|\omega|} \, d\omega < \infty$$

where $G(\omega)$ is the Fourier transform of $g(t)$, $L^2(R)$ is the set of all squares integrable or finite energy signals, and R denotes the real numbers. The constant c_g is the admissibility constant of the function $g(t)$, and the requirement that it is finite allows for the inversion of the wavelet transform. For a given function $g(t)$, the mother wavelet of the transform is assumed to be admissible. The mother wavelet may be orthogonal or non-orthogonal. The Wavelet Transform may be in the analog domain (CWT-Continuous Wavelet Transform) or in the digital domain (DWT-Discrete Wavelet Transform).

It is found that Wavelet Transform is applicable to the following:

(i) Power disturbance detection and localisation
(ii) Power disturbance data compression and storage
(iii) Power disturbance identification and classification
(iv) Power devices protection and power disturbance network/system analysis
(v) Identification of the transients on the power system

One way to use Wavelet Transform is through the implementation of MRA (Multi Resolution Analysis) by filter banks in which the analysed signal is filtered through a sequence of filter banks so that the result is a set of signals in the time domain. Each set represents a band pass filtered version of the original signal.

13.5
WIDE AREA PROTECTION

The present generation relays are based on the advancing technology of protective relays. New technology also has communication capabilities. This is why new relays are called Intelligent Electronic Devices (IED). The high-speed communication links allow the present generation relays to quickly update their status to each other or to a central control station. Wide area protection technology advancements and the synchronising capability of GPS are used in the Wide Area Back-up Protection Algorithms, which represent a hope for preventing blackout situations.

The Wide Area Current Differential Back-up Protection Scheme provides a higher selectivity than the conventional back-up protection which employs distance relays. ATM transmission networks and time synchronous systems may be utilised by the protection system configuration of the Wide Area Current Differential Back-up scheme. For long transmission line back-up protection, GPS synchronised current measurements may be taken at the ends of the line. The current measurements of the transmission lines, transformers and each busbar are required for operating the current differential protection scheme. The Wide Area Back-up Protection Scheme can also be very useful for preventing cascaded outages in power systems. This scheme may be employed to:

(i) Determine the precise location of a fault to facilitate the functioning of only circuit-breakers meant for isolating the fault, which results in tripping.

(ii) Prevent unnecessary trippings due to hidden failure or overloading, by blocking the trip signals of conventional back-up protection relays.

WABPES (Wide Area Back-up Protection Expert System) is used now-a-days for preventing cascading trips. WABPES, a wide area communication network based on a protection system, is designed to protect a region of the network by facilitating selective and secure back-ups. Its two modes are:

(i) Normal mode, and

(ii) Emergency mode.

WABPES monitors the closed/open states of circuit-breakers and the operational response of conventional protection relays on the network during the normal operation. In the event of occurrence of a fault, WABPES changes the concern relay to the emergency mode after which the system decides the best way to isolate the fault if the main protection has failed.

A conventional back-up protection relay may be blocked by WABPES, if necessary.

13.6
HYBRID PROTECTION

A hybrid relay is a combination of more than one relay concepts. It utilises both the impedance calculated from the fundamental components of the voltage and current signals, and the travelling wave information derived from high frequency transient signals. In the hybrid protection scheme, if the fault is too close to be detected by the travelling wave scheme, the impedance relay acts as a fast back-up. This enhances the reliability of the combined scheme. The hybrid protection algorithm operates in parallel and needs high-speed processing. This necessitates an inter-trip signal for faults, which occur on the line outside the maximum length of the protection zone. The maximum protection length depends on the accuracy of the distance estimate. Modern DSPs (Digital Signal Processors) and ADCs (Analog-to-Digital Converters) can handle various measurements and extensive calculations involved in the hybrid protection algorithm, which is obviously a complex algorithm. The accuracy of the relaying scheme depends on the processing power of the DSP.

The calculation of impedance and the derivation of the travelling wave information have to be carried out simultaneously on the basis of the measured voltage and current signals. The operation of the travelling wave relay depends on the high-frequency transient signal, which is why 50 Hz or 60 Hz main frequency components need to be filtered out from the measured waveforms. At the same time, since the impedance relay uses the fundamental frequency component, the high frequency noise has to be removed by filtering it out.

Since the travelling wave signals contain high-frequency transients, the voltage and current signals have to be sampled at a high sampling rate. The frequencies in the transient signals vary depending on the fault location and other line conditions.

For a distance relay based on impedance measurement, the impedance relay does not need such a high sampling rate because in this case, only the fundamental frequency component is worthy of interest. While cascading two algorithms, the sampled values used for the impedance algorithm may create noise in the travelling wave algorithm and vice versa. Thus, in order to prevent the sampling of identical signals at two different frequencies, the samples of voltage and current obtained for the travelling wave relay are decimated to achieve a lower sampling rate for the impedance relaying. In order to prevent aliasing, the sampled signals are specially filtered by using an anti-aliasing filter with a suitable cut-off frequency. This has to be carried out before decimation. One-cycle DFT (Discrete Fourier Transform) is capable of calculating the fundamental components, while rejecting any harmonics of the power frequency.

As far as the operating principle of the hybrid algorithm is concerned, the distance protection scheme based on impedance measurement moves rapidly into the trip zone for close-up faults. There is a need for speeding up towards the reach point of the relay. The travelling wave relay is capable of detecting faults close to the reach point of the impedance relay very quickly. It is also capable of detecting faults beyond the reach point depending upon the maximum distance setting of the travelling wave-based algorithm. The accuracy of the measuring transducers decides the maximum settings of the distance, which can also be calculated accurately. The output of the two algorithms combined in different configurations satisfies various protection requirements, but at the same time, there is a need for improving the reliability and speed of the overall scheme.

13.7
SCADA

13.7.1 Introduction

The acronym SCADA stands for Supervisory Control And Data Acquisition. This does not provide a full control over the system, but mainly focuses on the supervisory level. SCADA is a software package positioned on the top of the hardware modules to which it is interfaced. One of the popular hardware modules is the Programmable Logic Controller (PLC). The advantages of SCADA systems are their functionality, openness, performance and scalability.

Hardware architecture

The typical hardware architecture of a SCADA system consists of two basic layers, namely the *Client Layer* and the *Data Server Layer*. The client layer is intended to facilitate man-machine interaction. The data server layer handles the process data control. The data servers communicate with the field devices through process controllers, which are connected to the data servers in the network either via field buses or directly. The typical hardware architecture is shown in Figure 13.3. Data servers are generally connected to each other and to client stations via LAN (Ethernet).

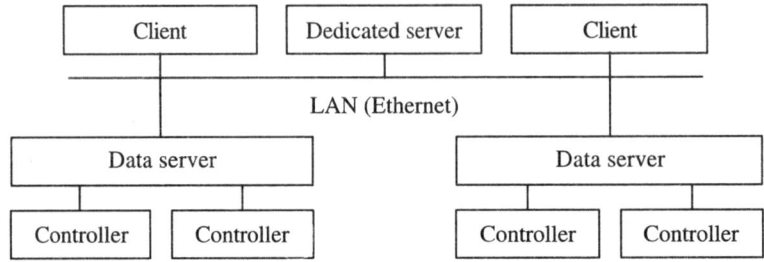

FIGURE 13.3   Typical hardware architecture.

Software architecture

SCADA software is based upon a real-time database (RTDB), which is available in different servers. Data acquisition and its handling on a set of parameters are undertaken by the servers. Data handling includes alarm checking, polling controllers, logging, archiving and calculations.

Functionality

In the SCADA system, the availability of functions like the access control of specific user groups, editing involving zooming, re-sizing, scrolling, on-line configuration, customisation and navigation makes it easy to operate the protection and control operators.

The other functionality of the SCADA system includes trending, alarm handling, logging, archiving and report generation. The majority of the SCADA products allow the action to be automatically triggered by events. Generally, SCADA products provide the scripting language whereby the actions to be taken are defined automatically.

Communication

Internal communication and access to devices are required in the SCADA system. Server-client and server-server communication, in general, fall under internal communication. A parameter owned by a particular server subscribes to the client application and only changes to the specified those particular parameters are then communicated to the client application.

Controllers are polled at a user-defined polling rate by the data servers during the process of access to communication devices. The parameters are polling rated differently. The time stampings typically performed in the controllers are taken over by the data server. The controllers pass the requested parameters to the data servers. Software providers provide communication drivers for the programmable logic controllers.

Scalability

Scalability implies the possibility to extend the SCADA-based control system by adding more process variables, more specialised servers and more clients. Each data server with a distinct configuration database and RTDB remains responsible for the handling of a sub-set of the process variables such as acquisition, alarm handling and archiving.

Interfacing

The required configuration data are stored in a logically centralised and physically distributed database. The real-time database (RTDB) resides in the memory of the servers. The following inputs are required for the interfacing:

(i) An open database connectivity (ODBC) interface to the data in the archive/logs

(ii) An ASCII import/export facility for configuration data

(iii) A library of API supporting C, C++ and Visual Basic to access data in RTDB, logs and archive

Evolution

The widespread development of industrial applications has led to the advent of revised versions of SCADA for handling devices and systems, even as vendors release new versions of SCADA with additions and modifications once a year. Facilities for on-line modifications to the configuration database and the graphics are generally available on SCADA systems. The development tools provided in these systems include a graphic editor, a database configuration tool, a scripting language, an Application Program Interface (API) supporting C, C++, VB and a driver development toolkit to develop drivers for hardwares.

13.7.2 Use of SCADA in Interconnected Power Systems

The rapid increase in electrical power demand worldwide, is compelling power suppliers to increase the size of power generating units. This necessitates the interconnection of different load centres located in different areas. Since the power system network is a highly complicated network, power engineers are also faced with a number of challenging tasks including real-time data recording, communication networking and man-machine interfacing. In the present competitive international scenario, electric power suppliers are also being forced to operate the system economically within the required safety and security limits.

Instability and voltage collapse constitute major concerns for electric energy providers, which can be resolved by stabilising the multi-area interconnected power system network. Additional advance features available in the Energy Management System (EMS) and SCADA systems are also used to provide system-based protection and control.

A block diagram showing a structure of the EMS/SCADA system is shown in Figure 13.4. This consists of different modules including *Generation Control and Scheduling Module, Network Analysis Module and Operator Training module.*

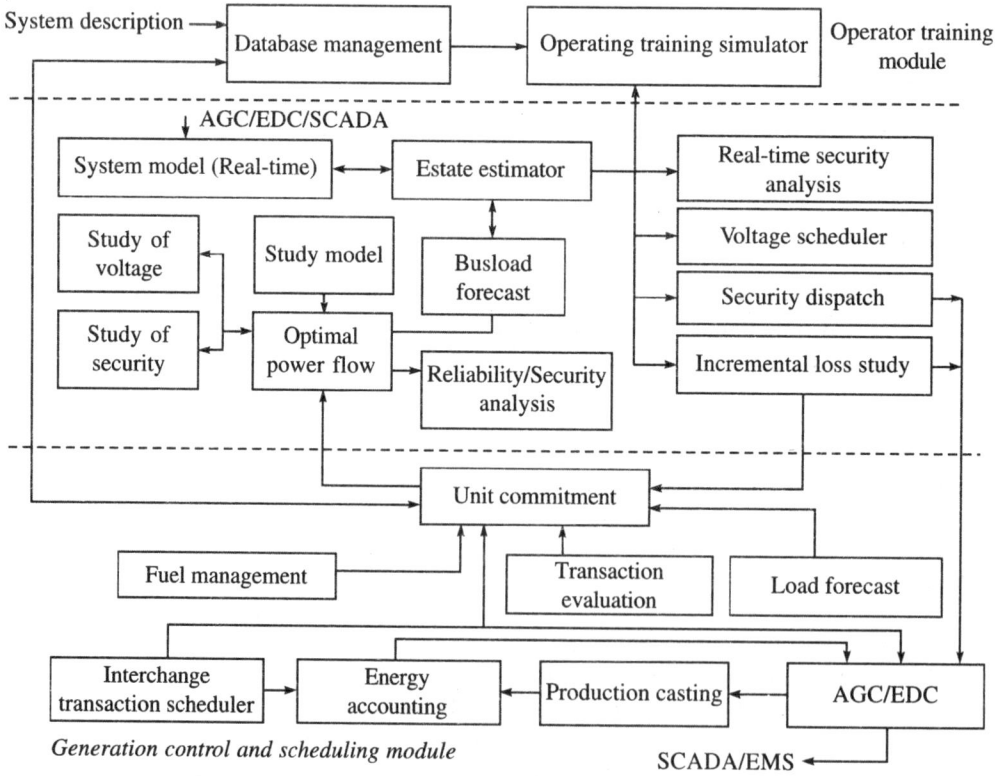

FIGURE 13.4 Various modules of the SCADA system.

The key to managing and scheduling the economic generation of an interconnected power network is the control of generation. Here, system frequency and tie line power flow are required to be maintained within prespecified limits as per system capacity. The generation control and schedule module deals with the determination of the most economic output of each generating unit for a given load, ON/OFF scheduling and generator commitment in order to meet the varying load demand. This module also deals with the pricing determination and the purchase and sale transactions with other power system networks.

Several probable faults may lead to outages in generating units, and in the transmission and distribution networks of a highly complex power system network. Proper analysis is essential to keep the system reliable and secure. The EMS (Energy Management System) must be capable of analysing the number of possible outages. Proper analysis of events in the systems ensures that the information is automatically transmitted to the operator who can then devise the best possible strategy to prevent losses in generation and voltage dipping problems. This analysis is carried out through the Network Analysis Module by monitoring several telemetered values and estimation of the effect of any possible outage on the operation of the power system network.

13.7.3   SCADA System Architecture

The data acquisition section of SCADA gathers the telemetered data, which are used by all the other functions within the Energy Management System (EMS). The SCADA system provides support to the system operator while controlling the locally or remotely placed equipments. The process of controlling of the equipments includes the closing and opening of the circuit breaker with security features like authorisation and "Select-Verify-Execute". The SCADA systems perform the following three critical functions:

   (i) Data acquisition
  (ii) Supervisory control
 (iii) Alarm display and control

These functions are discussed in detail below.

Data acquisition

SCADA has a sub-module called the *Data Acquisition Function*, which collects data in either the raw form or the processed form from remotely placed terminal units at definite time intervals. The five functional units of data acquisition are detailed below.

1. The *data collection unit* periodically collects data from Remote Terminal Units (RTUs) at the appropriate rate and monitors various scans to ensure that they operate during the proper time period.

2. The *data processing unit* converts raw data (analog value) into digital data. It also converts the system status points like '1' for CLOSED state and '0' for OPEN state. This unit also handles the data received from data links to other computer systems.

3. The *data monitoring unit* interfaces with the alarm processor and notifies it when the devices change their state values exceeding operating limits. This unit also provides the dead-band and return-to-normal features.

4. The *special computation unit* calculates various parameters like MVA from MW and MVAR, MVA from kV and amperes, and ampere from MVA and kV, and performs the other requisite calculations as and when required. These values are derived from the scanned data in the known database.

5. The *scan configuration control unit* works when a sustained communication error occurs. It removes the terminal unit from the scan or switches the channel assignment. This unit also re-establishes the terminated communication with the remote terminal units within a specified time interval.

Supervisory control function

The operations of the supervisory control unit follow multi-step procedures. The first step is the selection of the device to be operated. During the second step, visual verification is undertaken. The third or final step involves operator execution or cancellation. In this way, the supervisory control function allows the operator to control the remote devices and to replace or recondition the values in the database.

Conditioning the data involves various operations including the replacement of the telemetered data, enabling of the alarm inhibitor, reversing the state of the device, bypass entry, etc.

Alarm display and control function

This unit gives out the alarm signals to the operator. The alarm presentation includes the following:

- Construction of the alarm message
- Organisation of the alarms in categories
- Alarm summary display and its maintenance
- Maintenance of the console logs
- Initiation of audio/visual enunciators and interfacing with the other functions
- Assigning of priorities to the alarm messages
- Identification of the points which are inhibited from the alarm or originally replaced by the operator
- Acknowledgement of the alarm

In order to be able to perform some of the above functions, the EMS/SCADA system requires following sub-systems:

(i) User interface sub-system
(ii) Communication sub-system
(iii) Information management sub-system

13.8
COMPUTER-BASED PROTECTION

The invention of microprocessors during the beginning of the 1970s revolutionised the working of the technical world. Soon after the invention of microprocessors, researchers started developing microprocessor-based power system relaying schemes. These relays started working on numbers representing the instantaneous values of the fault signals. the advancement of research to the next stage during the late 1990s led to the advent of extremely fast relaying schemes based on travelling waves and microwaves. These relays work with the assistance of software specifically designed for the purpose. Thus computer-based relays have evolved from a torque-balancing device to a programmable information processor.

The power system fault signals are of an analog nature. After taking the fault signals through the CTs and PTs, the signals are passed through an analog type low pass filter. Filters are required to ensure that the fault signal does not contain the frequency component which has a frequency greater than one half of the sampling frequency. This is required because digital processing can take place only after the frequency spectrum of the signal is properly shaped.

During the conversion of the signal from the analog to the digital form, the analog signal must be kept constant. For this purpose, the analog signal is put to the sample and hold circuit after filtering. Without the sample and hold circuit, it is difficult for the analog-to-digital converters to handle the range of frequency. During the next stage, the sample and hold value is passed on to the analog-to-digital converter. Multiplexers are used before the analog-to-digital converter in order to accommodate a number of input signals. The computer processor controls the action of the sample and hold circuit along with the analog-to-digital conversion with the help of the end-of-conversion signal issued by the analog-to-digital converter. The output signal of the analog-to-digital converter represents the digital form of the instantaneous value of the fault signal, which is suitable for computer processors. The resolution of the analog-to-digital converters depends on the number of bits used to constitute the width of the output port. It may be 16-, 32-, or 64-bit wide or even wider.

The digital values of the signals are then stored in the RAM of the processor. The relay software processes these digital signals representing the fault signals in accordance with the relaying algorithm. The processor responsible for issuing the trip signal issues the trip signal on one of the bits of its output port. This trip signal is kept compatible with the trip coil of the circuit-breaker. The computer processor is also used to facilitate communication with the other relays or other interconnected supervisory computers. Computer-based relaying schemes are user-friendly and enhance the operator's confidence. This also improves the overall efficiency and reliability of the protection system. A block diagram showing the algorithm for computer-based protection is shown in Figure 13.5.

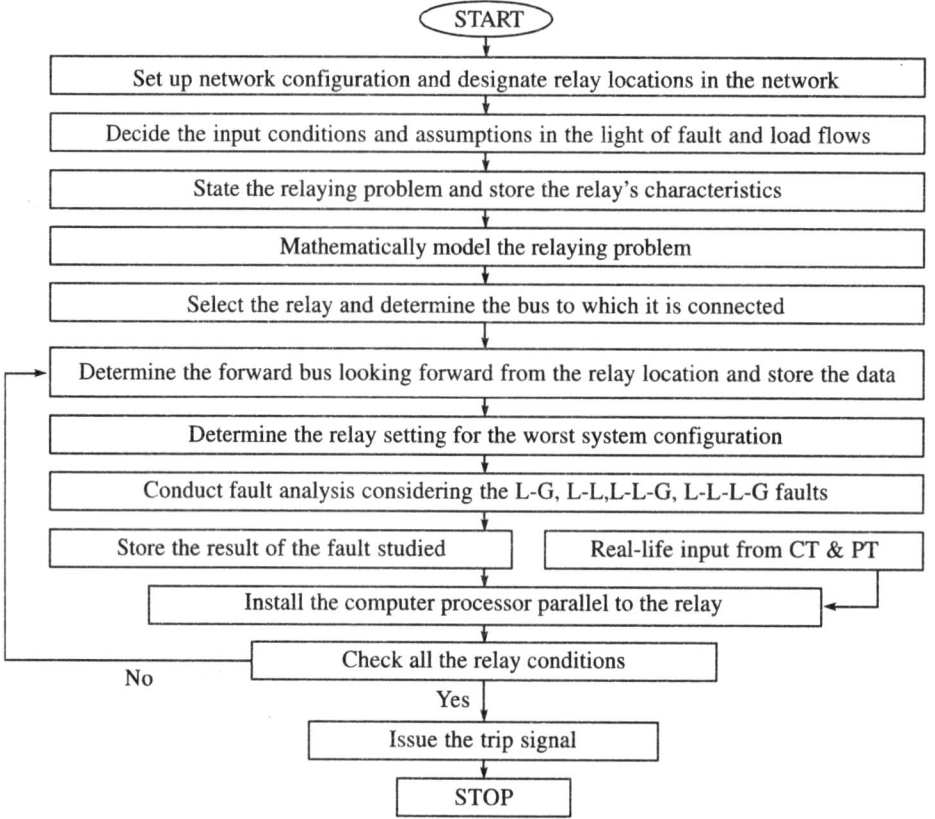

FIGURE 13.5 Algorithm for computer-based protection.

EXERCISES

1. How important is high-speed relaying in the present power system network?

2. Describe how the present adaptive protection schemes differ from the older ones.

3. Write a note on protection based on artificial intelligence.

4. Write a note on protection based on wavelet transform methods.

5. Write in detail about the need for the wide area protection scheme in the present scenario.

6. How is hybrid protection useful in improving the speed and reliability of the modern power system?

7. Describe the configuration and facilities available in the EMS/SCADA systems.

8. Explain how the SCADA system is used for power system application.

9. Write a note on computer-based protection of the power system.

Chapter 14

Surge Protection and Insulation Coordination

14.1
INTRODUCTION

Overvoltages can be of small amount and show adverse effects on the performance of the electrical equipment, apparatus and installations. Overvoltages can cause insulation failure in the power system network and can be appeared across any component of the network such as alternators, transformers, circuit-breakers, protective instruments, measuring instruments, control devices, insulators, transmission lines, etc.

It is very important for designers and power system engineers to take suitable measures and make proper arrangement for limiting the overvoltages and keeping within acceptable limits. Following are the causes of overvoltages:

Internal causes

    (i) Insulation failure
   (ii) Resonance
  (iii) Arcing ground
  (iv) Switching surges

External causes

    (i) Lightning
   (ii) Tree fall on transmission lines, etc. causing short circuits

There are three types of overvoltages which appear in the power system networks:

(i) Power frequency overvoltage

(ii) Switching overvoltage

(iii) Lightning overvoltage

14.1.1 Power Frequency Overvoltage

Power frequency overvoltages does not have damaging effects like switching or lightning surges, but is harmful, if sustained for longer duration. This occurs due to:

(i) Ground faults

(ii) Sudden load rejection

(iii) Ferranti effect

(iv) Loose connections

(v) Sudden changes of load

(vi) Resonance and ferro-resonance

Generally power frequency overvoltage does not exceed the double of rated system voltage.

14.1.2 Switching Overvoltage

Switching overvoltage is also known as *switching surge* or *overvoltage transient*. Sudden rise of voltage for a very short duration in the power system network is known as transient voltage or voltage surge.

Inductors and capacitors are energy storage devices. Instant change in energy stored in these devices is not possible. Actually, redistribution of stored energy requires finite time. The intermediate period between two steady states is known as *transient period*.

An electrical transient appears, if there is sudden change in the state of energy in the power system network. This sudden change may take place due to the following:

(i) Closing a switch

(ii) Opening a switch

(iii) Occurrence of fault in the system

Transient takes place to maintain the continuity in the stored energy in the storage elements. This happens for very short duration but the current through an inductor and voltage across a capacitor cannot change abruptly. Transients are for short duration but carry excessive current or voltage which can damage the system component. Mainly switching overvoltage occurs in the power system because of:

(i) Ferro-resonance phenomenon

(ii) Disconnection of an unloaded transformer

(iii) Opening of a circuit-breaker to clear a fault

   (iv) Line energisation and re-energisation

   (v) Fault clearing

   (vi) Load rejection

General practice to control the switching overvoltage is to insert a resistor between the contacts while switching off the circuit. Resistor is not required after few cycles and withdrawn.

Draining off the magnetic and electric energy before switching off the circuit is also advisable as an effective mechanism to control the switching overvoltages.

14.1.3   Lightning Overvoltage

Hitting by the lightning is explained as return of the stroke, which carries high current and originates from the earth, if stepped leader terminates at the earth during discharging process of the cloud charges. Stepped leader may also terminate at a tall object. The tall object is known to be hit by lightning. Conductor or transmission line tower experience a direct lightning stroke, if return stroke starts from the line conductor or tower. Whenever there is an induced stroke or direct stroke on the line, it causes direct connection to the transmission line or completes a closed mutually coupled circuit with the line. Following actions of the lightning stroke generate transients:

   (i) Direct stroke to phase conductor

   (ii) Back flashover

   (iii) Stroke to earth very close to line

Figure 14.1 shows waveform of a lightning surge.

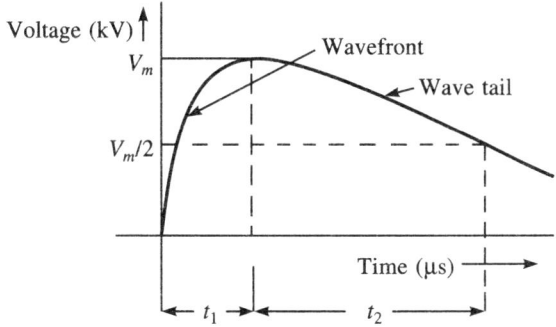

FIGURE 14.1   Waveform of a lightning surge.

A considerable number of outages in power system are caused by direct lightning stroke on the transmission lines or associated components. The main aim of a power system engineer should be:

   (i) To design the system in such a way so that chances of direct stroke can be minimized.

   (ii) To keep the amplitude and steepness of the lightning overvoltages minimum.

14.2
CLASSIFICATION OF INSULATIONS AND THEIR FAILURE MODES

External insulation

- Line insulators
- Bushings
- Tower air clearance, etc.

Internal insulation

- Insulation used in transformers, reactors, machines, etc.

Self-restoring insulation

- Insulation which completely recovers insulating properties after a disruptive discharge.

Non-self-restoring insulation

- Insulation which looses insulating properties after a disruptive discharge.

14.3
IMPACT OF CONTAMINATION ON ELECTRICAL INSULATION

With the industrial growth, on one side power demand is increasing day by day and on the other side pollution is becoming severe and severe throughout the world. The increase in demand of power necessitates going in for high voltage transmission due to economic reasons. As the transmission voltage level increases, the insulation contamination poses a major problem to maintain the reliability of power supply. Earlier, it was felt that insulation performance in higher voltage system was dictated by its switching surge behaviour. However, it has now been realized that the constraints are mainly imposed by pollution.

Insulation protects against atmospheric as well as switching overvoltages and works as a component at first line of defense to protect the costly generators, transformers and other installed equipment in substations, and other types of factories using electrical machineries and electricity.

Generally, the pollution on a solid insulation surface, i.e. housing of the surge arresters or body of the insulators does not cause serious problem under fair weather because in that condition, layers are substantially non-conducting. Lightly wet layers start conducting significantly. Due to Joule heating, there will be uneven drying up of the wet layers. Thick layers can have, under wet condition, a very different conductivity than the superficial layers. So, it is obvious that drying of the wet conducting layer will also be non-uniform. Due to non-uniform drying of layers, current gets constricted to a few narrow paths, where, there may be intense heating giving rise to hot moisture, formation and growth of dry bands. The voltage across dry bands may rise to a very high value leading to arcing. Continuous arcing sometimes causes complete flashover across the insulation and loss of its insulation property.

Insulations near a cement factory face a little different types of problems. The cement particles, when come in contact with the humid air, start setting in air itself. Before, they are permanently set, these cement particles are carried by winds and thrown on insulation surface and moist cement particles deposited get permanently set on the insulators themselves. These deposits are not soluble in water and rain.

14.3.1   Causes of Pollution

The worldwide power generation largely depends on the combustion of fossil fuel, i.e. coal, which produces major pollutants during its combustion like fly ash, SO_x, NO_x, CO and CO_2. These are the gases responsible for acid rain.

According to IEC Publication 71-2: *Insulation Coordination Application Guide*, pollution severity of sites has been divided into four levels given below. For each level of pollution, a rough description of some typical corresponding environments is made available. These pollutants and contaminants become conductive in the presence of fog, dew, rain and snow. With concentrating effects of pollutants on electrical transmission system, corona also becomes more effective on the line in the presence of conductive pollutants. These pollutants speed up the ionization processes around the conductor resulting in increased loss of energy. Early corrosion on malleable iron caps, forged ball pins and stem of insulators, surge arresters and other equipment take place in the presence of pollutants under adverse atmospheric conditions like rain, fog, ice, etc.

Following are the pollution levels and examples of a typical environment:

Light

- Areas without industries and with low intensity of houses equipped with heating plants.
- Areas with low density of industries or houses but subjected to frequent winds and/or rainfall.
- Agricultural areas[*].
- Mountainous areas.

All these areas must be situated far from the sea (10–20 km) and must not, in any case, be exposed to winds from the sea.

Medium

- Areas with industries not producing particularly polluting smokes and/or with average density of houses equipped with heating plants.

---

[*]Use of fertilizers sprayed by gun or by plane and burning of crops can lead to a higher pollution level.

- Areas with high density of houses and/or industries but subjected to frequent clean winds and/or rainfall.
- Areas exposed to winds from the sea, but not too close to the coast (at least a few km).[**]

Heavy

- Areas with high density of industries and suburb of large cities with high density of the heating plants producing pollution.
- Areas close to sea or, in any case, exposed to relatively strong winds from the sea.

Very heavy

- Areas generally of moderate extension subjected to conductive ducts and to industrial smoke producing particularly thick conductive deposits.
- Areas generally of moderate extension, very close to the coast and exposed to the sea sprays or to very strong and polluting winds from the sea.
- Those desert areas that are characterized by no rain for long periods are exposed to strong winds carrying sand and salt and subjected to regular condensation.

14.3.2   Remedial Measures to Avoid Surface Contamination

The main objective of the field engineers and researchers has been to know what economically and technically sound measures should be taken to avoid insulation failure. The common way of determining the behaviour of external insulation is to carry out artificial pollution tests, as representative as possible of the actual pollution. During the pollution test methods based on total flashover need a costly power source and lead to a considerable maintenance cost. So, the most economic method of assessment of pollution is by investigating the physical processes of the pollution flashover, either on model insulators or using computer modelling techniques. At the same time on the principle of 'precaution is better than cure' remedial measures should be adopted to prevent flashover/failure.

The extension of industrial areas and cities can contribute to increase in the different types of pollution for sites. Transmission and distribution lines have to pass through areas of this pollution. The following primary remedial measures to minimize accumulation of the contaminants and flashover on the surface of insulators, surge arresters and bushings are generally adopted.

Manual wipe cleaning

In this method, first a wipe cleaning with a non-fluffy cloth is done, then the same insulator is cleaned with water and finally wipe is cleaned with dry cloth at regular intervals. This is recommended for very light pollution only.

[**] These distances from the seacoast depend on the topography of the coastal area and on the wind condition.

Recoating of silicon greasing

The silicon greases are composed of a liquid phase, a solid phase and additional chemical agent to protect the coating from ultraviolet rays. The mechanism of action of these compounds has been divided into following three steps:

(i) Oil exudation

(ii) Encapsulation of the particles of contaminant

(iii) Gradual saturation of coating

Ultraviolet rays make the exuded oil hard on the surface and decrease the trapping efficiency, which can stop before the grease is saturated.

Since there is no self-cleaning on coated insulators, so after some time, leakage current can decay the silicon grease. This decay appears as whitish trails on the coated surface. The coating must be renewed before long whitish trails are observed. RTV (Room Temperature Vulcanized Silicon Rubber) coating is easier to apply as compared to silicon grease and can be hot washed.

Fixed hotline washing

The direction, intensity and duration of the wind influence the rate of building of pollutants on the insulator surface to a large extent. If rainfall of sufficient intensity occurs at intervals less than the period in which the built-up of contaminants on the insulators reaches the critical values, the rain acts like natural hotline insulator washing system.

The only way to prevent the building-up of the contaminants to the critical limits is to clean the insulators before these limits are reached. It can be done by using automatic hotline insulator washing system. The housing of energized conventional and metal oxide arresters can be cleaned without service interruption by applying a suitable flow of high resistance hot water. The system of high resistive water supply consists of pumps, distribution pipes and nozzles. Pollution sensors, differential anemometers and sequential control systems make it an automatic system. The two commonly used fixed hotline washing systems are as follows.

The *Spray System* uses fixed nozzles around each insulator. It is efficient only if the wind velocity is low and does not divert the spray water.

The *Screen System* consists of a number of nozzles arranged in a line producing a vertical water curtain in front of the equipment. The wind blowing from the opposite direction diverts the water jets towards the insulator. At the same time, the water screen dissolves part of the salts brought by wind and reduces the pollution intensity. During long-lasting heavy winds, fast building-up of contaminants takes place on the insulator surface. For such situations, the screen washing method is more effective. The capital and operational costs of hotline washing installations are very high.

Mobile manual hotline washing

This system comprises a stainless steel tank of required capacity in which demineralized water of conductivity of 0.5 to 1.0 micro mho per cubic centimetre and carbon dioxide gas of commercial quality are mixed. Out of the five nozzles on the top of the tank, one nozzle is connected to the CO_2 cylinder through a flexible hose and a regulator. A pressure gauge, a relieve valve set and a drain valve are each mounted on the separate nozzles. To the remaining nozzle, a long flexible hose is connected and the other end of this is connected to an insulated spray stick. This stick has a controlling lever at the bottom from where the flow of spray on the other end can be controlled.

The CO_2 gas is partially dissolved in water under pressure. Due to acidic nature of the fluid ($H_2O + CO_2 = H_2CO_3$), the conductivity of the fluid that passes through the hot spray stick increases. The dissolved CO_2 tries to escape from the water at the outlet of nozzle and hence the spray is in the minute droplets and not as a continuous jet, thus increasing the resistance of the washing column of jet. CO_2 is a very good fire-extinguishing agent also. The conductivity of the fluid at the outlet of the nozzle should not be greater than safe values such that leakage current passing through the operator while doing hotline washing is limited to five microamperes.

Wind deflectors

Wind deflectors are used near critical equipment. They do not provide effective and adequate protection to the insulators of ground equipment such as isolators, circuit-breakers, potential transformers, etc. due to the fact that the wind direction keeps on changing continuously. The height of the structure (erected in M.S. angles and asbestos sheets) normally does not increase from four metres as it may pose a clearance problem with energized lines. It has not been found feasible and cost-effective.

Green belts

Green belts means land area specified for plantation of long live trees. Green vegetation of sufficient height along the factories and coastlines avoids deposition of contaminants on the electrical insulation near the chemical and smoke-producing factories and salt deposition on the insulators near the seashore. Green belts also minimize deposition of contaminants and salt on the switchyard equipment.

Semi-conducting glazing

In order to attain the necessary power frequency insulation strength during wet contamination, insulator is provided with semiconducting glaze. Power dissipated in semiconducting glaze raises the surface temperature several degrees above the ambient. It prevents the fog condensation on the surface. In wet and contaminated conditions, it helps in the formation of very large dry bands or in the process of drying of the insulator surface.

Dielectric glaze in which barium titanate is used as a major constituent is also in use. Barium titanate increases the capacitance of the insulator as a whole, which as a result linearizes the voltage distribution across the leakage path, increases withstand/flashover voltage suppresses the transients during flashover, reduces corona loss and radio noise. This also improves glaze stability.

14.4
INSULATION COORDINATION

14.4.1   Objectives

Selection of electric strength of the equipment must be made for any system voltage so that any failure would be confined to the place on the system where:

- It would result in the least damage
- Be the least expensive to repair
- Cause the least hazard to the continuity of service

Definition of the insulation coordination as per the International Standard (IEC) is:

Insulation coordination refer to the selection of dielectric strength of equipment in relation to the voltages which can appear on the system for which the equipment is intended and taking into account the service environment and the characteristics of the protective devices.

So, insulation coordination is correlation between insulation provided to the electrical equipment and characteristics of the protective device so that protective device protects insulation against overvoltages.

A definite common insulation level for all the equipment in an electrical station is first established and insulation level for all the equipment and components is brought above this level. This insulation level is known as *Basic Insulation Level* (BIL). So, for proper insulation coordination, following three criteria should be satisfied:

1. Selection of suitable BIL
2. Breakdown voltage of all the insulation should always be higher than BIL
3. Selection of suitable protective device capable to provide protection

Margin assessment between BIL of the applied insulation to be protected and the protective level depends on the following:

(i) Shielding of the electrical substation
(ii) Location of the lightning arresters
(iii) Discharge voltage of the lightning arresters
(iv) Breakdown voltage of the lightning arresters
(v) Insulation level of the transmission/distribution level

Generally, a margin of 15% of BIL is provided for switching surge protection and 30% for lightning surge protection. Volt–time characteristics of insulation coordination are shown in Figure 14.2.

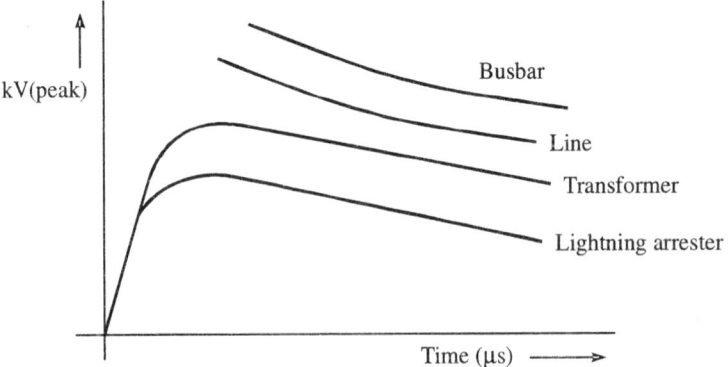

FIGURE 14.2   Volt–time characteristics of insulation coordination.

14.5
INSULATION PROTECTION

Overvoltage either power frequency overvoltage or transient overvoltage damage the insulation. One can say that insulate the system and equipment to withstand any voltage that can ever appear in the system. Adapting such view makes the system economically unviable. According to the economic view point, accepting all outages due to transient overvoltage, insulate the system for normal system voltage only. This reduces reliability of the supply and increases repair and need for replacement of the equipment. So, power system designers and planners make a judicious compromise between these two extreme unrealistic proportions. Reliability requirements and importance of the system decides to what extent does the system need to be insulated.

The protective device is generally installed parallel to the insulation to be protected as shown in Figure 14.3.

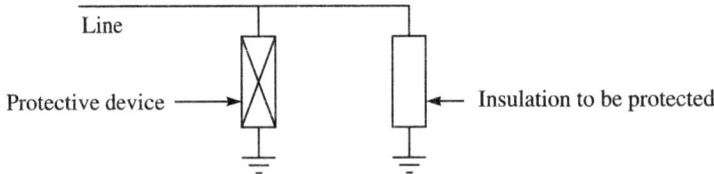

FIGURE 14.3   Schematic circuit of protective device.

The voltage rating of the protective device should always be less than the dielectric strength of the insulation so that the surge can bypass through the protective device before reaching the breakdown voltage of the insulation. It can be understood with the help of desired volt–time characteristics shown in Figure 14.4.

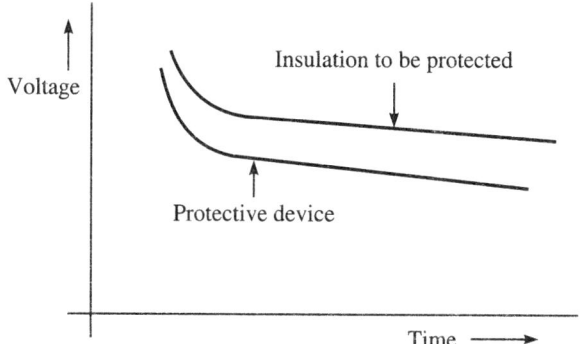

FIGURE 14.4   Desired volt–time characteristics.

14.6
UNDERSTANDING LIGHTNING PHENOMENON

Lightning is the discharge of charges contained in clouds to ground or air or to a nearby cloud.

In recent years, surge protection has been an important topic in electrical power distribution engineering. Transient overvoltages in power transmission and distribution system arise from two sources. One source is lightning strokes and the other source is switching operations. Overvoltage of these two types differs in terms of their energy level and rate of rise. Overvoltage due to switching operations has lower energy levels and is slower than that of those caused by lightning strokes.

For EHV lines, ground wires are arranged above the overhead phase conductors and grounded at every tower. Putting the ground wire above the line conductors acts as shield to the phase conductors. Lightning first hit the ground wire and gets grounded.

Lightning masts consisting of a large number of ground wires with various dispositions are used at substations. If the overvoltages across the line insulators or across the static and rotating machine insulation, transformer bushings or cable insulation exceeds the withstand value, the direct protective schemes are essentially required. These protective devices are installed parallel to the equipment insulation and their job is to keep the voltage across the insulation to an acceptable level by allowing a bypass to the surge to be grounded.

An ideal overvoltage protective device must have following characteristics:

(i) During normal operation, i.e. healthy conditions of the power system network protective device should act as a perfect insulator means no leakage current flows through it.

(ii) As soon as line voltage exceeds the rated voltage of the protective device, it should offer zero impedance so that overvoltage surge immediately passes to earth and the voltage across it becomes zero.

(iii) Just after the sharp rise in surge bypasses, the line voltage returns to its normal value. The protective device should again behave like a perfect insulator.

(iv) Protective device must have fast arc quenching component or provision.

(v) To avoid insulator failure the protective device must maintain the voltage across the insulation of the device at a level below its withstanding capability all the time.

(vi) Surge impedance of the protective device should be such that when surge current is bypassed, the voltage across it must not exceed the safe value set for the device being protected.

(vii) Protective device must be capable either to dissipate or to store energy release during bypassing the surges quickly and efficiently so that the protected equipment could not be damaged.

How electric charges spread through a system in the form of travelling electromagnetic waves which redistribute their energy is explained in the following topics.

14.7
CHARGE FORMATION

14.7.1 Wilson's Theory

This theory is based on the following assumptions:

(i) A large number of ions exist in the atmosphere. Ions, positive and negative attach themselves to minute particles of dust and extremely small drops of water called 'Ailken nuclei' to form large ions.

(ii) An electric field directed downward, towards the earth exists in the earth's atmosphere during fair weather. The magnitude of this electric field is 1 V/cm at the surface of earth, which gradually decreases with altitude. At 30,000 feet, it is only 0.02 V/cm.

Relatively large raindrop (0.1 cm radius) falling in this field becomes polarized by induction; the upper side acquires negative charge and the lower side acquires positive charge because weight of the positive ions is more than the weight of the negative ions. Lower part of the drop attracts negative charge from the atmosphere, which is available in abundance in the atmosphere leaving the positive charges in the air which got repelled.

The upward motion of air current tends to carry up the top of the cloud, the positive air and smaller drops that the wind can blow against gravity. Capturing negative charge is only possible at lower surface. No such selection occurs at the upper surface. This way both the positive and negative charges get mixed up and produce neutral space. So, finally according to the Wilson's theory larger negatively charged drops settle on the base of the cloud and the smaller positively charged drops settle on the upper portion of the cloud. Figure 14.5 shows a polarized water drop and Figure 14.6 shows the charge, separated according to the Wilson's theory.

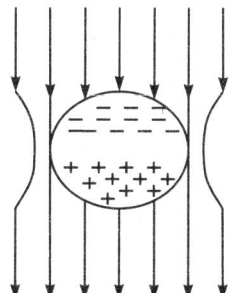

FIGURE 14.5   A polarized water drop.

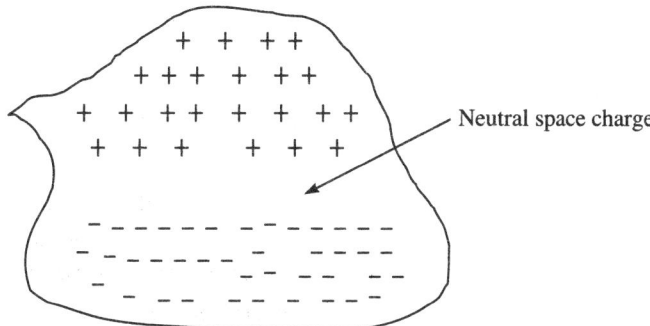

FIGURE 14.6   Charge separated according to the Wilson's theory.

14.7.2   GC Simpson's Theory

This theory is based on the temperature variations in the various regions of the cloud. In laboratories it has been shown that a water drop of radius greater than 0.25 cm becomes unstable, when it falls through still or ascending air. When water droplets are broken due to the air currents, water droplet acquires positive charge and the air is negatively charged. Also, when the ice crystals strike with the air, the air is positively charged and the crystals are negatively charged.

The steady state velocity of drops 0.25 cm in diameter is 8 m/s. Let the cloud move in the direction from left to right as shown by the arrow in Figure 14.7. If the velocity of air current is more than 8 m/s, in the base of the cloud water drops are broken and carried upwards unless they combine together and fall down in a pocket. Broken water particles get positively charged and are in the cloud and air is negatively charged.

These negative charges in the air are immediately absorbed by the cloud particles which are carried away upwards with the air currents. The air currents go still higher in the cloud where the moisture freezes into ice crystals at (–20°C temperature). The air currents when collide with ice crystals, air gets positively charged and it goes in the upper region of cloud whereas negatively charged ice crystals drift gently down in the lower region of the cloud. Once the charge separation is complete, the conditions are now set for a lightning stroke.

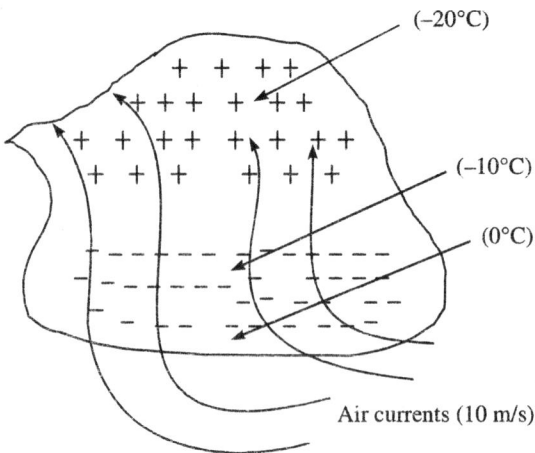

FIGURE 14.7   Charge separated according to the GC Simpson's theory.

14.7.3   Cloud Height

Simpson and Scarse from a limited number of observations in England estimate charge centres as occurring as low as 1500 feet and as high as 30,000 feet. The origin of the stroke may not coincide with the charge centres but may lie between charge centre and base. Altitude of the base of low-level thunder clouds lies between 500 feet and 100 feet.

14.7.4   Mechanism of Stroke

Lightning phenomena is the discharge of the cloud to the ground. The cloud and ground form two plate of capacitor and the dielectric medium is air. Since the lower part of cloud is negatively charged, the earth is positively charged by induction. Lightning discharge requires the puncture of the air between the cloud and the earth. For break down of air at STP condition the electric field required is 30 kV/cm peak. In the presence of moisture in the air, break down voltage of 10 kV/cm may be sufficient for air.

After a voltage gradient of approximately 10 kV/cm is set up in the cloud, the air surroundings get ionized. At this voltage gradient a negatively charged streamer starts from the cloud to the earth which can not be detected from the naked eyes. If the discharge potential gradient within the cloud is greater than at earth (100 V/cm), the discharge is initiated at the cloud rather than the earth.

Initial leader

The first component of stroke named as *Pilot Streamer* which represents propagation of the discharge into virgin air having very low ionization is associated with the current of 100 A and speed of 1.5×10^7 cm/s. Luminosity is likewise very less which can not be registered on the photographic plate.

Depending on the state of ionization of the air surrounding the streamer, it is accompanied by points of luminescence which travel in jumps and giving rise the term *Stepped Leader*. Velocity of these steps is 5×10^9 cm/s. And the distance travelled in one step is 50 m. The path of each step is straight in general but each fresh step takes a different direction. This zigzag path of stepped leader is because of the reasons:

- Variation of the head itself, or
- Variation in space ionization

A portion of charge in the centre from which the stroke originated is lowered and distributed over this entire system of temporary conductors. This process continues until one of the leaders strike the earth.

Return stroke

As the leader strikes the ground an extremely bright return streamer propagates upwards from the earth to cloud following the same path as the main channel of downward leader. The charge distributed along the leaders thus is discharged progressively to ground giving rise to very large currents varying from 1000 A to 200,000 A. The rate of propagation is about 3×10^9 cm/s.

With the development of a high-conducting arc-path between the charge centre and the ground, the potential of the charge centre is lowered considerably. This process may develop higher potential difference between this charge centre and another charge centre within the cloud resulting in the continued progress of the streamers into the cloud and the formation and attraction of the streamers from the other charge centre. With the meeting of two such approaching streamers a relatively low-conducting path to the ground for the new charge centre is formed. The resulting discharge traverses the same path blazed by the first stroke. This is known as *Dart Leader* and has no branching. Its velocity is 9×10^8 cm/s. Figure 14.8 (a, b and c) describes the lightning processes.

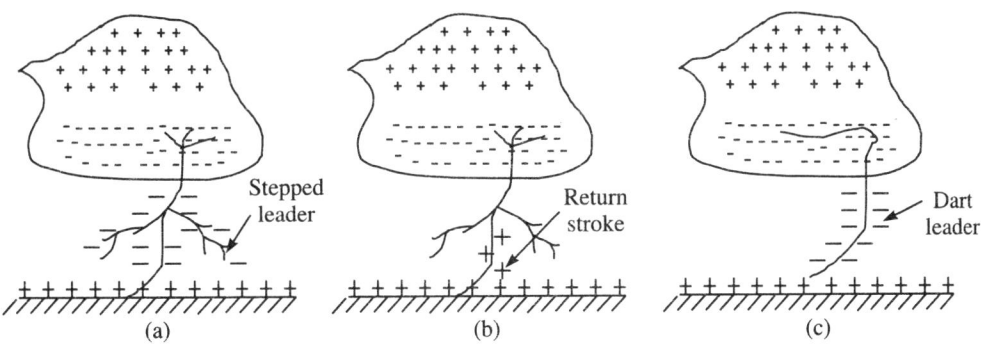

FIGURE 14.8 Lightning processes.

After the completion of the initial high current discharge, a smaller current continues to flow for some time. The magnitude and duration depend on the propagation of the streamers within the cloud body.

Note: Initially, ionization of air is very less hence pilot streamer is slowest. Stepped leader and dart leader follow the path blazed by the pilot streamer so these are faster because these leaders find easily less-resistive path. Pattern of the current flow due to lightning is shown in Figure 14.9.

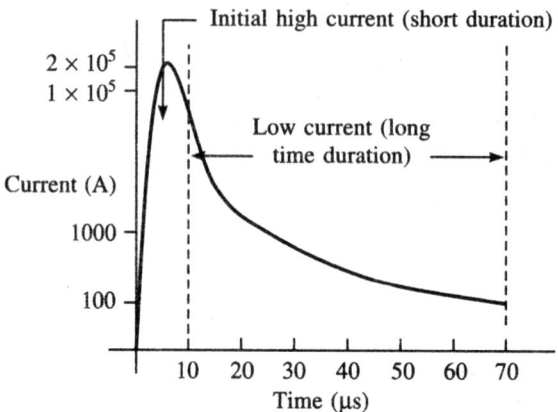

FIGURE 14.9   Current flow due to lightning.

Cold lightning

This is a type of lightning stroke which is not hot enough to start a fire. Return streamer is an example of cold lightning. A cold lightning stroke lasts for only a few microseconds and contains less energy.

Hot lightning

This is a type of lightning stroke which is hot enough to start a fire. Dart streamer is an example of hot lightning because even though the current in the streamer is relatively small, but it contains enough energy since continues for few milliseconds and even more.

The slower hot lightning stroke simply delivers more energy because of the longer duration of the current flow. In fact, the strokes travel so fast that they heat up the path to about 50,000 degrees. This temperature is no big deal, if it lasts for few microseconds, but in a full second this electron highway can cause some trouble especially, if it goes through a sunny dry weather.

14.7.5   Effects of Lightning

'Explosive and incendiary' effects of different strokes vary widely. One stroke might blow a trees, a part and still have little burning tendency, while another might have little burning effect and still result a fire.

- Those that come with dry flash do not cause fire but an explosion.
- The smoky one do not burn but blacken.

- The bright thunder draws casks dry without damaging their lids. It is Marcia a lady of high station in Rome was struck by lightning while pregnant and though the child was killed, she herself survived without being otherwise injured.

- *Stroke to tall buildings:* Because of the high altitude of high building, it acts much a large needle extending up from a plane. The gradients developed at the tower become so large that lost of the discharges develop from the tower propagate upwards. These discharges usually begin with small currents and may or may not develop into distinct discharges of high current value.

- *Stroke to aeroplanes:* Usually airplanes run above the cloud height 30,000 ft. But during ascending and descending period it may be struck by lightning so for protection purposes solid state control circuit is provided in airplanes.

14.8
EFFECT OF LIGHTNING ON POWER SYSTEM

The lightning voltage could build up to enormous values. Voltages of 4500 kV and 5000 kV were recorded on a 132 kV wood-pole line, and approximately 3000 kV on a 220 kV steel-tower line. The voltage depends on the current involved and the impedance of the circuit through which the current travels.

Suppose a cloud is positively charged. The transmission line will be charged to a negative potential by the electrostatic induction. This negative charge will, however, be present right under the cloud and the portion of the line away from this point will be charged positively as shown in Figure 14.10. But the positive charge on the far-ends of the line will not remain bound to cloud and it leaks to the earth slowly through the insulator's metallic parts thus leaving only the negative charge on the line. The charge on the line will not flow since it is a bound charge.

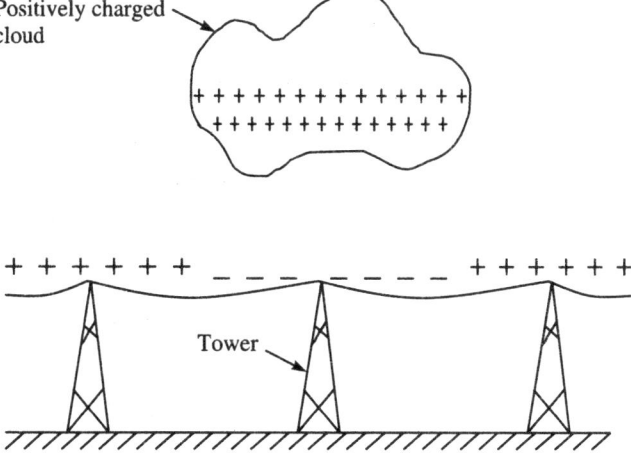

FIGURE 14.10   Effect of charged cloud on the line.

Now, consider that due to a direct discharge occurring between this cloud and another passing by cloud the charge on the cloud in question is neutralized then the charge on the line is no more a bound charge and is free to travel in both the directions in the form of travelling waves. These travelling waves will be of quite high magnitude (10 kV to 15 kV) and shall have very steep wavefronts which can damage the unprotected equipment connected to the line. So, it is required that these must be passed to the earth.

Suppose a line having surge impedance Z_s and let the discharge current be I_d, then the overvoltage due to a direct stroke is

$$V_d = I_d \times Z_s$$

If travelling waves travel in both the directions, current will be half for one direction and hence overvoltage

$$V_d = I_d \times \frac{Z_s}{2}$$

When, we consider that lightning stroke is on the earth wire or on top of the tower, then the overvoltage is

$$V_d = I_d \times Z_e + L_c \times \frac{dI_d}{dt}$$

where

Z_e = impedance of earth conductor

L_c = inductance of earth conductor

14.9

PROTECTION AGAINST LIGHTNING

14.9.1 Indirect Protection

Design of transmission and distribution lines must be such that it protects itself from lightning stroke. Following cares and concepts should be taken into consideration while designing and commissioning the transmission and distribution lines.

Ground wire

Ground wires must be of sufficient mechanical strength. Non-corrosive ground wires or earth wires are placed high above the conductor and they are so located that they are well out on the towers and not exactly over the conductors in order to avoid any possibility of short circuit occurring in the event of conductor swinging under the ice loading and thermal loading, etc.

The overhead ground wire performs two main functions:

- It intercepts the direct stroke, keeping it off the conductor
- It distributes the current in two or more paths and thereby reduces the voltage drop.

If the ground wire is struck in midspan, the current divides and flows toward both towers, and at the tower, the current divides again between the tower and the outgoing ground wire. If the tower is struck and there is one overhead ground wire, the current divides into three paths, one in the tower and the other two in the branches of the ground wire. A third function of the overhead ground wire, of a minor nature, is to reduce voltage induced on the conductors from nearby strokes.

Angle of safety

It is defined as the angle formed between the normal passing through the ground wire and line joining the supported centre point of the outer conductor and the ground wire. This angle should not be more than 30° for sufficiently good shielding of the line conductor. This is shown in Figure 14.11.

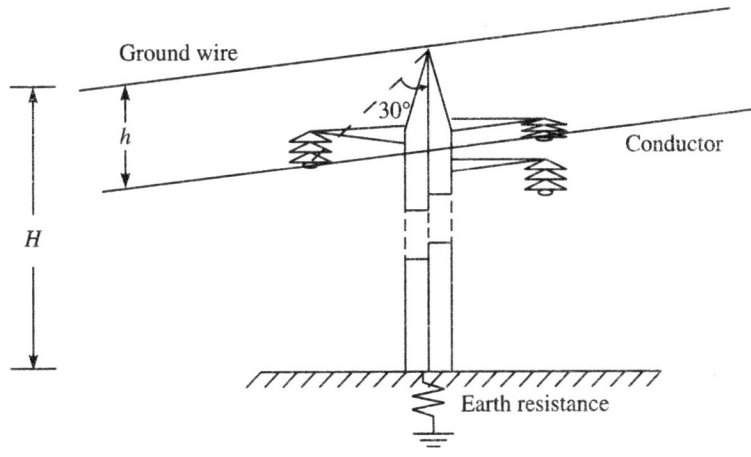

FIGURE 14.11 Configuration of tower and conductors with safety angle 30°.

Coupling factor

When lightning strikes at the ground wire, lightning current flows to ground. Due to the presence of earth resistance, voltage drop is there and the ground wire does not remain at earth potential. It is at higher potential.

Let this potential is V_g, so

$$V_g = IR_E$$

Now, the coupling factor is defined as, *the ratio of voltage between conductor and earth and voltage at ground wire when lightning discharge occurs.* See Figure 14.12.

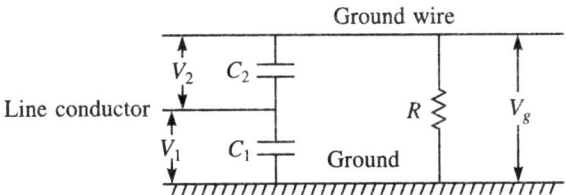

FIGURE 14.12 Equivalent circuit diagram of Figure 14.11.

So,
$$K = \frac{V_1}{V_g}$$

where

V_1 = voltage between conductor and earth

Clearance between wires

Considering an arrangement of conductors and earth wire and its equivalent circuit diagram as shown in Figures 14.11 and 14.12 respectively.

$$V_2 = V_g - V_1$$
$$= V_g - KV_g = V_g(1 - K)$$

Coupling factor K can also be defined as

$$K = \frac{\log_e (h/h_1)}{\log_e (2H/r)}$$

where

h = distance between conductor and ground wire

h_1 = distance between conductor image and ground wire

H = height of the ground wire above the ground

r = radius of the ground wire

If h is less, it implies that K is less and hence V_2 increases. If V_2 is high, it can cause flashover between the line conductor and ground wire. So, we must keep sufficient clearance between conductor and ground wire as well as between line conductor and tower to avoid flashover between the conductors and between the conductors and tower during lightning.

Resistivity of tower

It is very important to keep tower footings at low resistance so as to allow lightning discharge current to flow to earth easily. By using the ground rods and counter poises, tower footing resistance gets reduced.

Ground rods: Thin and long rods are put into the ground surrounding the tower structure. Rods are in parallel to reduce the overall resistance of tower footing.

Counterpoise: A counterpoise is galvanized steel wire run in parallel or radial or a combination of the two with respect to the overhead line as shown in Figure 14.13. The lightning stroke as is incident on the tower, discharges to the ground through the tower first and

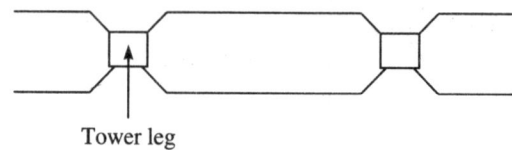

Tower leg

FIGURE 14.13 Double parallel continuous counterpoise.

then through the counterpoises. Leakage resistance of counterpoise should always be smaller than the surge impedance of counterpoise, otherwise positive reflections of the surge will take place and hence instead of lowering the potential of tower it will be raised.

14.9.2   Direct Protection

Direct protection refers to the protection of electrical system by using apparatus or protecting devices namely:

- Surge diverters
- Surge absorbers
- Surge arresters

Surge diverters

Surge diverters are of following types:

Condenser as surge diverters: If a condenser is placed in series with a resistance, the combination dissipates energy in addition to diverting it from the apparatus. Impedance of a condenser is inversely proportional to the frequency ($X_C = 1/j^2\pi fC$). At high frequency, its impedance is low and since combination of condenser and resistance is connected direct to earth, surge energy passes to earth. At line frequency impedance is so high and practically no current passes through it. Figure 14.14 shows circuit diagram of the use of condenser as a surge diverter.

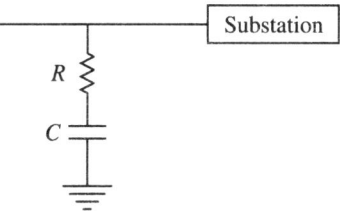

FIGURE 14.14   Circuit diagram of a surge diverter, using condenser.

Inductor as surge diverter: Another type of surge diverter consists of an inductance across which is placed a resistance. This combination is placed in series with the line. Impedance of an inductor is directly proportional to the frequency ($X_L = j^2\pi fL$). Steep wavefront or high frequency discharge finds the series inductance a high impedance path and is forced to get diverted through the resistance where it is dissipated. Figure 14.15 shows circuit diagram of the use of inductor as a surge diverter.

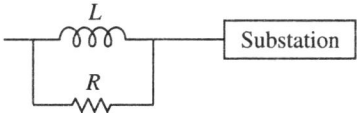

FIGURE 14.15   Circuit diagram of a surge diverter, using inductor.

Surge absorber

Surge absorber are of following types:

Ferranti surge absorber: Ferranti surge absorber consists of an air core inductor which is connected in series with the line and surrounded by an earth metallic sheet. The earth metallic sheet is known as *dissipater*. The dissipater is insulated from the inductor by the air as shown in Figure 14.16. This surge absorber acts like an air-cored transformer whose primary is the low inductance inductor and the dissipater as the single-turn short circuit secondary.

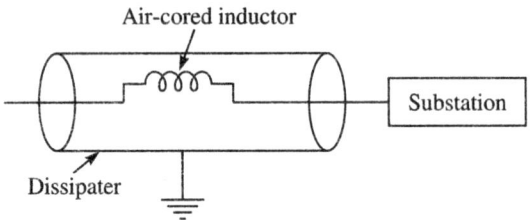

FIGURE 14.16   Ferranti surge absorber.

Whenever a travelling wave is incident on the surge absorber, energy is transformed by mutual inductance between the coil and dissipater. Because of the series inductance the steepness of the wave is also reduced.

ERA surge absorber: An improved form of the surge absorber is the Electrical Research Association (ERA)-type surge filter as shown in Figure 14.17 incorporated a gap G and expulsion gap E. When a wave reaches the inductor L, a high voltage is induced across it causing the gap G to break down putting the resistor R and expulsion gap E into circuit. An incoming wave is thus flattened by the inductor and the resistor and its amplitude is reduced by the expulsion gap.

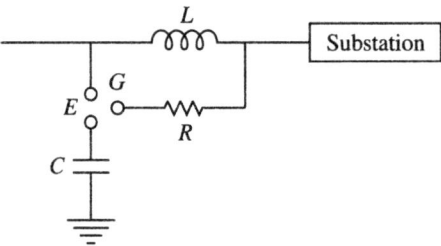

FIGURE 14.17   ERA surge absorber.

Surge arresters

Lightning flashover on transmission lines can be reduced by increasing the insulation or by reducing the tower footings resistance. Neither of these methods may be practical to apply, if tower clearances are not available or if lines are located in high resistance mountainous terrain. An alternate approach is to install transmission line surge arresters in parallel with the insulator strings to prevent insulator flashover.

Requirements for surge arresters: Following are the basic operational requirements for arresters:

  (i) If lightning strikes within a protected section, it should not create flashover either inside or outside the protected section.
 (ii) If lightning strikes outside a protected section, it should not create flashover inside the protected section.
(iii) Lightning flash can inject several strokes in less than a second so arresters must have sufficient energy capability.
(iv) Arresters which are for lightning protection should not conduct during switching or during low frequency overvoltage due to line to ground fault, etc.
 (v) It should be able to keep the system voltage as possible as normal.

Types of surge arresters: There are mainly following five types of surge arresters:

  1. Rod gap-type surge arrester
  2. Expulsion-type surge arrester
  3. Valve-type surge arrester
  4. Metal oxide surge arrester
  5. Zinc oxide surge arrester

Rod gap-type surge arrester.   It consists of two rods which are bent at right angles as shown in Figure 14.18(a). One rod is connected to line circuit, while the other rod is connected to earth. Distance between the rod end and the surface of the insulator should be at least one third of the main gap between the rods to avoid arc being blown over bushing insulator.

While lightning surges reach the arrester and attain the prefixed voltage, a spark is formed across the gap as shown by point P in Figure 14.18(b). Rod gap provides a low impedance path and air gap is punctured, and surge energy is sent towards ground. The surge impedance of the line limits the amplitude of current flowing to earth, it is necessary in order to protect the insulation of the equipment. The rod gap should be so adjusted that the breakdown should occur at 80 per cent of sparkover voltage in order to avoid very steep wavefront across the insulator.

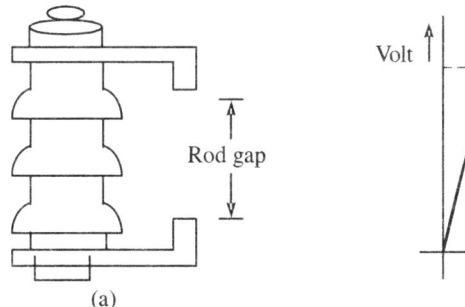 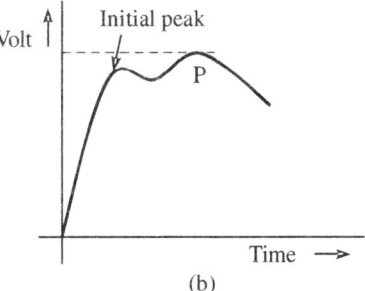

FIGURE 14.18   (a) Rod gap-type surge arrester. (b) Volt–time curve for rod gap.

These arresters have following limitations:

  (i) The arrester is not capable of stopping power frequency current to earth because of the ionization of air within the gap. Ones the gap is ionized by travelling surge in the presence of moisture, power frequency current continues to flow by getting lower resistance path.

(ii) When the arc is struck between the rods gap, large amount of heat is produced which can melt the rod and the rod can be damaged.

(iii) The climatic variations such as temperature and pressure also affect the performance of the rod gap arrester.

(iv) The polarity of the surge also affects the arrester.

Expulsion-type surge arrester. An improvement of the rod gap is the expulsion-type surge arrester, which consists of the following:

(i) In order to avoid the possibility of corona loss or leakage current through tube, a series gap external to the tube good enough to withstand the normal system voltage is used.

(ii) A tube made up of fibre (highly gas evolving material) lining on the inner side.

(iii) A spark gap in the tube.

(iv) An open vent at the lower end for expelling the gases.

It is desired that breakdown voltage of a tube must be lower than that of the insulation for which it is used. During the application of lightning overvoltage wave on it, both gaps get brokendown. Since discharge distance along the external surface is much greater than the length of the internal gap flashover can not take place along the external surface. Impulse current is carried to earth. After the impulse current is over, power frequency current continues to flow through arrester. An extensive emission of gas takes place in the tube under the action of high temperature. The neutral gas produced creates a lot of turbulence within the tube and is expelled out from the open bottom vent of the tube and extinguishes the arc at the first current zero.

The volt–time characteristic of the expulsion tube is some what better than the gap-type and has the ability to interrupt power voltage after flashover. A line diagram of an expulsion-type arrester is shown in Figure 14.19(a) and its inside view is shown in Figure 14.19(b).

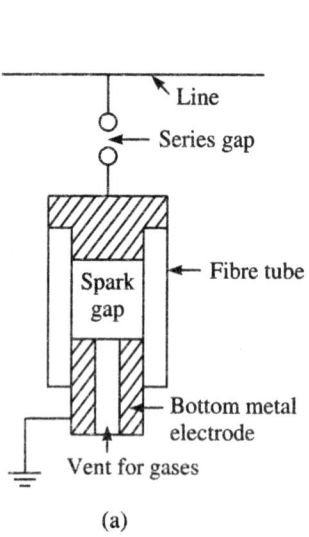
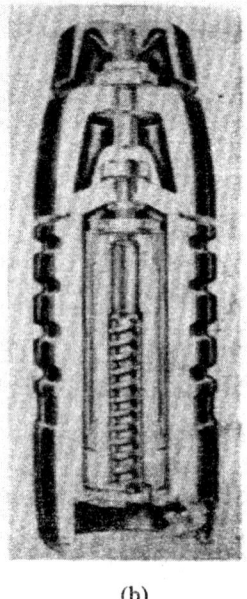

(a)   (b)

FIGURE 14.19   (a) Expulsion-type surge arrester (line diagram). (b) Expulsion-type surge arrester (inside view).

Valve-type surge arrester. Nonlinear resistor made up of silicon carbide is the basic element of this type of arresters. The other important component is the multiple spark gaps unit. During application of lightning overvoltage waves on the arrester, its multiple spark gaps unit is brokendown and overvoltage wave gets chopped the impulse current passing immediately through the nonlinear resistor. High resistance is connected across the disc so that system voltage is evenly distributed over the discs. High resistance helps keeping the inner assembly dry by generating some amount of heat. It is desirable to have residual voltage constant and equal in magnitude to the breakdown voltage of spark gap during the passage of lightning currents of different magnitude through the arresters. The use of nonlinear resistance permits to keep the residual voltage of the arrester at a level close to breakdown voltage of spark gap with only a small increase of the residual voltage with increase of current. A volt–ampere characteristic of nonlinear resistor is shown in Figure 14.20(a).

When a surge voltage is incident at terminal of the arrester it causes two gap units to flashover thereby a path is provided to the surge to the ground through the coil element and the nonlinear resistance element. Because of the high frequency of the surge, the coil develops sufficient voltage across its terminal to cause bypass gaps to flashover. With this the coil is removed from the circuit and the voltage across the arrester is IR drop due to the nonlinear element. This condition continues till power frequency currents follow the pre-ionized path.

For power frequency impedance of the coil ($j^2\pi fL$) is very low and, therefore, the arc becomes unstable and the current gets transferred to coil. A follow-up current in the coil develops magnetic field which reacts with the transferred current in the arcs of the gap assemblies causing them to be driven into arc quenching chambers of the gap unit. A line diagram of a valve-type surge arrester is shown in Figure 14.20(b).

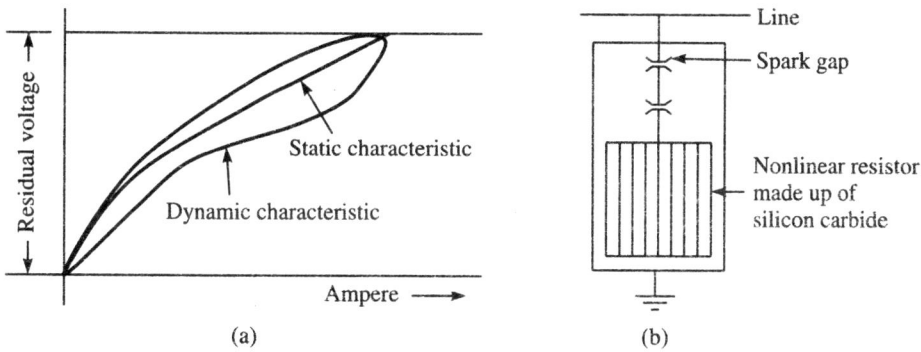

FIGURE 14.20   (a) Volt–ampere characteristic of nonlinear resistor. (b) Line diagram of valve-type surge arrester.

Metal oxide surge arrester. The methods of predicting the operation and performance of the metal oxide arresters in power system applications deal with:

(i) the low current *V–I* characteristics of the arrester for non-adiabatic thermal behaviour

(ii) the high current *V–I* characteristics for response to surges and system disturbances

The metal oxide technology opens many opportunities for arrester's applications in adverse operating and atmospheric conditions where gap-type arresters could not be used.

Based on the metal oxide technology metal oxide arresters are now being used extensively for protection of equipment on distribution and sub-transmission system. The applications of the metal oxide arresters include:

- Suppression of switching surge voltages
- Absorption of large amounts of energy
- Application in the presence of large temporary overvoltage

Thermal stability criteria for metal oxide surge arresters. The thermal stability of a metal oxide surge arrester is defined by its heat loss–input balance criteria. In this the AVR (Applied Voltage Ratio) is defined as the ratio of the peak 50/60 Hz applied voltage to the dc voltage that produces a direct current of mA through the entire valve element at a temperature of 20°C to 25°C.

Generally, two intersections of the heat loss and power dissipation curves are obtained. One occurs at low temperature slightly above the ambient and is referred to an operating temperature. The other intersection of the two curves occurs at higher temperature creating a stability limit of the valve element temperature. Temperature in excess of the stability limit will result in a catastrophic failure of the arrester due to the thermal run away. Thermal stability limit temperature determined by this method is only valid for conditions where the valve element and the housing have reached thermal equilibrium.

The mechanical damage energy absorption capability of the material is limited by the uniformity of surge current density and the shape of valve element.

The low current *V-I* characteristics show very significant variation with respect to temperature and wave shape. These characteristics determine the response of the arrester to non-adiabatic processes.

Zinc oxide (ZnO) surge arrester. Zinc oxide (ZnO) surge arrester has been developed first among the metal oxide surge arresters. The ZnO arrester design has many unique characteristics like:

(i) Because of the requirement for both high tensile and compressive strength, along with light weight for ease of installation, the housing is made from fiber-glass-epoxy with weather sheds made up of EPDM rubber.

(ii) End fittings are sealed to prevent moisture ingress and interior is kept dry by desiccant.

(iii) The arrester length can be changed by the use of pipe extension which is incorporated between tower articulation fitting and the fibrous glass housing.

(iv) The external series gaps uses two different diameter corona rings for voltage grading consistent voltage spark-over protection for both positive and negative polarities.

Cross-sectional view of ZnO surge arrester is shown in Figure 14.21(a). Line diagram of ZnO surge arrester assembled with insulator is shown in Figure 14.21 (b and c). The zinc oxide column can be represented by an RC circuit with a nonlinear variable resistor in parallel with a capacitance. Because of the high degrees of non-linearity in the resistance of ZnO with voltage RC, the time constant is short above 200 kV and longer at voltage below 200 kV. For example, at 195 kV the time constant is 60 ms and for 213 kV time constant is 0.6 ms. ZnO column, therefore, acts like a 200 kV bus voltage with regard to the impulse sparkover characteristics of

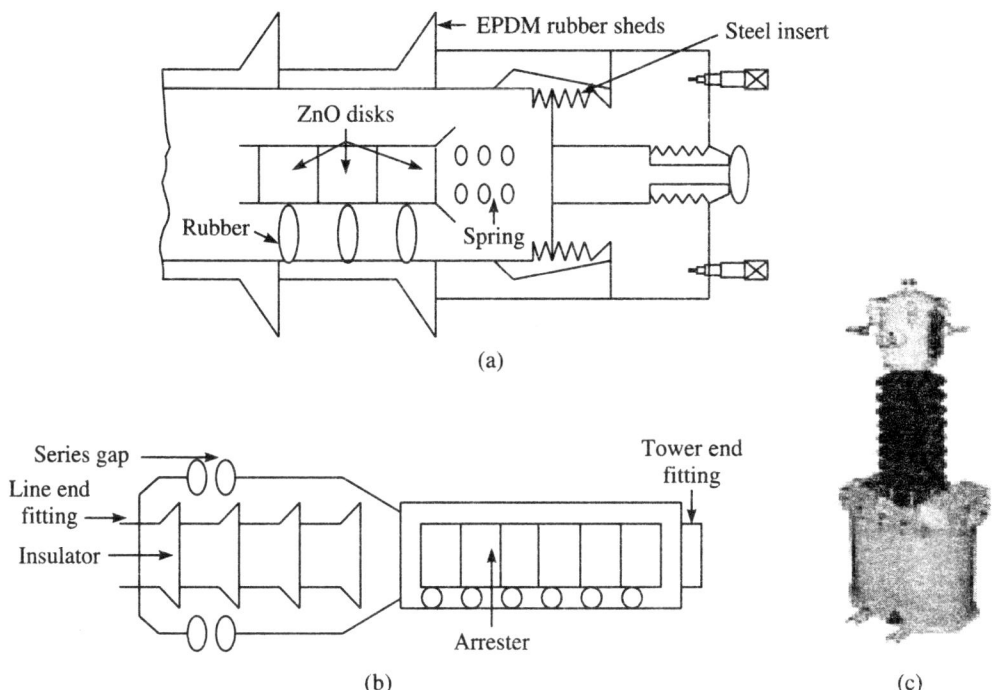

FIGURE 14.21   (a) Cross-sectional view of ZnO surge arrester, (b) Line diagram of ZnO surge arrester assembled with insulator, (c) ZnO surge arrester assembled with insulator.

series gap. It was found that a more non-uniform gap configuration reduces the gap capacitance and provides volt–time flashover characteristics approximately parallel to those of the insulator string.

The negative flashover voltage is too high to protect the insulator string as well as being 50% above the positive flashover voltage.

Advantages of zinc oxide surge arrester:

   (i) By eliminating series gaps, a number of parts are eliminated which can cause malfunctioning. Inherent simplicity of ZnO makes it reliable and cost-effective.

  (ii) Elimination of series gaps avoids variations in protective levels caused by erratic gap sparkover.

 (iii) ZnO energy discharge capability is in order of twice that of its SiC arrester.

 (iv) ZnO arresters are immune to pollution up to some level.

14.9.3   Protective Characteristics of Lightning Arrester

Volt–time curve

The breakdown voltage for a gap is a function of both the magnitude and the time of application of the voltage. Relationship between the crest flashover voltage and the time to flashover for a series of impulse applications of a given wave shape is determined by the volt–time curve. For the construction of volt–time curve, following procedures are adopted:

(i) Waves of the similar shape with different peak values are applied to the insulation or gap whose volt–time curve is required.

(ii) Flashover occurrence at front of the wave the flashover point gives one point on the curve.

(iii) The other possibility is that the flashover occurs just at the peak value of the wave; this gives another point on the curve.

(iv) The fourth possibility is that the flashover takes place on the tail side of the wave. In this case to find the point on volt–time curve, draw a horizontal line from the peak value of this wave and also draw a vertical line passing through the point where the flashover takes place. The intersection of horizontal and vertical lines gives the point on the volt–time curve.

The procedure of construction of volt–time curve is shown in Figure 14.22.

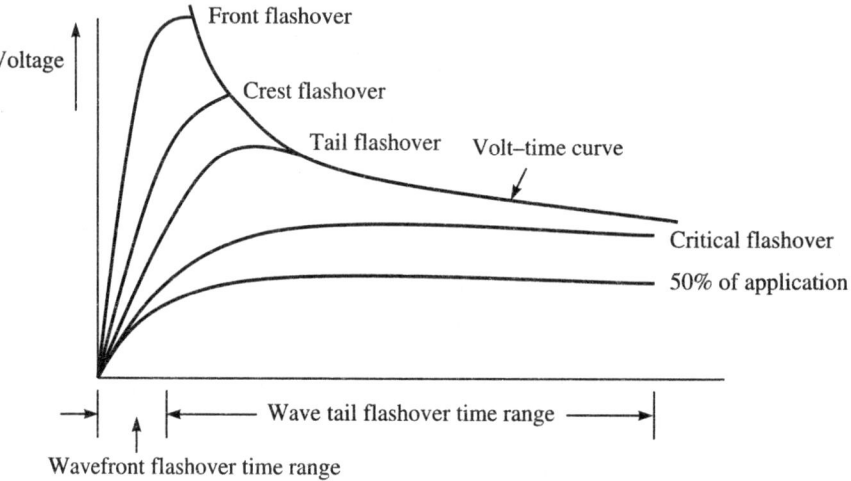

FIGURE 14.22   Construction of the volt–time curve.

Basic impulse insulation level (BIIL)

Basic impulse insulation level is reference level, expressed in impulse crest voltage, with a standard wave not longer than 1.2/50 μs. Apparatus insulation demonstrated by suitable tests shall be equal to or greater than the basic insulation level.

Protective level

Protective level is the maximum discharge voltage of lightning arrester. With a suitable margin, satisfactory protection is assured, if the arrester discharge voltage is below the BIIL of the equipment to be protected.

Voltage rating

Having known the system highest voltage and the coefficient of earthing, the arrester voltage rating should not be less than the product of these two values. For example, on 220 kV effectively earthed system the voltage rating of arresters should not be less than 245 (system highest voltage) × 0.8 = 196 kV.

Impulse ratio

It is a ratio of the breakdown voltage of a wave of special duration to the breakdown voltage of a 60/50 Hz wave.

14.9.4   Substation Protection

An electrical installation or a generating station is shielded against the direct lightning stroke by suitably placing the overhead ground wires, as shown in Figure 14.23. But protection from direct strokes is provided in most cases by lightning masts. If the construction of substation does not allow to put the masts or overhead ground wires, then, the lightning arresters and surge absorbers are used in each line, at entry, as shown in the figure.

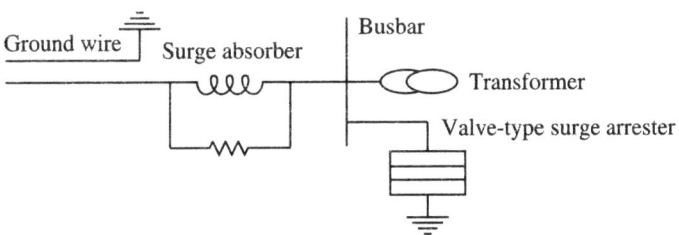

FIGURE 14.23   Protection of substation.

Lightning mast with lightning rod is described below with help of Figure 14.24(a and b). The lightning rod is clamped to the highest point on the mast. The ground wire is connected to the lightning rod and should be one continuous piece all the way to the earth ground rod.
Referring Figure 14.24(b)

$$\text{Radius of the cone } x = \frac{1.6(h - h_x)}{1 + (h_x/h)}$$

where

> h = height of the mast from the ground
>
> h_x = height of the object to be protected
>
> x = radius of the cone or umbrella

One single set of lightning mast with lightning rod creates protection zone like a cone or umbrella whose radius can be determined by the above formula. Similarly, requirement of the number of lightning masts can be calculated according to total area needs to be protected.

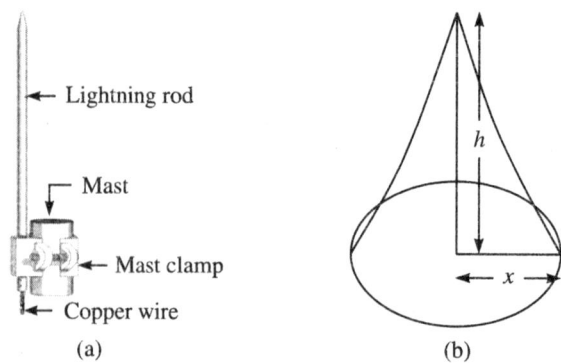

(a)   (b)

FIGURE 14.24   (a) Lightning mast with lightning rod, (b) Protected area by lightning mast and rod.

14.9.5   Protection of Three-phase Power Transformer

It is necessary to protect power transformer which is a very costly and important equipment of power system from lightning discharge. Surge absorbers and diverters are installed to protect it, as shown in Figure 14.25.

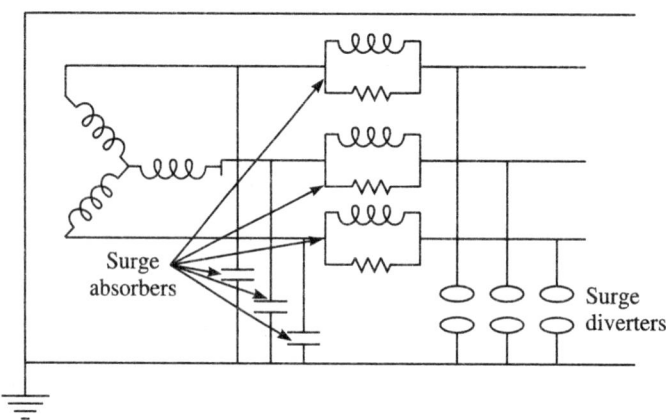

FIGURE 14.25   Connection diagram for protection of a winding of three-phase transformer.

Protective devices are connected between the primary lines and the transformer's tank. In similar manner the devices are installed between the secondary winding and the transformer's tank. Surge absorber avoids sudden voltage and current rise whereas surge arrester diverts the surges to the earth.

14.9.6   Location of Lightning Arresters

Normal practice is to locate the arresters as close as possible to the equipment to be protected. The reasons are:

(i) Lightning arrester reduces the chances of surge entering the circuit between the protective equipment and the equipment to be protected.

(ii) Distance between two steep fronted waves after being incident on the lightning arrester which sparks over corresponding to its spark overvoltage facilitates entry of the transients into the circuitry of the equipment to be protected after travelling over the lead between the two points. The transient wave suffers reflections at the terminal of the equipment and, therefore, total voltage at the terminal is the sum of two voltages which is twice the incident voltage.

(iii) If L is the inductance of lead between two points and IR is the residual voltage of lightning arrester, the incident voltage of the equipment terminal will be:

$$V = IR + L\left(\frac{dI}{dt}\right)$$

where dI/dt is the rate of change of surge current.

Note: Connecting a capacitor at the terminals of the equipment to be protected reduces the steepness of the wave and hence the rate dI/dt. Installation of capacitor also reduces the stress distribution over the winding of the equipment.

EXERCISES

1. Describe the different types of overvoltages and mention their causes of initiation.

2. Specify the classifications of HV insulations and their protection against contamination.

3. Describe with neat sketches the lightning phenomena, explaining the different theories of charge formation in clouds.

4. How lightning strokes develop and induce overvoltages on the overhead power lines.

5. Explain the effects of lightning in general and in particular on the power system.

6. What is meant by insulation coordination? How are the protective devices chosen for optimal insulation level in a power system?

7. Explain a surge diverter using its functions as a shunt protective device.

8. What are ground rods and counterpoises? Explain their uses and show how these are used to improve grounding conditions.

9. Explain the principle of operation of the following:
   (a) Ferranti surge absorber
   (b) ERA surge absorber

10. What are the basic requirements of surge arresters? Differentiate between a surge absorber and a surge arrester.

11. Write short note on the following:
    (a) Rod gaps used as protective devices
    (b) Ground wires for protection of overhead lines
    (c) Lightning rod with mast
    (d) Hot lightning
    (e) Cold lightning
    (f) Location of the lightning arresters

12. Discuss the construction and principle of working of a valve-type arrester.

13. Discuss the construction and principle of working of an expulsion-type arrester.

14. What are the merits of metal oxide arresters over other types of arresters?

15. Discuss the construction and principle of working of a ZnO lightning arrester.

16. Discuss in brief the following:
    (a) Protection of a transformer against lightning
    (b) Protection of substation against lightning

17. Describe the requirements of protective characteristics of lightning arrester.

Chapter 15

Power System Management

15.1

LOAD DISPATCH CENTRE

The occurrence of all kinds of changes in power systems including the upgradation of the power system network, rise in the voltage level of transmission, interconnections within grids, consumer population growth, categorisation of consumers, and the use of fuel variety in power generation have made load dispatching operations more complicated than before. The responsibilities of the load dispatchers have also become multi-dimensional because the integration of various grids and utilities leads to security problems in the system and creates the need for economic load dispatching.

The main duties of the load dispatch centres are as follows:

    (i) Data processing and system studies
    (ii) Load forecast and system behaviour analysis
    (iii) Load generation balance and maintenance of the quality of supply
    (iv) Economic load dispatch
    (v) Energy distribution and load pattern study
    (vi) Integration with headquarters and the regional load dispatch centre
    (vii) Maintenance of grid discipline and co-ordination with neighbour grids
  (viii) Devising of a strategy for system operation
    (ix) Event analysis and preventive measures
    (x) Black start preparedness
    (xi) Clearing of outages for generating units and EHV lines
    (xii) Maintenance of scheduling of generating units and transmission lines
  (xiii) System security and islanding facility
  (xiv) Communication and SCADA management
   (xv) Public relations and consumer interaction

It is necessary to ensure automatic, effective and timely control on generating units and load centres directly from the load dispatch centre in order to prevent the occurrence of any major blackout. Reliable interlocks and back-up systems are also required to ensure the safe working of the grid and foolproof arrangements for preventing blackouts.

It is also imperative to set up a reliable speech and data communication facility with each load centre and substation with generating units, and a load dispatch centre to facilitate effective power system management. A reliable stand-by network equipped with the full capacity to handle data and speech channels is also desirable in a dedicated communication network. A block diagram of a load dispatch centre is shown in Figure 15.1.

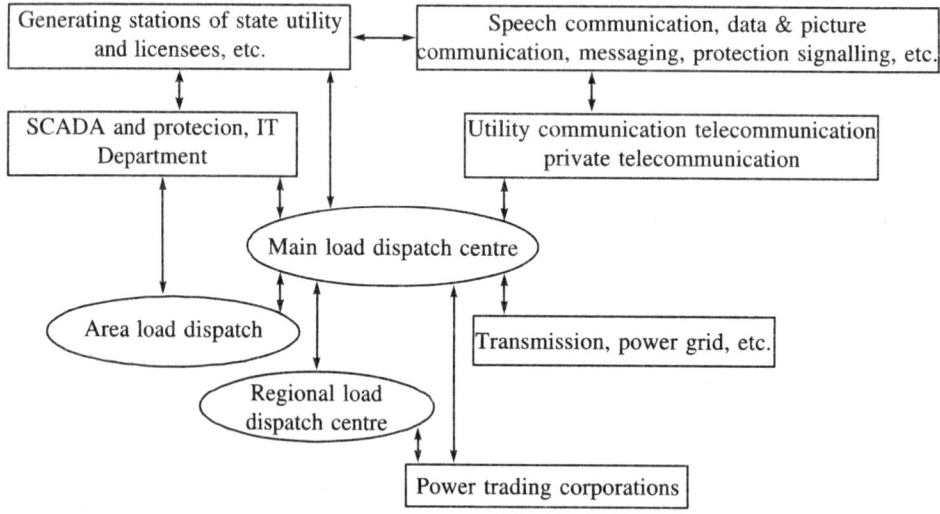

FIGURE 15.1   Block diagram of load dispatch centre.

15.2
REPORTING AND DATA MANAGEMENT

Futuristic power demand projections and planning for upcoming generating units necessitate proper data management with historical perspectives. In order to attain this objective, desk engineers working at the load dispatch centre need to carefully and systematically record the following data:

- Hourly recording of power flow on different sections
- Tie line power flow of inter-state lines
- Power generation of different central generating units
- Power generation of different captive power plants
- Power generation of different private generating units
- Voltage levels at load centres
- Voltage levels at central grid stations
- Voltage levels at inter-state grid stations
- Outage details on the transmission lines

- Line outage duration
- Generating unit outage duration
- Reactive power flow on inter-state tie lines
- Reactive power flow on transmission lines within the grid
- Hourly grid frequency
- Sudden change in system frequency
- Warning from the meteorological department
- Details of generating units under outages with reasons
- Power swing or hunting in the system with reasons
- SCADA or communication system failure
- Switchgear or any major equipment failure
- Details of the fault in the system
- Major fault in the neighbouring grid system

It is thus said that the load dispatch centre is the nerve centre of any power system network as it not only reports the various changes in the power system on a regular basis but also gathers data and maintains a complete record of the various system parameters along with climatological changes. Data exchange on a daily and periodic basis among the various sections of the power system network including power stations, commercial departments, transmission and distribution sub-stations, and fuel support departments, ensures the more effective management of electric power supply. The load dispatch centre plays a vital role in this regard. Data supplied by the load dispatch centre on a periodical basis helps in preparation of the future strategy and demand forecast.

15.3
LOAD DISPATCHER IN THE CONSUMER SET-UP

A user of electricity who pays for it is known as a consumer while the one who provides the electricity is known as a supplier. As in other fields, the consumer-supplier relationship is also important in the field of power system management and consequently, the rule of correct and prompt payment by the consumer against the receipt of electric power and services is also applicable here.

A consumer expects to obtain clarifications (as and when required) regarding all aspects of the power system like generation, power trading, power transmission, power corporate, etc. The load dispatcher must thus be able to answer the queries of the consumers. Any divergence between the indent and the actual power availability in terms of the quantity as well as the duration must be declared and the load dispatcher must apprise the consumer of this to avoid probable bitterness on the part of the latter.

In the present competitive deregulated and complicated power system network, the load dispatcher must himself be very clear and precise about the size and nature of the data and the quantum of information to be extracted from any related agency or person. This helps to minimise inconvenience to any person and prevents duplication of efforts that may arise at any stage. It is thus also important to maintain proper documentation of data at the load dispatch centre.

The load dispatcher must also be capable of carrying out proper formatting and recordings of the outcomes of the regular interactions with the authorities responsible for transmission, distribution, generation, control and trading. If detailed information is available with the load dispatcher, it becomes easy for him to take prompt and corrective action to prevent any trouble in the system. In a majority of the cases, a communication gap or miscommunication often creates bitterness between the supplier and the consumer, leading to adverse consequences. A block diagram of the load dispatcher's relationships in a consumer set up is shown in Figure 15.2.

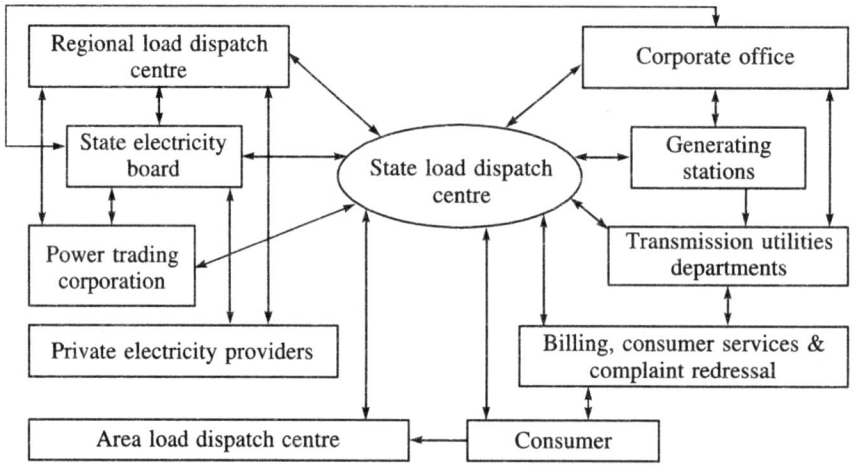

FIGURE 15.2   Load dispatcher in the consumer set-up.

15.4
LOAD CONTROL CENTRE

In order to ensure the power quality, it is essential to maintain the voltage and frequency of the power supply within permissible limits. Any change in the system frequency is a very sensitive issue as it affects the performance of the system. The system frequency of present day complex interconnected power system network may change, if there is a change in the demand or generation of the load in any part of the interconnected power system. Thus, it is essential to regulate the system frequency in order to maintain it close to the nominal value. During the commonly used process of load frequency control, the generation is controlled in such a way that the load demand is exactly met, the system frequency becomes close to the nominal value and the tie lines flows are also maintained at the scheduled values. Marginal load variations are automatically adjusted at the regulator stations.

The main job of the control centre of an interconnected power system is to handle any emergency during the failure of a transmission line or loss of generation by restoring first the emergency loadshedding to eliminate overloads and then the normal operation of the system.

In order to ensure effective emergency control, loadshedding is spread throughout the system, which is generally initiated in steps. Once the corrective measures have been taken, loadshedding is automatically done by the relays at regular intervals to restore the system. Load control centres are also set up with the load dispatch centres. In fact, these are a part of the load dispatch centres and possess SCADA systems for data collection and monitoring. Another important function of the control centres is to limit the duration and extent of the repercussions of the faults. A block diagram depicting control hierarchy in the power system is shown in Figure 15.3.

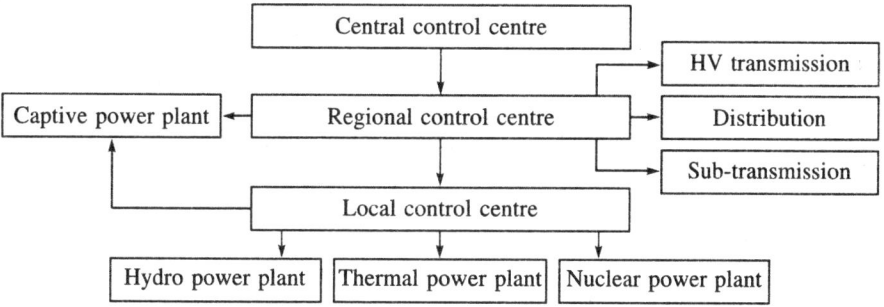

FIGURE 15.3   Block diagram of control hierarchy in a power system.

15.5
COMPUTERISED POWER SYSTEM CONTROL

The main objective of computerising the power system functions is to ensure secure and economic operations by reducing the error factors, especially human errors. In computerised operations, human operators are needed only to interact with consumers for any feedback action, if required. Computerised control and management of the power system also facilitates the storage of information and records of minute-to-minute tasks carried out by the human operators.

Various functional modules can be installed in the power system depending on the size and nature of the power system network. These modules include the following:

(i)   Main computer system placed at the main station
(ii) RTU (Remote Terminal Unit) system, which controls the power stations and grid sub-stations
(iii) Data storage modules
(iv) Output mode module

The task of the main computer system is to handle the computation of data, input data acceptance, storage of the desired information, processing of the operations and issuing of the requisite instructions to the respective RTUs. The output mode module consists of the printer, console or dynamic map board. The RTU transmits data and receives the processed information.

Generally, in order to ensure high reliability and prevent failures in the power system network, two computer sets are used for remote supervision, data acquisition, control, energy management and system security. Figure 15.4 shows the computer configuration in which one is known as the On-line computer while the other is known as the Back-up computer. Each of the computers has its own memory and drive system, and input and output devices. The On-line computer monitors and controls the power system. This unit periodically updates the disk memory shared between the two computers. The Back-up computer executes off-line programmes like load forecasting or hydrothermal allocation. New digital codes used to control the system are generally compiled and tested in the Back-up computer.

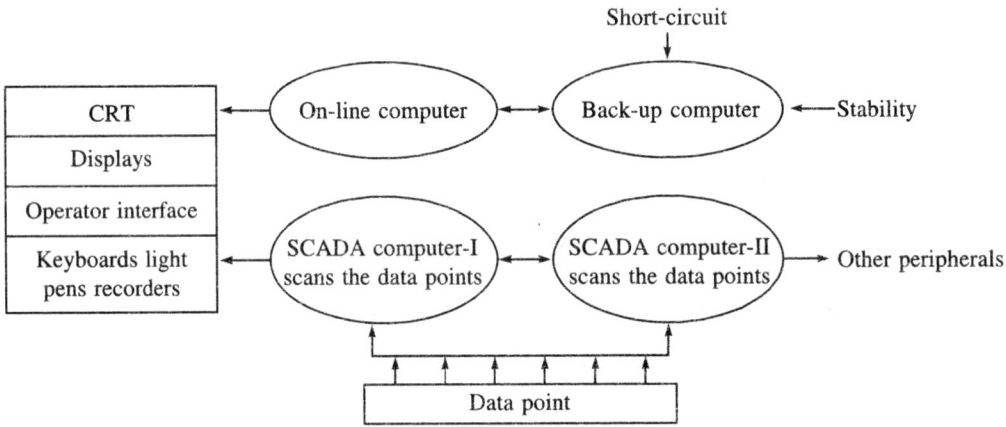

FIGURE 15.4 Computer configuration.

After they are tested, the new digital codes are switched to the On-line status. The scanning of the various status points including the transformer tap position, switchgear contact position, tie-line interchange schedule, and telemetry verification to detect failures and errors, generator loads, voltage, operating limits and boiler capacity, is done at two-second intervals. The time scale of various controls at the load control and dispatch centre is shown in Table 15.1.

TABLE 15.1 Time Scale of Various Controls at the Load Control and Dispatch Centre

Time scale	Controls
Milliseconds	Relaying, system voltage control and excitation control
1–10 seconds	Automatic generation control
1–15 seconds	Economic load dispatch
Few hours to one week	Unit commitment
1–6 months	Maintenance scheduling of generation
1–5 years	System planning (renovation/extension) maintenance

EXERCISES

1. Write in brief about the need for power system management.

2. Describe the main functions of load dispatch centres.

3. How does a control centre function?

4. Describe the importance and merits of computerised power system control.

5. Explain the role of a load dispatcher in the consumer set-up.

6. How significant are reporting and data management at load dispatch centres?

Test Your Knowledge

FILL IN THE BLANKS

1. According to the Indian Electricity Rules, the maximum margin between the declared voltage and the voltage available at the consumer end should not be more than ....

2. .... are required at the receiving end to get a good voltage profile under load conditions for a long transmission line.

3. Modern high voltage lines are designed with a safety factor of about ....

4. At natural temperature and pressure, the breakdown strength of air is ....

5. The frequency of resonant vibrations lies in the range of ....

6. .... fault among the transmission line faults causes the highest electromagnetic interference between the power line and a nearby communication line.

7. In the .... fault, zero sequence fault currents are absent.

8. .... sequence network is affected by the methods of neutral grounding.

9. Electromagnetic waves are propagated on the overhead transmission lines with a speed equal to ....

10. Voltage wave denoted as +200/1/50 means that ....

11. When a wave reaches the short circuit point in the transmission line, the voltage at the termination is ....

12. The recommended shielding angle for an HV line is ....

13. The recommended shielding angle for an EHV line is ....

14. Power systems are operated with the power angle being approximately ....

15. In the HV and EHV transmission lines, the ground wires are placed ....

16. A high-speed circuit-breaker improves the ....

17. The outstanding feature of the fuse action is that it breaks the circuit ....

18. .... is the process of interrupting the continuity of the circuit in which the electric current is flowing.

19. The possible overvoltages on power systems are .... and ....

20. The cost of a circuit-breaker is strongly influenced by .... and ....

21. The fuse element rated current is the maximum current that the .... carries continuously.

22. The voltage on fuse is the value indicated on them to show the .... for which they are to be used.

23. The maximum value of the fault current reached to a point at which it is to be cut off is known as ....

24. The difference between a circuit-breaker and a fuse is that a fuse breaks the circuit automatically while a circuit-breaker breaks and .... the circuit.

25. Practically, the speed of the break of a circuit-breaker should be ....

26. .... effect is an important phenomenon in switchgears at higher currents.

27. An increase in fault .... and .... time is the main consideration for the breaking operation of a circuit-breaker.

28. The cut-off current for a fuse is its .... value actually reached.

29. The arc between two electrodes of a circuit-breaker is divided into three distinct parts, i.e. ...., ... and ....

30. The .... of the circuit decides the instantaneous voltage in a circuit-breaker.

31. Bushings are used in electrical apparatus to prevent an .... between the enclosed conductor and the surrounding earthed metal work.

32. The circuit-breaker must not trip during the .... switching operations.

33. .... implies the opening of the circuit-breaker under pre-specified conditions without the intervention of the operator.

34. If the rate of rise of the building up of dielectric strength is higher than the rate of voltage rise in the circuit-breaker, the arc does not ....

35. The rate of rise and the .... of the re-striking voltage characterise it.

36. .... is the value of rms voltage that reappears across the poles of a circuit-breaker before the final arc is extinguished.

37. The relationship among the zero sequence, negative sequence and positive sequence impedances of the solidly grounded system under a steady state condition is ....

38. Carrier current protection schemes are normally used for .... lines only.

39. An Mho relay is normally used for protecting the .... transmission lines.

40. A reactance relay is normally used for protecting the .... transmission lines.

41. The shape of the disc of the induction disc relay is a ....

42. The reactance relay is a directional restraint ....

43. The biasing for generator protection is kept between ....

44. .... wires are required for the translay protection scheme of three-phase transmission lines.

45. In the event of a shunt fault, the current ...., the frequency .... and the power factor ....

46. The vacuum circuit-breaker .... generate a problem in capacitor switching.

47. Vacuum circuit-breakers are preferred for interrupting the .... current when the voltage is high.

48. In order to minimise the current chopping tendency in the SF_6 circuit-breaker, the gas should be kept at .... pressure and .... velocity.

49. Because of its low gaseous viscosity and higher molecular weight, SF_6 gas, which is useful in an SF_6 circuit-breaker, has an excellent .... transfer property.

50. RRRV is most severe in the case of .... length transmission lines.

51. An SF_6 circuit-breaker is most suitable for a short length line fault if the .... resistor is absent.

52. Temperature sensitive relays are employed to protect large size alternators against ....

53. The .... voltages form the basis of insulation design for EHV transmission lines.

54. The .... is the basis of deciding the conductor size of the EHV transmission lines.

55. An Mho relay is a .... restrained directional relay.

56. Arc interruption in the subsequent current zero .... the chances in case of an oil circuit-breaker.

57. Arc interruption in the subsequent current zero .... the chances in case of Air Blast Circuit-breakers (ABCBs).

58. The RRRV depends upon the .... and .... of the system.

59. In case the combination of a circuit-breaker and an HRC fuse is employed, the circuit-breaker will operate at .... overload currents.

60. The Buchholz relay protects the transformer from all types of .... faults.

61. In order to provide mechanical strength to the cables, ... are employed.

62. In order to fill up less volume of oil in Oil Circuit-breakers .... insulation is applied for insulating the .... from the earth.

63. The making current of a circuit-breaker is specified in terms of .... value.

64. In case the system frequency rises from 50 to 52 Hz, the most adversely affected machine would be the ...., among the alternator, power transformer and turbine.

65. The wave-trap unit in the carrier current relaying offers .... impedance to carrier frequency signals but .... impedance to power frequency signals.

66. An .... protection scheme is generally employed for detection of loss of excitation of a very large generating unit feeding power into a grid.

67. The CT secondaries connection in the primary and secondary windings of the transformer would be star-delta in the Merz Price percentage differential protection of a .... transformer.

68. As far as the RRRV is concerned, a fault occurring on an end-supplied transmission line is more severe for a .... fault.

69. Filtering the .... harmonic content of the inrush current flowing through the operating coil and passing through the restraining coil may prevent mal-operation of differential protection of transformers due to the magnetising inrush current.

70. Among the reactance, impedance and mho relays, the operation of the .... relay is most affected by arc resistance.

71. The .... relay among the reactance, impedance and mho relays is most sensitive to power swings.

72. Auto reclosing is used in case of ....

73. A characteristic on R-X diagram for a distance relay is called inherently directional if it is ....

74. Electromagnetic relays are generally rated as .... and ....

75. The voltage rating of the electromagnetic relay is generally kept ....

76. The unit protection scheme provides .... protection.

77. In order to prevent mal-operation of the differentially connected relay while energising a transformer, the relay's restrained coil is biased with the .... current.

78. A protection scheme consists of both the protective relay and ....

79. Relay contacts are preferably made up of .... and enclosed in a ....

80. Air-actuated, i.e. a .... relay is employed to protect the .... against an .... fault.

81. An overcurrent relay is a .... comparator.

82. A distance relay is a .... comparator.

83. The threshold characteristic of an Mho relay is .... from the origin on the real axis in the complex β-plane.

84. .... circuit-breakers are preferably used for long EHV and UHV transmission lines.

85. The .... rating of the relay means the burden of the relay.

86. The carrier current pilot scheme requires the frequency of the carrier in the range of ....

87. The value of the bias factor (S) in the unit protection scheme of an alternator is in the range of ....

88. A short-circuit between turns of the same phase is known as an .... fault in the winding.

89. Infinite impulse response filters are .... type filters.

90. Impedance relaying is used for the protection of ....

91. A circle with the centre at the origin represents the threshold characteristics of a plane ....

92. A .... is used for protection against a line-to-ground fault.

93. A ....…................ is preferred for transformer protection.

94. The inrush magnetising current in a transformer is rich in ....…................ components.

95. The inrush magnetising current in a transformer is ....…................ times the normal rated current.

96. A relay function is a ....…................ device in the protection scheme.

97. The carrier transmitted by a microwave pilot has a frequency ranging between ....…................ .

98. The reactance relay characteristic is realised by setting the set impedance angle equal to ....…................ .

99. Phase faults, which involve more than one phase, are protected by ....…................ .

100. An ....…................ is preferred for protection against a ground fault.

101. For more than one phase fault in a loaded long line, an ....…................ is used.

102. The second harmonic component current induced in the field circuit of a synchronous generator is due to the ....…................ sequence stator current.

103. The wave trap connected at the end of the transmission line consists of a ....…................ circuit.

104. Due to momentary overloading, a ....…................ phenomenon is caused.

105. The travelling wave-based protective relay operate in less than ....…................ cycle.

106. A line trap or coupling capacitor is not required in a ....…................ pilot scheme.

107. The coupling capacitor needed for injecting carrier wave to transmission line is about ....…................ μF.

108. The most severe fault on a three-phase transmission line is a ....…................ .

109. The main advantage of employing a digital relaying scheme is that it provides ....…................ and ....…................ .

110. A ....…................ study needs to be conducted for designing an adequate protection scheme.

111. When the switch is closed at voltage zero, the inrush current in a transformer is the ....…................ .

112. When three-phase quantities are unbalanced but the three-phase excitation is balanced, ...…................ components treatment is applied.

113. The zero sequence quantity is used in if the ....…................ relay is applied.

114. An overcurrent relay having a current setting of 125 per cent is connected to a supply circuit through a CT of ratio 400/5. The pick-up value shall be ....…................

115. Earth relays have ....…................ current settings.

ANSWERS

1. ±6%

2. Shunt capacitors

3. 2

4. 30 kV peak/cm

5. 50-100 Hz

6. Single line-to-ground

7. line-to-line

8. zero

9. speed of light = 3×10^8 m/s

10. Crest of the positive wave = 200 kV, Rise time of the wave = 1 micro seconds, Fall time of the wave = 50 micro seconds

11. Zero

12. 30°

13. 20°

14. 30°

15. above the phase conductors

16. system stability

17. before the fault

18. circuit breaking

19. power frequency overvoltage, high frequency overvoltage

20. the rated voltage, the normal load current rating

21. fuse element

22. maximum service voltage

23. cut-off current

24. makes

25. slow

26. Skin

27. magnitude, time

28. maximum

29. anode fall, cathode fall, plasma

30. power factor

31. electrical breakdown

32. normal

33. Automatic tripping

34. re-struck

35. crest value

36. Recovery voltage

37. $Z_1 > Z_2 > Z_0$

38. HV transmission

39. EHV/UHV long

40. medium

41. spiral

42. overcurrent relay

43. 5 to 10%

44. Pilot

45. increases, reduces, reduces

46. does not

47. low

48. low, low

49. heat

50. short

51. switching

52. overloads

53. switching

54. corona

55. voltage

56. increases

57. decreases

58. capacitance, inductance

59. low
60. internal
61. sheaths
62. solid, contacts
63. peak
64. turbine
65. high, low
66. offset Mho relay
67. delta-star
68. short line
69. second
70. impedance
71. reactance
72. ABCB
73. a circle that passes through the origin
74. 1A, 5A
75. 110 V and 220 V
76. primary
77. second harmonic
78. circuit-breaker
79. silver, moisture-proof enclosure
80. Buchholz, transformer, internal
81. single-input
82. dual-input
83. circle offset
84. Air blast type
85. volt-ampere
86. 50 kHz to 500 kHz
87. 0.05 to 0.1 p.u.
88. inter-turn
89. recursive
90. transmission lines
91. impedance relay
92. ground relay
93. unit protection scheme
94. odd harmonic
95. 10 to 15
96. sensing
97. 900 MHz to 6000 MHz
98. 90°
99. phase relays
100. overcurrent relay
101. Mho relaying scheme
102. negative
103. parallel LC
104. power swing
105. a quarter
106. microwave
107. 0.001
108. three-phase fault
109. self-checking, flexibility
110. short-circuit
111. maximum
112. symmetrical
113. ground overcurrent
114. 6.25 A
115. lower

Bibliography

Computer Relaying Tutorial Text: IEEE Power System Engineering Society Special Publication No. 79 EHO148-7-PWR, 1997.

Date, M.A., B. Oza, and N.C. Nair, *Power System Protection*, Bharati Prakashan, 1999.

Electricity Council, *Power System Protection*, Vols. I, II, III, Macdonald & Co. Ltd., London, 1969.

El-Hawary, Mohamed E., *Electrical Power Systems: Design and Analysis*, IEEE Publication, 1995.

Elmore, Walter A., *Pilot Protective Relaying*, Marcel Dekker, 1999.

Fink, D.G. and H.W. Beaty, *Standard Handbook of Electrical Engineers*, McGraw-Hill, 1975.

Gers, Juan and Edward Holmes, *Protection of Electricity Networks*, IEE Publication, 2004.

Grigsby, Leo-L, *The Electrical Power Engineering Handbook*, CRC Press, 2000.

Johns, A.T. and S.K. Salman, *Power and Energy Digital Protection for Power Systems*, IEE Publication, 2005.

Kimbark, E.W., *Power System Stability*, Vol. II, John Wiley & Sons Inc., New York, 1965.

Mason, C.R., *The Art and Science of Protective Relaying*, John Wiley & Sons, 1956.

McDonald, *Intelligent Knowledge Based Systems in Electrical Power Engineering*, Springer, 1997.

Paithankar, Y.G., *Transmission Network Protection: Theory and Practice*, Marcel Dekker Inc., New York, 1998.

Paithankar, Y.G. and S.R. Bhide, *Fundamentals of Power System Protection*, Prentice-Hall of India, New Delhi, 2005.

Phadke, A.G. and J.S. Thorp, *Computer Relaying for Power Systems*, Research Studies Press, 1988.

Rabindranath, B. and M. Chander, *Power System Protection and Switchgear*, New Age International, 1993.

Ram, Badri and D.N. Biswakarma, *Power System Protection and Switchgears*, Tata McGraw-Hill, 1995.

Ramaswamy, Natrajan, *Computer-aided Power System Analysis*, Marcel Dekker, 2002.

Rao, T.S.M., *Power System Protection—Static Relays with Microprocessor Applications*, Tata McGraw-Hill, 1989.

Rehtanz, Ch., *Autonomous Systems and Intelligent Agents in Power System Control and Operation*, Springer, 2003.

Robertson, D., *Power System Protection Reference Manual*, Oriel Press, Stockfield, London, 1982.

Russel, B.D. and M.E. Council, *Power System Control and Protection*, Academic Press, 1978.

Sadat, H., *Power System Analysis*, Tata McGraw-Hill, New Delhi, 2002.

Shahidehpour, Mohamed and Yaoyu Wang, *Communication and Control in Electric Power Systems*, John Wiley & Sons, 2003.

Singh, L.P., *Advanced Power System Analysis and Dynamics*, 3rd ed., Wiley Eastern, New Delhi, 1992.

Singh, L.P., *Digital Protection—Protective Relaying from Electromechanical to Microprocessor*, New Age International, 1997.

Stagg, G.W., et al., *Computer Methods in Power System Analysis*, McGraw-Hill, New York, 1968.

Strauss, Cobus, *Practical Electrical Network Automation and Communication Systems*, Elsevier, 2003.

Van, A.R., Warrington, C., *Protective Relays*, Vol. 2, Chapman & Hall, London, 1974.

Warrington, A.R. Van C., *Protective Relays Their Theory and Practice*, Vol. I, Chapman & Hall, London, 1976.

Willis, H. Lee, *Power Distribution Planning Reference Book*, Marcel Dekker, 2004.

Research Papers (in chronological order)

Crichton, L.N., 'The Distance Relay for Automatically Sectionalising Electrical Networks', *AIEE Trans.*, Vol. 42, 1923, pp. 527–537.

McLaughlin, H.A. and E.O. Erickson, 'The Impedance Relay', *AIEE Trans.*, Vol. 47, 1928, pp. 776–782.

Crichton, L.N., 'High Speed Protective Relays', *AIEE Trans.*, Vol. 49, 1930, pp. 1232–1242.

George, E.E., 'Operating Experience with Reactance-type Distance Relay', *AIEE Trans.*, Vol. 50, 1931, pp. 288–293.

Goldborough, S.L. and W.A. Lewis, 'New High Speed Distance Relay', *Electrical Engineering*, Vol. 51, March 1932, pp. 157–160.

Warrington, A.R. Van C., 'A High Speed Reactance Relay', *Electrical Engineering*, Vol. 52, April 1933, pp. 248–252.

Sonnemann, W.K., 'A High Speed Differential Relay for Generator Protection', *AIEE Trans.*, Vol. 59, November 1940, pp. 608–612.

Warrington, A.R. Van C., 'Prospective Relaying for Long Transmission Lines', *AIEE Trans.*, Vol. 62, 1943, pp. 261–268.

McConnell, A.J., 'A Generator Differential Relay', *AIEE Trans.*, Vol. 62, January 1943, pp. 11–13.

Goldborough, S.L., 'A Distance Relay with Adjustable Phase-Angle Discrimination'. *AIEE Trans.*, Vol. 63, 1944, pp. 835–838.

Cordrey, and A.R. Van C. Warrington, 'The MHO Carrier Relaying Scheme', *AIEE Trans.*, Vol. 63, 1944, pp. 228–235.

Hutchinson, R.M., 'The MHO Distance Relay', *AIEE Trans.*, Vol. 65, June 1946, pp. 353–359.

Dewey, C.G. and J.R. McGlynn, 'A New Reactance Distance Relay', *AIEE Trans.*, Vol. 67, Part 1, 1948, pp. 743–746.

Mason, C.R., 'A New Loss of Excitation Relay for Synchronous Generator', *AIEE Trans.*, Vol. 68, 1949, pp. 1240–1245.

New Warren, C., 'Combined Phase and Ground Distance Relaying', *AIEE Trans.*, Vol. 69, Part 1, 1950, pp. 37–44.

Hampe, G.W. and B.W. Storer, 'Power-Line Carrier for Relaying and Joint Usage', *AIEE Power Apparatus and Systems*, No. 1, August 1952, pp. 661–670.

Barkle, J.E. and Frankvon Roeschaub, 'Application of Relays for Unbalanced Faults on Generators', *AIEE Trans.*, Vol. 72, Part III, 1953, pp. 277–282.

Moris, W.C. and L.E. Goff, 'A Negative Sequence Overcurrent Relay for Generator Protection', *AIEE Trans.*, Vol. 72, Part III, 1953, pp. 615–621.

Barkle, J.E. and W.E. Glassburn, 'Protection of Generators against Unbalanced Currents', *AIEE Trans.*, Vol. 72, Part III, 1953, pp. 282–286.

Adamson, C. and L.M. Wedepohl, 'A Dual-Comparator Mho-type Distance Relay Utilizing Transistors', *Proc. IEE*, Vol. 103-A, September 1956, pp. 509–517.

Adamson, C. and L.M. Wedepohl, 'Power System Protection with Particular Reference to the Application of Junction Transistors to Distance Relays', Proc. IEE, Vol. 103 A, 1956, p. 379.

Adamson, C. and E.A. Talkhan, 'The Application of Transistors to Phase-Comparison Carrier Protection', *Proc. IEE*, Vol. 106, Part A, February 1959, pp. 51–63.

Dewey, C.G. and M.E. Hodges, 'Transistorised Phase-Comparison Relaying: Principles and Circuits', *AIEE Trans.*, Vol. 79, Part III, August 1960, pp. 373–380.

Ahmad, H. El-Abiad, 'Digital Calculation of Line to Ground Short-circuit by Method', *AIEE Trans.*, Vol. 79, Part III, 1960, pp. 323–332.

Ahmad, H. EI-Abiad, et al., 'Calculation of Short-circuit Using a High Speed Digital Computer', *AIEE Trans.*, Vol. 80, Part III, 1961, pp. 702–707.

Skuderna, J.E., 'A Mathematical Basis for a Protective Relay with Conic Pick-up Characteristics', *AIEE Trans.*, Vol. 81, Part III, 1962, pp. 81–88.

Dewey, C.G., C.A. Mathews and W.C. Morris, 'Static Mho Distance and Pilot Relaying: Principles and Circuits', *IEEE Trans. on Power Apparatus and Systems*, Vol. PAS-82, June 1963, pp. 391–400.

Dalasta, D., F. Free and A.P. Desnoo, 'An Improved Static Overcurrent Relay', *IEEE Trans. on Power Apparatus and Systems*, No. 68, October 1963, pp. 705–716.

Mathews, P. and B.D. Nellist, 'Transients in Distance Protection', *Proc. IEEE*, Vol. 110, 1963, pp. 407–418.

Caleca, V., S.H. Horowitz, A.J. McConnell and H.T. Seeley, 'Static Mho Distance and Pilot Relaying: Application and Test Results', *IEEE Trans. on Power Apparatus and Systems*, Vol. PAS-82, August 1963, pp. 424–436.

Begian, S.S., et al., 'A Computer Approach on Setting Relay in a Network', *IEEE PICA Conference Record*, Vol. 31c 69, May 1964, pp. 447–447.

Rao, T.S.M., 'Behaviour of Rectifier Bridge Comparator as an Amplitude Comparator', *Journal of Institution of Engineers* (India), Vol. XLV, No. 10, June 1965, pp. 220–230.

J. Rushton, 'Pilot Wire Differential Protection Characteristics for Multi-ended Circuits', *Proc. IEE*, November 1965, p. 2095.

Wedepohl, L.M., 'Polarised Mho Distance Relay—A New Approach to the Analysis of Practical Characteristics', *Proc. IEEE*, Vol. 112, March 1965, pp. 525–535.

Humpage, W.D. and S.P. Sabberwal, 'Developments in Phase-Comparison Techniques for Distance Protection', *Proc. IEEE*, Vol. 112, July 1965, pp. 1383–1394.

Hoel, Hans, W.D. Humpage and C.P. Chapman, 'Composite Polar Characteristics in Multi-zone Systems of Phase-Comparison Distance Protection', *Proc. IEEE*, Vol. 113, October 1966, pp. 1631–1642.

K. Parthasarathy, 'Pilot Wire Differential Protection', *Electrical Times*, March 10, 1966, p. 320.

Anil Kumar, N.M. and K. Parthasarathy, 'A Mathematical Basis for Multi-Input Sine Comparators', *Journal of Institution of Engineers* (India), Vol. 48, Part EL 1, October 1967, pp. 9–21.

Anil Kumar, N.M., K. Parthasarathy and K.K. Thakkar, 'A Three-Step Transistorised Distance Relay with Quadrilateral Polar Characteristics', *Journal of Institution of Engineers* (India), Vol. 48, Part EL 2, December 1967, pp. 189–204.

Vitanov, A., 'Quadrilateral-Characteristic: Transistor Distance Protection', *CIGRE*, Paris, 1968, Paper No. 31–03.

Jackson, L., J.B. Patrickson and L.M. Wedephol, 'Distance Protection: Optimum Dynamic Design of Static Relay Comparators', *Proc. IEEE*, Vol. 115, February 1968, pp. 280–287.

Slemon, G.R., S.D.T. Robertson and M. Ramamoorty, 'High Speed Protection of Power Systems Based on Improved System Models', *CIGRE*, Paris, 1968, Paper No. 31–09.

Jackson, L., J.B. Patrickson and L.M. Wedephol, 'Distance Protection: Optimum Dynamic Design of Static Relay Comparators', *Proc. IEE*, Vol. 113, February 1968, pp. 280–287.

Maleev, A.M., 'Three-Phase Impedance Relay Responding to the Phase-Sequence of Four Electrical Quantities', *Electrihertvo*, USSR, 1968.

McLaren, P.G., 'Static Sampling Distance Relays', *Proc. IEEE*, Vol. 115, November 1968, pp. 418–424.

Gupta, S.C. 'Static Polyphase Distance Relay Schemes for the Protection of Transmission Lines', Ph.D. Thesis, Roorkee University, November 1969.

Rockefeller, G.D., 'Fault Protection with a Digital Computer', *IEEE Trans. on Power Applications and Systems*, Vol. PAS-88, April 1969, pp. 438–464.

Slemon, G.R., S.D.T. Robertson and M. Ramamoorty, 'High Speed Protection of Power System Based on Improved System Models', *CIGRE*, Paris, 1969, Paper No. 31–09.

Ramamoorty, M. and N.S. Wani, 'Static Distance Relay with Quadrilateral Characteristic', Presented at the IEEE Summer Meeting, 1970, Paper No. 70C571.

Anil Kumar, N.M., 'New Approach to Distance Relays with Quadrilateral Polar Characteristics for EHV Line Protection', *Proc. IEEE*, Vol. 117, October 1970, pp. 1986–1992.

Walker, L.N., et al., 'Special Purpose Digital Computer Requirements for Power System Sub-station Needs', *IEEE Winter Meeting*, Paper No. 70CP142-PWR, 1970.

Khincha, H.P., K. Parthasarathy and B.S. Ashok Kumar, 'Developments in Amplitude Comparator Techniques for Distance Relays', *Proc. IEEE*, Vol. 117, June 1970, pp. 1118–1124.

Khincha, H.P., K. Parthasarathy, B.S. Ashok Kumar and C.G. Arun, 'New Possibilities in Amplitude and Phase Comparison Techniques for Distance Relays', *Proc. IEEE*, Vol. 117, Nov. 1970, pp. 2133–2141.

Khincha, H.P., K. Parthasarathy and B.S. Ashok Kumar, 'Developments in Amplitude Comparator Techniques for Distance Relays', *Proc. IEE*, Vol. 117, June 1970, pp. 1118–1124.

Ramamoorthy, M., 'A Note on Impedance Measurement Using Digital Computers', *IEE-IERE Proc.* (India), Vol. 9, November/December 1971, pp. 243–247.

Mann, B.J. and I.F. Morrison, 'Digital Calculation of Impedance for Transmission Line Protection', *IEEE Trans. on Power Apparatus and Systems*, Vol. PAS-90, January/February 1971, pp. 270–279.

Mann, B.J. and I.F. Morrison, 'Relaying a Three-phase Transmission Line with a Digital Computer', *IEEE Trans. on Power Apparatus and Systems*, Vol. PAS-90, March/April 1971, pp. 742–750.

McInnes, A.D. and I.F. Morrison, 'Real-time Calculation of Resistance and Reactance for Transmission Line Protection by Digital Computer', *IEE Trans.*, Institution of Engineers, Australia, Vol. EE7, No. 1, 1971, pp. 16–23.

Poncelet, R., 'The Use of Digital Computers for Network Protection', *CIGRE*, Paris, 1972, Paper No. 32–08.

Choudhury, S., S.K. Basu and S.P. Patra, 'Polyphase Ground Distance Relaying by Phase Coincidence Principle', Transaction Paper No. T-72, 4281, Presented at *IEEE Summer Power Meeting*, 1972, *IEEE Trans.*, PAS-92, 1972, pp. 626–634.

Gilcrest, G.B., G.D. Rockfeller and E.A. Udren, 'High Speed Distance Relaying Using a Digital Computer: Part-I—System Description', *IEEE Trans. on Power Apparatus and Systems*, Vol. PAS-91, May/June 1972, pp. 1235–1242.

Johns, A.T., 'Generalised Phase Comparator Techniques for Distance Protection Basis of Their Operation and Design', *Proc. IEEE*, Vol. 119, July 1972, pp. 833–841.

Johns, A.T., 'Generalised Phase Comparator Techniques for Distance Protection—Theory and Operation of Multi-input Devices', *Proc. IEEE*, Vol. 119, November 1972, pp. 1595–1603.

Paithankar, Y.G. and M.V. Despande, 'Polyphase Distance Relay', *Proc. IEE*, Vol. 120(9), 1973, pp. 1013–1015.

Brown, P.G., 'Generating Requirement for System Faults', *IEEE Trans.*, Vol. PAS 72, July/August 1973, pp. 1247–1251.

Johns, A.T., 'Vibrable-Characteristic Generalised Techniques for Distance Protection: Theory and Initial Performance Studies', *Proc. IEEE*, Vol. 120, August 1973, pp. 891–899.

Gupta, S.C. and T.S.M. Rao, "Improved Static Overcurrent Relay with IDMT Characteristics', *Journal of Institution of Engineers* (India), Vol. 53, February 1973, pp. 138–140.

Paithankar, Y.G. and V.T. Ingole, 'New Techniques in Multi-input Amplitude Comparators to Generate Quadrilateral Distance Relay', *Journal of Institution of Engineers* (India), Vol. 54, Part EL 3, February 1974, pp. 79–85.

Hope, G.S. and V.S. Umamaheswaran, 'Sampling for Computer Protection of Transmission Lines', *IEEE Trans. on Power Apparatus and Systems*, Vol. PAS-93, September/October 1974, pp. 1522–1534.

Sanderson, J.V.H. and A. Wright, 'Protective Scheme for Series-Compensated Transmission Lines', *Proc. IEE*, Vol. 121, November 1974, pp. 1377–1384.

Basu, S.K., S. Chaudhuri and R. Chakrabarti, 'Poly-phase Distance Relaying by Phase Sequence Detection', *Journal of Institution of Engineers* (India), Vol. 54, Part EL-4, 1974, pp. 135–141.

Patra, S.P., S.K. Basu, S. Chaudhuri and R. Chakrabarti, 'A Single Unit Distance Relay for Three-phase Transmission Lines', *Electrical India*, March 1974.

Jackson, L., 'Distance-Protection Comparator with Signal Dependent Phase-Angle Criterion', *Proc. IEEE*, Vol. 121, August 1974, pp. 817–825.

Johns, A.T., 'Vibrable-Characteristic Generalised Techniques for Distance Protection: Double-Circuit Application Studies', *Proc. IEEE*, Vol. 121, December 1974.

Carr, J. and R.V. Jackson, 'Frequency Domain Analysis Applied to Digital Transmission Line Protection', *IEEE Trans. on Power Apparatus and Systems*, Vol. PAS-94, July/August 1975, pp. 1157–1166.

Ranjbar, A.M. and B.J. Cory, 'An Improved Method for the Digital Protection of Transmission Lines', *IEEE Trans. on Power Delivery*, Vol. PAS-94, March/April 1975, pp. 544–550.

Luckett, R.G., P.J. Munday and B.E. Murray, 'A Sub-station-based Computer for Control and Protection', Conference on Modern Developments in Protection, Institution of Electrical Engineers, London, 1975, Conference Publication No. 125, pp. 252–260.

Gilbert, J.G. and R.J. Shovlin, 'High Speed Transmission Line Fault Impedance Calculation Using a Dedicated Minicomputer'. *IEEE Trans. on Power Apparatus and Systems*, Vol. PAS-94, May/June 1975, pp. 872–883.

Danniel, J., 'Generator Protection with New Static Negative Sequence Relay', *IEEE Trans.*, Vol. PAS 94, 1975, pp. 1205–1213.

Berdy, John, 'Loss of Excitation Relaying', *IEEE Trans.*, Vol. PAS 94, 1975, pp. 1457–1463.

Rajbar, A.M. and B.J. Cory, 'An Improved Method for the Digital Protection of Transmission Lines', *IEEE Trans. on Power Apparatus and Systems*, Vol. PAS-94, March/April 1975, pp. 544–550.

Phadke, A.G., T. Hlibka and M. Ibrahim, 'A Digital Computer System for EHV Sub-stations: Analysis and Field Tests', *IEEE Trans. on Power Apparatus and Systems*, Vol. PAS-95, January/February 1976, pp. 291–301.

Verma, H.K. and T.S.M. Rao, 'Inverse Time Overcurrent Relay Using Linear Components', *IEEE Trans. on Power Apparatus and Systems*, Vol. PAS-95, No. 5, September/October 1976, pp. 1738–1743.

Hope, G.S., P.K. Dash and O.P. Malik, 'Digital Differential Protections of a Generation Unit', *IEEE Trans.*, PAS 75, 1976.

Hope, G.S., O.P. Malik and M.E. Rasmy, 'Digital Transmission Line Protection in Real-time', *Proc. IEE*, Vol. 123, December 1976, pp. 1349–1354.

Horton, J.W., 'Walsh Functions for Digital Impedance Relaying for Power Lines', *IBM Journal for Research and Development*, No. 10, 1976, pp. 530–541.

Brooke (Jr.), A.W., 'Distance Relaying Using Least-Square Estimates of Voltage, Current and Impedance', *Proceedings of PICA Conference*, IEEE Publication No. 77 Ch 1131-2 PWR, May 1977, pp. 394–402.

Imlof, J.A., et al., 'Out-of-Step Relaying for Generators', Working Group Report, *IEEE Trans. on Power Apparatus and Systems*, September/October 1977, pp. 1556–1564.

Phadke, A.G., M. Ibrahim and T. Hlibka, 'Fundamental Basis for Distance Relaying with Symmetrical Components', *IEEE Trans. on Power Apparatus and Systems*, Vol. PAS-96, March/April 1977, pp. 635–646.

Miki, Y., Y. Sano and J.I. Makino, 'Study on High Speed Distance Relay Using Microcomputer', *IEEE Trans. on Power Apparatus and Systems*, Vol. PAS-96, No. 2, March/April 1977, pp. 602–613.

Johns, A.T. and M.A. Martin, 'Fundamental Digital Approach to the Distance Protection of E.H.V. Transmission Lines', *Proc. IEE*, Vol. 125, May 1978, pp. 377–384.

Dommel, H.W. and J.M. Michels, 'High Speed Relaying Using Travelling Wave Transient Analysis', Paper presented at the IEEE PES Winter Meeting, January, IEEE Publication. No. 78CH1295-PWR, Paper No. A78, pp. 214–219.

Esztergalyos, J., M.T. Yee, M. Chamia and S. Liberman, 'The Development and Operation of an Ultra High Speed Relaying System for EHV Transmission Lines', *CIGRE*, Paris, 1978, Paper No. 34–04.

Chamia, M. and S. Liberman, 'Ultra High Speed Relay for EHV/UHV Transmission Lines-Development, Design and Application', *IEEE Trans. on Power Apparatus and Systems*, Vol. PAS-97, November/December 1978, pp. 2104–2116.

Gilbert, J.G., E.A. Udren and H. Sackin, 'The Development and Selection of Algorithms for Relaying of Transmission Lines by Digital Computers', *Power System Control and Protection*, edited by B.D. Russel and M.E. Council, Academic Press, New York, 1978, pp. 83–127.

Vitins, M., 'A Correlation Method for Transmission Line Protection', *IEEE Trans. on Power Apparatus and Systems*, Vol. PAS-97, September/October 1978, pp. 1607–1616.

Breingan, W.D., M.M. Chen and T.F. Gellen, 'The Laboratory Investigation of a Digital System for the Protection of Transmission Lines', *IEEE Trans. on Power Apparatus and Systems*, Vol. PAS-98, No. 2, March/April 1979, pp. 350–368.

Johns, A.T., 'New Discriminative Distance—Protective Relays for Selective Pole Auto-reclosure Applications', *Proc. IEE*, Vol. 126, February 1979, pp. 159–161.

Smolonski, W.J., 'An Algorithm for Digital Impedance Calculation Using a Single Pi Selection Transmission Line', *IEEE Trans. on Power Apparatus and Systems*, Vol. PAS-98, September/ October 1979, pp. 1546–1551.

Lee, D.C., 'Loss of Excitation Protection Scheme to Prevent Coupling of Sensitive Load Adjacent to Station with Only One Unit Operation', *IEEE Trans.*, Vol. PAS 98, 1979, pp. 1895–1897.

Roy, L., 'Generalised Poly-phase Fault Analysis Program: Calculations of Cross-country Fault', *Proc. IEE*, Vol. 126, No. 10, 1979, pp. 995–1001.

Sachdev, M.S. and M.A. Baribeau, 'A New Algorithm for Digital Impedance Relay', *IEEE Trans. on Power Apparatus and Systems*, Vol. PAS-98, May/June 1979, pp. 2232–2240.

Takagi, T. and Y. Yamakosi, 'Digital Differential Relaying System for Transmission Line Primary Protection Using Travelling Wave Theory—Its Theory and Field Experience', Paper presented at the *IEEE PES Winter Meeting*, 1979, Paper No. A79-096-9.

Ramamoorty, M. and S.N. Lall, 'A Versatile Phase Comparator Relay Using Digital Circuits', *Journal of Institution of Engineers* (India), Vol. 59, Part E16, June 1979, pp. 309–314.

Sachdev, M.S. and M.A. Baribeau, 'A New Algorithm for Digital Impedance Relay', *IEEE Trans. on Power Apparatus and Systems*, Vol. PAS-98, May/June 1979, pp. 2232–2240.

Kellog, A.J., L.P. Singh and G.K. Dubey, 'Protection of EHV Transmission Lines by Using Static Relays', Proceedings of Conference on Power System Protection, Madras, April 1980, Vol. 1, *Institution of Engineers* (India), Calcutta, p. B5.

Partasarathy, K., H.P. Khincha and B.L. Mathur, 'An Adaptive Distance Relay for the Protection of Transmission Lines', Conference Papers, Vol. 1: Conference on Power System Protection, Madras (India), April 16–19, 1980, Paper No. A3.

Ramamoorthy, M. and S.N. Lall, 'Digital Multi-Input Phase Comparator', Conference Papers, Vol. 1, Conference on Power System Protection, Madras (India), April 16–19, 1980, Paper No. A5.

Paithankar, Y.G. and A.S. Thoke, 'Variable Characteristic Earth-Fault Quadrilateral Distance Relay Suitable for Double-End-Infeed Lines: A New Technique', Conf. Paper: Vol. 1, Conf. Power System Protection, Madras (India), April 16–19, 1980, Paper No. A6.

Parthasarathy, K. 'New Static Three-Step Distance Relay', *Proc. IEEE*, Vol. 113, April 16–19 1980, Paper No. A6.

Ramamoorty, M., M.N. Gandhi, R.J. Phansalkar and P.J. Durkal, 'Static Distance Relay with Conic Characteristics', Conference Papers: Vol. 1, Conf. Power System Protection, Madras (India), April 16–19, 1980, Paper No. A7.

Weller, G.C., A. Newbould and R.B. Miller, 'New Principles for Distance Protective Relays', Second Int. Conf. on Developments in Power System Protection, London, June 10–12, 1980, Conference. Publication. No. 185, *Institution of Electronics Engineers*, 1980, pp. 284–288.

Sakaguchi, T. and K. Uemura, 'A New Directional Protection Based on Laplace Transformation with Special Reference to Computer Relaying', Second International Conference on Developments in Power System Protection, London, June 10–12, 1980, Conference Publication No. 185 Institution of Electrical Engineers, London, 1980, pp. 146–150.

Humpage, W.D., K.P. Wong, T.T. Nguyen and D. Sutanto, 'Z-Transform Electromagnetic Transient Analysis in Power System', *Proc. IEE*, Part C, Vol. 127, June 1980, pp. 370–378.

Johns, A.T., 'New Ultra-High Speed Directional Comparison Technique for the Protection of EHV Transmission Lines', *Proc. IEE*, Vol. 127, Part C, July 1980, pp. 228–239.

Vitins, M., 'A Fundamental Concept for High Speed Relaying', *IEEE Trans. on Power Apparatus and Systems*, Vol. PAS-100, January 1981, pp. 163–173.

Girgis, A.A. and R.G. Brown, 'Application of Kalman Filtering in Computer Relaying', *IEEE Trans. on Power Apparatus and Systems*, Vol. PAS-100, July 1981, pp. 3387–3395.

Wiszniewaski, A., 'How to Reduce Error of Distance Fault Locating Algorithm', *IEEE Trans. on Power Apparatus and Systems*, Vol. PAS-100, December 1981, pp. 4815–4820.

IEEE Committee Report, 'Review of Recent Practices and Trends in Protective Relaying', *IEEE Trans. on Power Apparatus and Systems*, Vol. PAS-100, August 1981, pp. 4054–4063.

Desikachar, K.V. and L.P. Singh, 'Protection of EHV/UHV Transmission Lines Using Travelling Wave Phenomena', Paper presented to the National Power Systems Conference (India), December 1981, pp. S4.4.1.

Nanda, S.P., S.N. Tiwari and L.P. Singh, 'Fault Analysis of Six-Phase System', *Electrical Power System Research*, Vol. 4, 1981, pp. 201–211.

Rehman, M.A. and P.K. Dash, 'Fast Algorithm for Digital Protection of Power Transformer', *Proc. IEE*, Vol. 129, Part C, March 1982, pp. 79–85.

Kellog, A.J., L.P. Singh and G.K. Dubey, 'General Purpose Static Relay Using Digital Techniques', *Journal of Institution of Engineers* (India), Vol. 63, pt. EL-1, August 1982, pp. 5–11.

Tiwari, S.N. and L.P. Singh, 'Mathematical Modelling and Analysis of Multi-phase Systems', *IEEE Trans.*, Vol. PAS-101, 1982, pp. 1784–1793.

Bernard, P. and J.C. Bastide, 'A Prototype of Multiprocessor-based Distance Relay', *IEEE Trans. on Power Apparatus and Systems*, Vol. PAS-101, February 1982, pp. 491–498.

Sriniwas Rao, B.G., H.P. Khincha, K. Parthsarathy, 'Algorithm for Digital Impedance Calculation Using Walsh Functions', *Journal of Institution of Engineers* (India), Vol. 62, February 1982, pp. 184–187.

Thirupathaiah, G., A. Varshney and L.P. Singh, 'On-line Digital Protection Using a microcomputer—A Decentralised Approach', Technical Proceedings, National Conference on Planning, Operation and Control of Large Scale Power Systems, Annamalai University, India, June 24–25, 1982, Paper No. S5.3.

Desikachar, K.V., Y. Asthana and L.P. Singh, 'New Relay Scheme for EHV/UHV Transmission Lines', *Journal of Institution of Engineers* (India), Vol. 63, Part EL 3, December 1982, pp. 159–164.

Girgis, A.A., 'A New Kalman Filtering-based Digital Distance Relay', *IEEE Trans. on Power Apparatus and Systems*, Vol. PAS-101, September 1982, pp. 3471–3480.

Bornard, P. and J.C. Bastide, 'A Prototype of Multiprocessor-based Distance Relay', *IEEE Trans. on Power Apparatus and Systems*, Vol. PAS-101, February 1982, pp. 491–498.

Thorp, J.S. and A.G. Phadke, 'A Microprocessor-based Three-phase Transformer Differential Relay', *IEEE Trans.*, Vol. PAS-101, February 1982, pp. 426–432.

Degan, A.J., 'Microprocessor-implemented Digital Filter for the Calculation of Symmetrical Components', *Proc. IEE*, Vol. 129, May 1982, pp. 111–118.

Dasgupta, K.S., O.P. Malik and G.S. Hope, 'Kalman Filtering Approach to Impedance Protection', *Trans. on Engineering and Operating Division*, Can. Elect. Assoc., March 1983, Paper No. 83-SP-171, pp. 1–14.

Desikachar, K.V. and L.P. Singh, 'Fault Analysis of Multi-node Power System for Designing Ultra Speed Protective Relays along with a Proposed Relaying Scheme', *Electrical Power System Research*, Vol. 6, 1983, pp. 13–25.

Desikachar, K.V., 'Ultra High Speed Protective Relaying Scheme for EHV/UHV Transmission Line Based upon Travelling Wave Phenomena', Ph.D. thesis submitted to IIT Kanpur, May 1983.

Ram, B. and B.B. Chakravarty, "Microprocessor-based Distance Relays", *Journal of Microcomputer Applications*, No. 6, 1983, Academic Press Inc. (London).

Dasgupta, K.S., O.P. Malik and G.S. Hope, 'Kalman Filtering Approach to Impedance Protection', *Transaction on Canada Electrical Association*, March 1983, pp. 1–14.

Basu, K.P., G. Mahboob and B.H. Khan, 'Distance Relay Characteristics for Protection of Series Compensated Transmission Lines', *Journal of the Institution of Engineers* (India), 63, Part EL 4, February 1983, pp. 203–205.

Tiwari, S.N. and L.P. Singh, 'Six-Phase (Multi-Phase) Power Systems: Some Aspects of Modelling and Fault Analysis', *Electrical Power System Research*, Vol. 6, 1983, pp. 193–202.

Desikachar, K.V. and L.P. Singh, 'A Novel Protective System for Transmission Line Protection Based on Travelling Wave Phenomena', Paper presented to the Second National Power Systems Conference, Hyderabad (India), September 1983.

Kellog, A.J., L.P. Singh and G.K. Dubey, 'Three-phase Poly-phase Distance Relay Using Digital Circuits', Paper presented to the Second National Power Systems Conference, Hyderabad (India), September 1983.

B. Jeyasuria and W.J. Smolinski, 'Identification of a Best Algorithm for Digital Distance Protection of Transmission Lines', *IEEE Trans. on Power Apparatus and Systems*, Vol. PAS-102, No. 10, October 1983, pp. 3358–3369.

Desikachar, K.V. and Singh L.P., 'Digital Travelling Wave Protection of Transmission Lines', *Electric Power System Research*, Vol. 7, 1984, pp. 19–28.

Fakhruddin, D.R., K. Parthsarthy and D. Tukaram, 'An Intregrated Scheme Based on Haar-Functions for Protection of Synchronous Generators', Proceedings of the Third National Power System Conference, 1984, Delhi.

Kellog, A.J., L.P. Singh and G.K. Dubey, 'A High Speed Three-Zone Distance Relay Using Digital Circuit', International IEEE Conference on Computers etc., Bangalore, December 1984, Proceedings III, pp. 1349–1353.

Desikachar, K.V. and L.P. Singh, 'A Proposed Ultra High Speed Relaying Scheme for the Protection of an Existing 400 KV Line', Proceedings of International (IEEE) Conference on Computers, etc., Bangalore, December 9–12, 1984, Proc. 3, pp. 1359–1363.

Sunak, V.P. and L.P. Singh, 'Microprocessor-based Protection Scheme for EHV/UHV Transmission Line', *Journal of Institution of Engineers* (India), (1) Vol. 65, Parts EL2 and EL3, October-November 1984, pp. 89–96.

Gokul, P., V.P. Sunak and L.P. Singh, 'Digital Protection of EHV Transmission Line', Proceedings of the 51st R&D Section of CBIP at Vadodara (India), January 1984, pp. 75–88.

Jeyasra, B.J., W.J. Smolinski, 'Design and Testing of a Microprocessor-based Distance Relay', *IEEE Trans.*, Vol. PAS-103, No. 5, May 1984, pp. 1104–1110.

Kellog, A.J., L.P. Singh and G.K. Dubey, 'A High Speed Three-Zone Distance Relay Using Digital Circuits', *Electrical Power System Research*, Vol. 8, 1984–85, pp. 187–195.

Kellog, A.J., 'Novel Distance Relaying Schemes for the Protection of EHV/UHV Transmission Lines Using Digital Techniques', Ph.D. Thesis, IIT Kanpur, July 1984.

Fakhruddin, D.B., K. Parthasarathy, L. Jenkins and B.W. Hogg, 'Application of Haar Functions for Transmission Line and Transformer Differential Protection', *Electrical Power and Energy System*, Vol. 6, No. 3, July 1984, pp. 169–180.

Rajendra, S. and P.G. Mc-Laren, 'Travelling Wave Techniques Applied to Protection of Teed Circuits: Principles of Travelling Wave Techniques', *IEEE Trans. on Power Apparatus and Systems*, Vol. PAS-104, 1984, pp. 3544–50.

McLaren, P.G. and S. Rajendra, 'Ultra High Speed Distance Protection Based on Travelling Wave', Third International Conference on Development in Power System Protection, *IEE Conference Publication No. 249*, 1985, pp. 106–110.

Jeyasurya, B. and M.A. Rohma, 'Application of Walsh Functions for Microprocessor-based Transformer', *Digital Protection of Power Transformer*, Vol. EMC-27, No. 4, November 1985, pp. 221–225.

Islam, K.K. and L.P. Singh, 'Microprocessor: Application to Power System Protection', 52nd R&D Session of CBIP, Aurangabad, February 1985, Vol. III, pp. 7–12.

Sachdev, M.S., J.C. Wood and N.G. Johnson, 'Kalman Filtering Applied to Power System Measurements for Relaying', *IEEE Trans. on Power Apparatus and Systems*, Vol. PAS-104, December 1985.

Rajendra, S, and P.G. McLaren, 'Travelling Wave Techniques Applied to the Protection of Teed Circuits: Multi-phase/Multi-conductor System', *IEEE Trans. on Power Apparatus and Systems*, Vol. PAS-104, 1985, pp. 3551–3557.

Paithankar, Y.G. and M.T. Sant, 'A New Algorithm for Relaying Fault Location Based on Travelling Waves', *Electrical Power System Research* (Switzerland), Vol. 8, 1985, pp. 179–185.

Umapal and L.P. Singh, 'Feasibility and Fault Analysis of Multi-Phase (12-Phase) Systems', *Journal of Institution of Engineers* (India), Vol. 65, Pt El-4, February 1985, pp. 138–146.

Johns, A.T. and M.A. Martin, et al., 'A New Approach to EHV Direction Comparison Protection Using Digital Signal Processing Techniques', *IEEE Trans. on Power Delivery*, (USA), Vol. PWRD-1, 1986, pp. 24–43.

Ram, B., 'Microprocessor-based Overcurrent and Directional Relays', *Journal of the Institution of Engineers* (India), Vol. 67, Part EL2, October 1986.

Jeysurya, B. and M.A. Rahman, 'Transmission Line Distance Protection by Spectral Estimation Using Rectangular Wave Transforms', *Electrical Machines and Power Systems*, Vol. 11, 1986, pp. 65–75.

Islam, K.K., S.K. Bose and L.P. Singh, 'On-line Microprocessor-based Relaying Scheme', Proceedings of the Platinum Jubilee Conference on Systems and Signal Processing, IISc, Bangalore, December 1986, pp. 241–244.

Mansour, M.M. and G.W. Swift, 'A Multiprocessor-based Travelling Wave Relay', *IEEE Trans. on Power Delivery* (USA), Vol. PWRD-1, 1986, pp. 272–279.

Desikachar, K.V. and L.P. Singh, 'Application of FET to the Computation of Electrical Transients', Proceedings of the National Power System Conference, Annamalai University, Chidambaram, August 1986.

Siyaram and L.P. Singh, 'Digital Protection of Transmission Line', Proceedings of the 53rd CBIP, R&D Conference, Bhubaneswar, Vol. 7, May 1986, pp. 85–98.

Chu, H.Y., S.L. Chen, C.L. Haung, 'Fault Impedance Calculation Algorithms for Transmission Line Distance Protection', *Electric Power System Research*, Vol. 10, 1986, pp. 60–75.

Singh, L.P. and V.P. Singh, 'Coordination and Setting of Relays: A Case Study of Chukha Transmission System', Proceedings of the International Symposium on EHV/UHV A.C. and HVDC Transmission System, Bangalore, 1987, SiV-27.

Islam, K.K., S.K. Bose and L.P. Singh, 'On-line Microprocessor-based Relaying Scheme for EHV/ UHV Transmission Line: An Existing 400 KV Transmission Line', *Electrical Machines and Power System*, Vol. 12, 1987, pp. 313–324.

Ram, B., 'Modelling of Distance Relays for Digital Protection', Proceedings of the International AMSE Conference, 29–31 October 1987, New Delhi.

Singh, L.P. and V.P. Singh, 'Coordination and Setting of Relays: A Case Study of the Chukha Transmission System', Proceedings of the CBIP International Conference, May 1987, Bangalore, pp. 27–32.

Islam, K.K., S.K. Bose and L.P. Singh, 'Digital Protection of EHV/UHV Transmission Line Based Upon Travelling Wave Phenomena', *Electrical Machines and Power System*, No. 14, 1988, pp. 413–431.

Viswakarma, D.N. and B. Ram, 'Microprocessor-based Offset MHO Relays", *Journal of Institution of Engineers* (India), Vol. 69, EL2, October 1988, p. 78.

Kudo, H., et al., 'Implementation of a Digital Distance Relay Using an Integral Solution of a Differential Equation', *IEEE Trans. on Power Delivery*, Vol. PWRD-3, No. 4, October 1988, pp. 1475–1484.

Islam, K.K., 'Digital Protection of EHV/UHV Transmission Line', Ph.D. Thesis submitted to IIT Kanpur, September 1988.

Singhal, Arun and L.P. Singh, 'Microprocessor-based Relaying Scheme for the Protection of Power Transformer', Proceedings of the National Conference on Power System Protection, Bangalore, January 11–13, 1988, pp. B: 2.1.

Singhal, Arun and L.P. Singh, 'On-line Protection of Power Transformer Using Microprocessor', *Journal of Institution of Engineers* (India), Vol. 69, EL-2, October 1988, pp. 52–58.

Dash, P.K. and D.K. Panda, 'Digital Impedance Protection of Power Transmission Line Using Spectral Observer', *IEEE Trans. on Power Delivery*, Vol. 3, No. 1, January 1988, pp. 121–129.

Horowitz, S.H., A.G. Phadke and J.S. Thorp, 'Adaptive Transmission System Relaying', *IEEE Trans. on Power Delivery*, Vol. 3, No. 4, October 1988, pp. 1436–1445.

Dash, P.K. and H.P Khincha, 'New Algorithms for Computer Relaying for Power Transmission Lines', *Electric Machines and Power Systems*, Vol. 14, No. 3–4, 1988, pp. 163–178.

Singhal, Arun and L.P. Singh, 'On-line Protection of Power Transformer Using Microprocessor', *Journal of Institution of Engineers* (India), Vol. 69, Pt El-2, October 1988, pp. 52–58.

Trivedi, Maj. V.I., 'Digital Protection of a Synchronous Generator', M.Tech. thesis submitted to IIT Kanpur, March 1989.

Chandra, Amulya and L.P. Singh, 'Fault Studies on Six-phase Systems and Proposed Ultra High Speed Relaying Scheme', Proceedings of National Seminar on High Phase Order System, Kanpur, March 1989, pp. 124–134.

Daneshdoost, M. and R. Shaat, 'A PC-based Integrated Software for Power System Education', *IEEE Trans. Power System*. Vol. 4, No. 3, August 1989, pp. 1285–1292.

D'Amoro, D. and A. Ferrero, 'A Simplified Algorithm for Digital Distance Protection Based on Fourier Techniques', *IEEE Trans. on Power Delivery*, Vol. PWRD-4, No. 1, January 1989, pp. 157–163.

Chandra, A., Sachchidanand and L.P. Singh, 'Proposed Ultra High Speed Relaying Scheme for High Phase Order (HPOT) System'. Proceedings of the 16th Annual Conference of the *IEEE Industrial Electronic Society* (IECON'90), Pacific Grove, California, USA, November 27–30, 1990.

Jain, Rajeev and L.P. Singh, 'Digital Protection of Transformers', Proceedings of the Sixth National Power System Conference, Bombay, June 1990, pp. 564–569.

Barnwal, Rajiv Kumar, P.K. Kalra and L.P. Singh, 'Walsh Transform-based Digital Protection of Power Transformer', Proceedings of the *IEEE Annual Convention and Exhibition*, Bangalore, January 1991, pp. 62–66.

Biswakarma, D.N. and B. Ram, "Microprocessor-based Quadrilateral Distance Relay for EHV/UH Transmission Line Protection, *Journal of Microcomputer Applications*, Academic Press (London), Vol. 15, No. 4, October 1992, pp. 347–360.

Gupta, A.K., L.P. Singh and G.K. Dubey, 'Design, Fabrication and Testing of a New Dynamic Test Bench', *Journal of Institution of Engineers* (India), November 1993, pp. 100–103.

Trivedi, V.I., Ravi Aggarwal and L.P. Singh, 'Integrated Digital Protection of a Synchronous Generator', Proceedings of the Power System Conference, Calcutta, February 1993, pp. 130–134.

Bhagat, Nitin Sachchidanand and L.P. Singh, 'Setting and Coordination of Distance Relaying Scheme Using Computer Graphics: A Case Study of the 400 KV UPSEB Sub-system', Proceedings of the Seventh National Power System Conference, February 1993, Calcutta, pp. 395–400.

Laway, N.A. and H.O. Gupta, 'A Method for Adaptive Coordination of Overcurrent Relays in an Interconnected Power system', Proceedings of the Fifth International Conference on Development in Power System Protection, 1993, pp. 240–243.

Sachdev, M.S., T.S. Sidhu and B.K. Talukdar, 'Topology Detection for Adaptive Protection of Distribution Networks', Proceedings of the International Conference on Energy Management and Power Delivery, Vol. 1, 1995, pp. 445–450.

Codling, J.D., et al., 'Adaptive Relaying: A New Direction in Power System Protection', *IEEE Potentials*, Vol. 15, No. 1, February-March 1996, pp. 28–33.

Singh, L.P. and K.V. Desikachar, 'Ultra High Speed Fault Locating Protective Relay for Three-phase Long Heavily Loaded EHV/UHV Transmission Line', *Journal of Institution of Engineers* (India), March, 1996.

Parthasarathy, K., 'New Static Three-Step Distance Relay', *Proc. IEEE*, Vol. 113, April 1996, pp. 633–640.

Stedall, B., et al., 'An Investigation into the Use of Adaptive Setting Techniques for Improved Distance Back-up Protection', *IEEE Trans. on Power Delivery*, Vol. 11. April 1996, pp. 757–762.

Phadke, A.G. and J.S. Thorp, 'Expose Hidden Failures to Prevent Cascading Outages', *IEEE Computer Applications in Power*, July 1996, pp. 20–24.

Parthasarathy, K., 'Three-system and Single-system Static Distance Relay', *Proc. IEEE*, Vol. 113, April 1996, pp. 641–651.

Faucon, O. and L. Dousset, 'Coordinated Defense Plan Protects against Transient Instabilities', *IEEE Computer Applications in Power*, Vol. 10, No. 3, July 1997, pp. 22–26.

Butler, K.L., et al., 'Shipboard Systems Deploy Automated Protection', *IEEE Computer Applications in Power*, Vol. 11, No. 2, April 1998, pp. 31–36.

Phadke, A.G., S.H. Horowitz and J.S. Thorp, 'Aspects of Power System Protection in the Post-structuring Era', Proceedings of the 32nd Hawai International Conference on System Sciences, 1999.

Damborg, M.J., et al., 'Adaptive Protection as Preventive and Emergency Control', Proceedings of the *IEEE PES Summer Meeting*, Seattle, WA, Vol. 2, July 2000, pp. 1208–1212.

Kezunovic, M., 'Intelligent Systems in Protection Engineering', *Power System Technology*, 2000, Proc. POWERCON 2000, Vol. 2, pp. 801–806.

Breslaua, L., et al., 'Advances in Network Simulation', *IEEE Trans. on Computers*, Vol. 33, No. 5, May 2000, pp. 59–67.

Coury, D.V., et al., 'Improving the Protection of EHV Teed Feeder Using Local Agent', Paper presented at the IEEE 2000 Summer Meeting, pp. 1196–1201.

Tan, J.C., et al., 'An Expert System for the Back-up Protection of a Transmission Network', *IEEE Trans. on Power Delivery*, Vol. 15, April 2000, pp. 508–514.

Li, G., et al., 'Experiments in Fast Restoration Using GMPLS in Optical/Electronic Mesh Network', Optical Fibre Communication Conference and Exhibit, 2001, OFC 2001.

Tan, J.C., et al., 'Intelligent Wide Area Back-up Protection and its Role in Enhancing Transmission Network Reliability', IEE Seventh International Conference on Developments in the Power System Network, 2001, pp. 446–449.

Wang, X.R., et al., 'Developing an Agent-based Back-up Protection System for Transmission Networks', Proceedings of the Conference on Power Systems and Communication Infrastructure for the Future, Beijing, September 2002.

Bertsch, J., C. Carnal, D. Karlsson, J. Mcdaniel and Khoi Vu, 'Wide Area Protection and Power System Utilization', *Proc. IEEE*, Vol. 93, No. 5, May 2005.

Zima, M., M. Larsson, P. Corba, Ch. Rehtanz and G. Andersson, 'Design Aspects for Wide Area Monitoring and Control Systems', *Proc. IEEE*, Vol. 93, No. 5, May 2005.

Index